GEOMETRY OF THE LAPLACE OPERATOR

PROCEEDINGS OF SYMPOSIA
IN PURE MATHEMATICS
Volume XXXVI

GEOMETRY
of the
LAPLACE OPERATOR

AMERICAN MATHEMATICAL SOCIETY
PROVIDENCE, RHODE ISLAND

1980

PROCEEDINGS OF THE SYMPOSIUM IN PURE MATHEMATICS
OF THE AMERICAN MATHEMATICAL SOCIETY

HELD AT THE UNIVERSITY OF HAWAII
HONOLULU, HAWAII
MARCH 27–30, 1979

EDITED BY
ROBERT OSSERMAN
ALAN WEINSTEIN

Prepared by the American Mathematical Society
with partial support from National Science Foundation grant MCS 78-27127

Library of Congress Cataloging in Publication Data

AMS Symposium on the Geometry of the Laplace Operator, University of Hawaii at Manoa, 1979.
Geometry of the Laplace operator.
(Proceedings of symposia in pure mathematics; v. 36)
"Proceedings of the Symposium in Pure Mathematics of the American Mathematical Society, held at the University of Hawaii, Honolulu, Hawaii, March 27–30, 1979."
Includes bibliographies.
1. Global analysis (Mathematics)–Congresses. 2. Manifolds (Mathematics)–Congresses. 3. Geometry, Differential–Congresses. 4. Laplacian operator–Congresses. I. Osserman, Robert. II. Weinstein, Alan, 1943– III. American Mathematical Society. IV. Symposium in Pure Mathematics, University of Hawaii, 1979. V. Title. VI. Series.
QA614.A18 1979 514′.74 79-26934
ISBN 0-8218-1439-7

1980 Mathematics Subject Classification. Primary 58–06; Secondary 53–06.

CONTENTS

FOREWORD

The AMS Symposium on the Geometry of the Laplace Operator was held at the University of Hawaii at Manoa (Honolulu) from March 27th through 30th, 1979, in conjunction with the 764th meeting of the American Mathematical Society. The organizing committee consisted of David D. Bleecker and Robert Osserman (co-chairmen), Victor W. Guillemin, Henry P. McKean, Jr., Karen Uhlenbeck, Joel L. Weiner, and Alan Weinstein.

The present volume contains expanded versions of most of the invited lectures and a few contributed papers; it represents quite faithfully the range of topics discussed in the symposium.

We would like to thank all those whose work on the arrangements made the symposium a success. These include David Bleecker and Joel Weiner of the University of Hawaii, as well as Dorothy Smith and Heather MacDonald of the AMS. We are also grateful to NSF for financial support. We hope that in publishing this volume we will allow a much wider audience to share in some of the work and enthusiasm of the participants in the Symposium.

Robert Osserman
University of California, Berkeley

Alan Weinstein
University of California, Berkeley

Proceedings of Symposia in Pure Mathematics
Volume 36, 1980

RIESZ MEANS ON RIEMANNIAN MANIFOLDS

Pierre H Bérard

ABSTRACT. In this paper we investigate convergence properties of the Riesz means of eigenfunction expansions associated with the Laplace operator on a compact Riemannian manifold. This is done by using the wave equation on the manifold.

A. INTRODUCTION.

1. Let (M,g) be a d-dimensional, compact, Riemannian (smooth) manifold (we shall also assume that it is connected). The manifold M is naturally equipped with a positive measure dvol and with an elliptic second order differential operator Δ. Because this operator is symmetric and nonnegative, it extends to a nonnegative selfadjoint operator (which we shall also denote by Δ) on the space $L^2(M)$ (by $L^p(M)$ we mean the L^p-space associated with the measure dvol on the manifold M). Let c be any positive number, and let $P = \Delta + c$. We denote by $\mathrm{Spec}(P) = \{u_j \mid j = 0,1,2,\ldots\}$ the spectrum of P. This spectrum is a nondecreasing sequence of positive eigenvalues with finite multiplicities (and written as such) tending to infinity. We denote by $\{f_j\}$ an orthonormal basis in $L^2(M)$ consisting of smooth real eigenfunctions of P, associated with the eigenvalues u_j.

2. Given a function f in $L^1(M)$, we define the generalized Fourier coefficients of f by

$$(3)\qquad c_j(f) = \int_M f(x)f_j(x)\,d\mathrm{vol}(x) .$$

Mimicking the case of multiple Fourier series, we define the Riesz means of order a (a nonnegative number) of f by

$$(4)\qquad s_R^a(f;x) = R^{-2a}\,\Sigma\,(R^2 - u_j)_+^a\, c_j(f)\, f_j(x)$$

where R is a positive number which will eventually tend to infinity (recall that x_+^a is defined as the function equal to x^a when x is positive and 0

1980 Mathematics Subject Classification: Primary 53C20, 58G99, 42C15.

otherwise). In this paper we investigate the convergence properties of the Riesz means $s_R^a(f)$ when R tends to infinity. Our first result is (Section B)

5. PROPOSITION. Assume that $a > d/2$ and let f be in $L^p(M)$, for $1 \leq p < \infty$. Then the Riesz means $s_R^a(f)$ converge to f in $L^p(M)$ as R tends to infinity.

6. The previous proposition shows the existence of an index, namely $d/2$, such that the Riesz means of order larger than $d/2$ converge in norm. In Section B, we shall also investigate convergence properties below the value $d/2$

7. The index which appears in the classical case of Riesz means of multiple Fourier series is $(d-1)/2$; furthermore this index is critical in the sense that some results fail to hold below $(d-1)/2$: see [S-W] Chapter 7. The most general case of a selfadjoint elliptic operator on a manifold is studied in [HR1] and [HR2] ; the index which appears is then $(d-1)$. The case of compact Lie groups is examined in [CC1], that of compact symmetric spaces of rank one in [B-C] ; in both instances $(d-1)/2$ appears as critical index. In Section C we shall show how we can improve the results of Section B in the case in which the manifold has periodic geodesic flow. This enables us to generalize part of the results of [B-C] to that case ; this also gives a method of dealing with the different cases of symmetric spaces of rank one simultaneously. At the end of Section C, we shall comment further upon the index $d/2$ found in B.

8. Let us now give more notations. We let $e(u,x,y)$ denote the spectral function of the operator $\sqrt{P}$

$$(9) \qquad e(u,x,y) = \sum_{\sqrt{u_j} \leq u} f_j(x) f_j(y) \ .$$

We recall that the operator $\sqrt{P}$ is defined by

$$(10) \qquad \sqrt{P}\, f_j = \sqrt{u_j}\, f_j$$

and it is well known that $\sqrt{P}$ is a pseudodifferential operator by a theorem of Seeley. One also defines the operators $\exp(-it\sqrt{P})$ (resp. $\cos t\sqrt{P}$). We shall denote by $U(t)$ (resp. $C(t)$) the kernels of these operators. We recall that $C(t)$ is the fundamental solution of the Cauchy problem for the wave equation

$$(11) \qquad \begin{aligned} & C''(t,x,y) + P_y C(t,x,y) = 0 \ ; \\ & C(0,x,y) = \delta(x-y) \ ; \quad C'(0,x,y) = 0 \ ; \end{aligned}$$

where $'$ denotes differentiation with respect to t.

12. REMARK. The Riesz means which we defined above are associated with the operator P. We point out that the Riesz means associated with the operator Δ are essentially the same (they differ from the above ones by a multiplicative factor and by a change of scale on the parameter R, as is easily seen).

B. RIESZ MEANS : THE GENERAL CASE.

1. Let us first introduce the kernels of the Riesz means

(2) $s_R^a(x,y) = R^{-2a} \Sigma (R^2 - u_j)_+^a f_j(x)f_j(y)$.

3. REMARK. The fact that R^2 and not R appears in (2) as well as in A(4) is due to the fact that Δ is a second order operator (compare with the classical case of multiple Fourier series).

One can also write the kernel of the Riesz means in the following manner

(4) $s_R^a(x,y) = R^{-2a} \int_{\mathbb{R}} (R^2-u^2)_+^a \, de(u,x,y)$.

The general idea is now to study the growth properties of these kernels and then to apply general theorems in order to obtain convergence results as R tends to infinity. To do so we shall express the kernels of the Riesz means in terms of the fundamental solution of the Cauchy problem for the wave equation (indeed, the kernel $C(t)$ contains information on the geometry of the manifold as shown in [D-G] for example). For this purpose we write $U(t)$ as the Fourier transform of the spectral measure $de(u)$ of the operator $\sqrt{P}$

(5) $U(t) = \int_{\mathbb{R}} \exp(-iut)\, de(u)$.

6. Now, notice that

$$R^{-2a} \int_{\mathbb{R}} (R^2-u^2)_+^a \exp(itu)\, du = \sqrt{\pi}\, \Gamma(a+1)\, (Rt/2)^{-a-1/2} J_{a+1/2}(Rt)$$

(see [G-S] p. 363) where J_m is the Bessel function of order m . In the sequel we shall write $j(t) = t^{+1/2} J_{a+1/2}(t)$. Now a formal application of Parseval's formula gives

(7) $s_R^a(x,y) = b(a)\, R^{-a} \int_0^\infty t^{-a-1} j(Rt)\, C(t,x,y)\, dt$

where $b(a)$ is a numerical constant. We shall now show that (7) makes sense, for fixed x and y with $x \neq y$, if we interpret the integral in the right hand side in the distribution sense.

8. PROPOSITION. Let x and y be two given points ($x \neq y$) in M and let $f_{x,y}$ be the application

$$f_{x,y} : \mathbb{R} \longrightarrow \mathbb{R} \times M \times M \qquad ; \qquad f_{x,y}(t) = (t,x,y) .$$

Then the inverse image $f_{x,y}^*(U)$, which we shall denote by $U(t,x,y)$ for short, is a well defined <u>tempered</u> distribution on $\mathbb{R}$. Furthermore, its wave front set is contained in the set of (t,τ) in $\mathbb{R} \times \mathbb{R}_-$ such that there exists a geodesic of length $|t|$ from x to y .

PROOF. Write

$$N(f_{x,y}) = \{(t,\tau;x,\xi;y,\eta) \in T^*(\mathbb{R} \times M \times M) \mid {}^*Df_{x,y}(\tau,\xi,\eta) = 0 \};$$

(here $*$ denotes transposition, or cotangent objects). Then $N(f_{x,y})$ is given by $\tau = 0$. On the other hand U is a Fourier integral operator associated with the canonical relation

$$C = \{(t,\tau;x,\xi;y,\eta) \in T^*(\mathbb{R} \times M \times M) \mid (x,\xi) = \phi_t(y,\eta) \ , \ \tau + ||\xi|| = 0\};$$

(see [D-G] p.43) where ϕ_t is the geodesic flow of M. Since $x \neq y$ by assumption, we have $N(f_{x,y}) \cap C = \emptyset$. From this we deduce (see [DT] p. 29) that $f^*_{x,y}(U)$ exists. The assertion on the wave front set follows from the calculus of wave front sets. In order to prove that $f^*_{x,y}(U)$ is a tempered distribution, we use the same method as that of [D-G] p.42. The application $V : S(\mathbb{R}) \longrightarrow C^\infty(M \times M)$ defined by (S means Schwartz space)

$$V(h) = \int_{\mathbb{R}} \exp(-it\sqrt{P})\, h(t)\, dt$$

is continuous (we identified smoothing operators and their smooth kernels) : indeed, it suffices to perform integration by parts and to use the fact that the operator $\exp(-it\sqrt{P})$ is unitary in $L^2(M)$. We then compose with the evaluation map of a smooth function at the point (x,y). To finish the proof we remark that

$$V(h)(x,y) = \int_{\mathbb{R}} f^*_{x,y}(U)(t)\, h(t)\, dt$$

for any function h with compact support.

9. Since we have $C(t) = \text{Real } U(t)$ we can also apply Proposition 8 to $C(t)$. Recall that

$$\text{Supp}(C) \subset \{(t,x,y) \in \mathbb{R} \times M \times M \mid |t| \geq \overline{xy}\}$$

where $\overline{xy}$ denotes the Riemannian distance from x to y. Assuming that the points x and y are distinct, we deduce that the integral (7) exists near 0 at least when $x \neq y$: indeed the distribution $t^{-a-1} C(t,x,y)$ vanishes near the point $t = 0$. Now remark that the integral (7) is the Hankel transform of $t^{-a-1}C(t,x,y)$ and deduce that the right hand side of the equality (7) is well defined in $S'(\mathbb{R})$ (the space of tempered distributions) .We have proved

10. LEMMA. The equality (7) makes sense, for $x \neq y$, provided that the integral in the right hand side be interpreted in $S'(\mathbb{R})$.

Let us now fix a smooth function h such that the support of h be contained in the closed interval $[-r/2 , +r/2]$ and such that h be equal to 1 on the interval $[-r/3 , +r/3]$, where r denotes the injectivity radius of the manifold M (this is a positive number because M is compact). Denote by $\hat{h}$ the inverse Fourier transform of h, and set

$$(11) \quad s^a_{R,h}(x,y) = b(a)\, R^{-a} \int_0^\infty t^{-a-1}\, j(Rt)\, h(t)\, C(t,x,y)\, dt \quad , x \neq y .$$

12.LEMMA. There exists a constant D such that for any nonnegative a, for any x and y in M the following inequality holds

$$|s^a_{R,h}(x,y)| \leqq D\, R^d$$

(the proof also shows how to give a meaning to $s^a_{R,h}(x,y)$ when $x = y$).

PROOF. For $x \neq y$ one has

$s^a_{R,h}(x,y) = R^{-2a} \int_{\mathbb{R}} (R^2-u^2)^a_+ * \hat{h}(u)\, de(u,x,y)$ (which makes sense even for $x = y$) ;

$$s^a_{R,h}(x,y) = R^{-2a} \iint (R^2-s^2)^a_+ \hat{h}(u-s)\, ds\, de(u,x,y)$$
$$= R^{-2a} \int \Big(\Sigma \check{\hat{h}}(s-\sqrt{u_j})\, f_j(x) f_j(y) \Big) (R^2-s^2)^a_+\, ds \quad .$$

As is shown in [D-G] p. 46 one has

$$\Sigma\, \check{\hat{h}}(s-\sqrt{u_j})\, f_j^2(x) \sim s^{d-1} \;\Sigma\; a_k(x)\, s^{-k} \qquad \text{(summation in } k \geqq 0) .$$

The inequality of the lemma follows by first applying Cauchy-Schwartz inequali--ty and then by integrating the above asymptotic expansion against $(R^2-s^2)^a_+$.

13. LEMMA. Let us denote by $\overline{xy}$ the Riemannian distance from x to y .

Then, there exists a constant D such that for all x and y in the manifold M, for all nonnegative a and for all positive R, the following inequality holds

$$|s^a_{R,h}(x,y)| \leqq D\, R^{-a+(d-1)/2} \quad (\overline{xy} + 1/R)^{-a-(d+1)/2} \quad .$$

PROOF. Let us first notice that if $R\,\overline{xy} \leqq 1$, then the estimate in the lemma reduces to that of Lemma 12 . We can therefore assume that $R\,\overline{xy} > 1$. By our choice of the function h we have $h(t)\, C(t,x,y) = 0$ as soon as we have $\overline{xy} > r/2$. We can use for C the expression given in [BD] p.258. We write

$$C(t,x,y) = \sum_{k=0}^{N} U_k(x,y)\, |t| \left. \frac{(\overline{xy}^{\,2} - t^2)^{k-m}_-}{\Gamma(k-m+1)} \right|_{m\, =\, (d+1)/2} + |t|\, C_N(t,x,y)$$

where we choose N so large that C_N be enough regular . Since the function $h(t)\, C_N(t,x,y)$ vanishes when $\overline{xy} > |t|$ the integral

$R^{-a} \int_0^\infty t^{-a}\, j(Rt)\, h(t)\, C_N(t,x,y) dt$ is bounded by a constant times $\overline{xy}^{\,-a} R^{-a}$

and since $R\,\overline{xy} > 1$ the integral is bounded by a constant. One can estimate in a similar way the integrals involving the functions $(\overline{xy}^2 - t^2)^{k-(d+1)/2}_-$ when k is larger than the integer part of $(d+1)/2$. In order to estimate the main contributions we divide the problem into two cases.

1st case : $d = 2p + 1$; this is the simpler case. One has to estimate integrals of the following form (see [G-S] p. 56-58)

$$I = R^{-a} \int_0^\infty t^{-a}\, h(t)\, j(Rt)\; \delta^{(p-k)}(\overline{xy}^{\,2} - t^2)\, dt \qquad \text{for} \quad 0 \leqq k \leqq p$$
$$= R^{-a} \int_0^\infty t^{-a}\, h(t)\, j(Rt)\, (\overline{xy} + t)^{-p+k-1}\; \delta^{(p-k)}(\overline{xy} - t)\, dt \quad .$$

Leibnitz' formula shows that one obtains a sum of terms of the following form

$J = R^{-a} R^{q} \, \overline{xy}^{\,-a-m} \, \overline{xy}^{\,-p+k-1-r}$, with $m+q+r \leq p-k$; this gives

$J = R^{p-a-k} \, \overline{xy}^{\,-p-1-a+k} \, (\overline{xy}^{\,-m-r} R^{q-p+k})$, the third term is bounded

because $R\,\overline{xy} \geq 1$ and $m+q+r \leq p-k$. It then follows that

(14od) $I = O(R^{-a+(d-1)/2}) \, \overline{xy}^{\,-a-(d+1)/2}$ (here $p = (d-1)/2$) .

2nd case : $d = 2p$. One has integrals of the form

$I = R^{-a} \int_0^{\infty} t^{-a} \, h(t) \, j(Rt) \, (\overline{xy}^{\,2} - t^2)_-^{k-p-1/2} \, dt$, $0 \leq k \leq p$;

Performing the changes of variables $t = \overline{xy}u$ and $u = v + 1$ gives

$I = R^{-a} \, \overline{xy}^{\,-d-a+2k} \int_0^{\infty} v_+^{-p+k-1/2} (v+2)^{-p+k-1/2} (v+1)^{-a} h(\overline{xy}(v+1)) j(R\overline{xy}(v+1)) \, dv.$

If we now write, with obvious notations, $I = I_h + I_{1-h}$ we see that

$R\,\overline{xy} \geq 1$ implies that $I_{1-h} = O(R^{-a+(d-1)/2}) \, \overline{xy}^{\,-a-(d+1)/2}$. On the other hand, the asymptotic expansions for Bessel functions (see [M-S] p.139) yield the same estimate for I_h . Hence we have proved

(14ev) $I = O(R^{-a+(d-1)/2}) \, \overline{xy}^{\,-a-(d+1)/2}$. The lemma follows.

Let us now define the following kernel

(15) $r^a_{R,h}(x,y) = s^a_R(x,y) - s^a_{R,h}(x,y)$.

Then, for $x \neq y$, one has (in $S'(\mathbb{R})$)

(16) $r^a_{R,h}(x,y) = b(a) \, R^{-a} \int_0^{\infty} t^{-a-1} (1 - h(t)) \, j(Rt) \, C(t,x,y) \, dt$.

17. LEMMA. Given any positive a , any (arbitrarily small) positive number e , there exists a constant D such that for any positive R, any x and y in M , the following inequality holds

$|r^a_{R,h}(x,y)| \leq D \, R^{-a+e+d/2} \, k(x,y)$

where k is a positive function on $M \times M$ which is in $L^1(M)$ with respect to each variable, uniformly with respect to the other one.

PROOF. Given e we write $2d' = (e + d/2)$, and we define a distribution $F(t)$ on $M \times M$ by $F(t) = P_y^{-d'} U(t)$ (i.e. we let $P^{-d'}$ act on $U(t)$ with respect to the second variable). Then, we can write (in $L^2(M \times M)$ for example)

$F(t) = \Sigma \exp(-it\sqrt{u_j}) u_j^{-d'} \, f_j \otimes f_j$.

By [D-G] p.53 (see also [BE]Chap.8) we deduce that $F(t,x,\cdot)$ is in $L^2(M)$, with L^2-norm uniformly bounded with respect to $(t,x) \in \mathbb{R} \times M$ .Now, notice that

$D^{2d'} F(t) = U(t)$ where $D = \left| \dfrac{d}{dt} \right|$.

Since the function h takes value 1 near $t=0$ the integral in (16) also makes sense when $x = y$. One can write

$$r^a_{R,h} = b(a)\, R^{-a} \int_0^\infty t^{-a-1} (1-h(t))\, j(Rt)\, D^{2d'} F(t)\, dt$$

(the integral converges in $L^2(M \times M)$ in the norm sense if a is positive). By the asymptotic properties of the Bessel functions (see [M-S] p.139) one has

$D^{2d'} (t^{-a-1} (1-h(t))\, j(Rt)) = O(R^{2d'})\, (1 + t)^{-a-1}$. The lemma follows from the facts that $(1+t)^{-a-1}$ is integrable when a is positive, and that one has the inclusion $L^2(M) \subset L^1(M)$ because M is compact.

18. REMARK. In proving Lemma 17, we have used brute force ; this explains why the index $d/2$ appears here instead of $(d-1)/2$ as in Lemma 13 (see comments at the end of Section C).

19. LEMMA. ([CC1] Lemme 4 p. 161) Let g be a nonincreasing positive function on $\mathbb{R}_+$. Then for any integrable function f

$$\int_M |f(y)|\, g(\overline{xy})\, dvol(y) \leqq D \left(\int_0^\infty g(r)\, r^{d-1}\, dr \right) Mf(x)$$

where $Mf(x)$ is the maximal function of Hardy-Littlewood; and the constant D does not depend on f or g.

Recall that the maximal function of Hardy-Littlewood is defined by

$$Mf(x) = \sup_{r>0} \left(\int_{\overline{xy} \leqq r} |f(y)|\, dvol(y) \right) \left(\int_{\overline{xy} \leqq r} dvol(y) \right)^{-1}$$

(see [S-W] for details about maximal functions). Let us set

$g(r) = (r + 1/R)^{-a-(d+1)/2}$, then the integral in the right hand side of the inequality which appears in the above lemma converges if $a > (d-1)/2$.

20. THEOREM. (a) Let f be in $L^p(M)$, $1 \leqq p < \infty$, then the Riesz means of order $a > d/2$ converge to f in $L^p(M)$ as R tends to infinity ;

(b) Let $a > d/2$, and let k be the function described in Lemma 17 (e.g. with $2e = a-d/2$), then there exists a constant D such that for any function f in $L^1(M)$ the following inequality holds

$$\sup_{R>0} \left| s^a_R(f;x) \right| \leqq D \left\{ Mf(x) + \int_M k(x,y)\, |f(y)|\, dvol(y) \right\}$$

PROOF. Notice that assertion (a) is clear for smooth functions. On the other hand, it follows from Lemma 13 and Lemma 17 that the kernels s^a_R are uniformly bounded in $L^1(M)$ with respect to each variable. Assertion (a) follows by a density argument. Assertion (b) is an easy consequence of Lemmas 12, 17, 19.

22. COROLLARY. Let $a > d/2$, then there exists a constant D such that for any function f in $L^1(M)$ the following inequality hold

$$\mathrm{Vol}(\{ \sup_R |s^a_R(f)| > s \}) \leqq \frac{D}{s} \| f \|_1 .$$

The Riesz means of the function f of order $a > d/2$ converge almost everywhere to f as R tends to infinity .

PROOF. It suffices to apply Theorem 3.12 p. 60 of [S-W] after making use of the properties of the maximal function of Hardy-Littlewood.

23. THEOREM. Let $1 < p \leq 2$. Then, if $a > (d/2)(2/p - 1)$ there exists a constant D_p such that, for any function f in $L^p(M)$ the following inequality holds

$$\| \sup_R |s_R^a(f)| \|_p \leq D_p \|f\|_p \quad ;$$

In particular, the Riesz means of f of order $a > (d/2)(2/p - 1)$ converge in norm and almost everywhere to f as R tends to infinity.

PROOF. We remark that the theorem is true when $a > d/2$ and $p > 1$. The case $p = 2$ is Kacmarcz' lemma (see [HR1] Lemma 3.4 p;171). In order to prove the theorem, one uses the method of interpolation of Stein between the cases $p > 1$ and $p = 2$ as is done in [HR1] and [B-C] (for the method of interpolation see [S-W] Chapter 7).

REMARK. This result is interesting in the sense that it gives convergence properties below the index $d/2$.

One can also investigate localization properties of the Riesz means ; we have the following result (it is rather weak : compare with [HR1] and [B-C])

24. THEOREM. Let e be a (small) positive number, and let x_0 be a fixed point in M. Assume that $a > d/2$. Then, if the function f is in $L^q(M)$, with $q \geq 1$ and $q > d/a$, and if f vanishes in the ball $\{\overline{x_0 y} < e\}$, the following holds

$$\lim_{R \infty} s_R^a(f; x_0) = 0 \quad .$$

PROOF. First of all notice that the theorem is obviously true if the function f is a smooth function (or a finite linear combination of eigenfunctions) Hence, it suffices to prove that there exists a constant D such that for any positive R the following inequality holds

$$|s_R^a(f;x_0)| < D\|f\|_q \quad .$$

Apply Theorem 20(b). Notice that the part involving $Mf(x_0)$ disappears because the function f vanishes near x_0. Now one has

$$\int_M k(x_0,y)\ |f(y)|\ dvol(y) \leq D\|f\|_q$$

as soon as $\|k(x_0,.)\|_p \leq D$ where $1/p + 1/q = 1$. Using the lemma of Sobolev-Krylov (see [SZ] Proposition XIV.3,2) we see that the inequality on the kernel k holds if $a > d/2$ when $p \leq 2$ and $a > d - d/p$ when $p \leq 2$. This gives the condition on q required in the theorem.

C. RIESZ MEANS ON MANIFOLDS WITH PERIODIC GEODESIC FLOW.

1. The purpose of this section is to improve the results which were obtained in Section B , in the case in which the geodesic flow of the manifold is periodic. Typical examples of such manifolds are given by the symmetric spaces of rank one (compact).

2. Let us set (we shall choose m later on)

$$G(t) = \Sigma \exp(-it\sqrt{u_j})\, u_j^{-m}\, f_j \otimes f_j .$$

Since the kernel $U(t)$ belongs to the space $I^0(M \times M , C_t)$ of Fourier integrals of order 0 associated with the graph of the geodesic flow at time t (see [BE] Chap.8 for example), it follows that $G(t)$ belongs to the space $I^{-2m}(M \times M , C_t)$.

3. Given a point x in M , consider the application (we shall assume that t is away from 0)

$$i_x : M \longrightarrow M \times M \quad ; \quad i_x(y) = (x,y) .$$

The pullback operator i_x^* is a Fourier integral operator of order $d/4$ (see [DT] p; 56-60). It is easy to check that $i_x^*(G(t))$ is well defined and belongs to the space of Fourier integral distributions of order $-2m+d/4$ (see [D-G] p. 63 and notice that in our case the excess is 0 , for example).

4. LEMMA. The distribution $i_x^*G(t)$ is in $L^p(M)$, $1 \leqq p < \infty$ if

(5) $d/2 - (1/p - 1/2) < 2m$ when $p \leqq 2$;

(6) $d(1/2 - 1/p) + d/2 < 2m$ when $p \geqq 2$;

Furthermore, the L^p-norm of $i_x^*G(t)$ is uniformly bounded with respect to x in M and to t in any compact subset of $\mathbb{R}^\cdot$ .If q is the dual exponent to p then inequality (6) reads $q \geqq 1$ and $q > d/2m$.

PROOF. Remark that the projection of the wave front set $WF(i_x^*G(t))$ on M has rank at most $(d-1)$ and at least 1 ; now apply Theorem 10.5 p. 561 of [C-F] .

7.REMARK. For $p = 1$ the previous lemma is sharper than the embedding lemma of Sobolev-Krylov which we used in Section B. However, using Lemma 4, we loose the uniformity of the L^p-norm with respect to t in $\mathbb{R}$.

8. In the sequel, we shall assume that the geodesic flow of M is periodic with period 2π . Given such a manifold M , there exist a number c (see Introduction) and a number T ($T = 4\pi$, or $T = 8\pi$) such that

$\exp(-iT\sqrt{\Delta + c}\,) = Id + C$, where C is a pseudodifferential operator of degree -1 (see [D-G] p. 53 , [BE] Chap. 8). Furthermore, if M is a symmetric space of rank one, then $C \equiv 0$. Let h be the smooth function defined in B.10 and let g be a smooth function such that

$h(t) + g(t) = 1$ when t is positive and such that $g(t) = 0$ when t is nonpositive. As in Section B, we write

(9) $s_R^a(x,y) = s_{R,h}^a(x,y) + b(a)\, R^{-a} \int_0^\infty t^{-a-1}\, g(t)\, j(Rt)\, C(t,x,y)\, dt$.

In order to study the "integral" in the right hand side, we choose $2m = a$, and we use the distribution $G(t)$ defined in n°1 with that spacial value of m:

$$s_R^a(x,y) - s_{R,h}^a(x,y) = s_{R,g}^a(x,y) = \text{Real}\{ R^{-a} \int_{\mathbb{R}} t^{-a-1} g(t) j(Rt)\, D^a G(t)\, dt \}$$

where $D = |d/dt|$. It follows that we can write

$$s_{R,g}^a = \text{Real}\{ \int_{\mathbb{R}} t^{-a-1}\, \hat{g}(t,R)\, j(Rt)\, G(t)\, dt \} ;$$

where the function $\hat{g}$ vanishes for $t \leqq r/3$ (see B.10), and is bounded with respect to t and R . Choose a smooth function u with compact support in the interval $(-T, +T)$ and such that

$\Sigma\, u(t - kT) = 1$ (summation with respect to k in $\mathbb{Z}$)

(the existence of such a function is guaranteed by Lemma 3.21 p.265 of [S-W]).

It follows that

(10) $s_{R,g}^a = \text{Real} \{ \int_{\mathbb{R}} (t+kT)^{-a-1}\, \hat{g}(t+kT,R) j(Rt+kRT) G(t+kT)\, dt \}$.

Using the assumption that the geodesic flow is periodic, we have

$$G(t+ kT) = G(t) + H_k(t) \quad , \text{ with } \quad H_k(t) = C \sum_{p=1}^{|k|-1} G(t+\text{sgn}(k)pT) ,$$

and when M is acompact symmetric space of rank one $H_k \equiv 0$.

Using Lemma 5 we conclude that $G(t)(x,\cdot)$ is in $L^1(M)$ when $a > (d-1)/2$, uniformly with respect to $(t,x) \varepsilon [-T,T] \times M$. On the other hand, by Lemma17 of Section B, we have that $C\, G(t)(x,\cdot)$ is in $L^1(M)$ when $a > (d-1)/2$, uniformly with respect to $(t,x) \varepsilon \mathbb{R} \times M$. We conclude from this fact that the L^1-norm of $H_k(t)(x,\cdot)$ is bounded by a constant times $|k|$ uniformly with respect to $(t,x) \varepsilon \mathbb{R} \times M$.

11. THEOREM. Assume that $a > (d-1)/2$ ($a > 1$ if $C \not\equiv 0$). Then

(a) the Riesz means of order a , $s_R^a(f)$, of a function f in the space $L^p(M)$, $1 \leq p < \infty$, converge in norm to f ;

(b) there exists a constant D such that for any function f in $L^1(M)$ the following inequality holds

(12) $\sup_R |s_R^a f;x)| \leqq D \{ Mf(x) + \int_M k(x,y)|\, f(y)\, |dvol(y) \}$,

where k is a positive function in $M \times M$ which belongs to the space $L^1(M)$ with respect to each variable, uniformly with respect to the other one.

PROOF. It is the same as that of Theorem B.20 , once we have noticed that the series with general term n^{-a-1} (resp. n^{-a}) converge when a is positive (resp. $a > 1$) .

As in Section B, one can deduce the following corollaries

13. COROLLARY. Assume that $a > (d-1)/2$ ($a > 1$ if $C \not\equiv 0$), then there exists a constant D such that for any function f in $L^1(M)$

the following inequality holds

$$\mathrm{Vol}\,(\{\sup_R |s_R^a f| > s\}) \leqq \frac{D}{s} ||f||_1 \;;$$

the Riesz means of order a of a function f converge almost everywhere to the function f.

14. COROLLARY. Assume that $1 < p \leqq 2$. Then, if $a > ((d-1)/2)(2/p - 1)$ there exists a constant D_p such that for any function f in $L^p(M)$ the following inequality holds

$$||\sup_R |s_R^a(f)| \,||_p \leqq D_p ||f||_p \;;$$

the Riesz means of order a of f converge to f in $L^p(M)$.

As in Section B we have the following localization result (using (5) & (6))

15. PROPOSITION. Assume that $a > (d-1)/2$, $q \geqq 1$ and $q > d/a$. If the function f in $L^q(M)$ vanishes in a neighborhood of x_0

$$\lim_{R\,\infty} s_R^a(f;x_0) = 0 .$$

SOME COMMENTS.

(a) In Section C, we have been able to improve the index $d/2$ found in Sect. B to the index $(d-1)/2$. We did so by estimating the growth of the kernel $C(t)$ as t tends to infinity ; for this purpose, we used the assumption that the geodesic flow of the manifold is periodic. This assumption is only a technical one. As a matter of fact, we could prove part of the results on the convergence of the Riesz means of multiple Fourier series by our method (indeed in the case of flat tori, the kernel $C(t)$ is known explicitly so that we can conclude ;

(b) Given any smooth function g with compact support in $\mathbb{R}$, we can give estimates for the growth of the functions $s_{R,g}^a(x,y)$ as R tends to infinity (these estimates depend on whether the points x and y are conjugate along a geodesic or not). Such estimates are unfortunately not uniform as g varies. We point out that precise localization results rely on good pointwise estimates for the kernels s_R^a ;

(c) It might be worthwhile to notice that a way of getting the index $(d-1)/2$ in the general case, would be to have explicit expressions for the weight functions controlling the Lipschitz regularity of the solutions of the wave equation : see [C-F] and the references in that paper ;

(d) Finally, we point out that "good" control on the Riesz means of eigenfunction expansions also lead to "good" multiplier theorems : we refer to [B-C] and [CC2] .

ADDENDUM.

I have been informed that B. Randol (Trans. Amer. Math. Soc. 168 (1972), 265-271) obtained results concerning Riesz means convergence and localization for elliptic operators on tori.

BIBLIOGRAPHY

[B-C] Bonami A et Clerc J.L, Sommes de Cesàro et multiplicateurs des développements en harmoniques sphériques, Trans. Amer. Math. Soc., 183 (1973), 223-263.

[BD] Bérard P, On the Wave Equation on a Riemannian Manifold without Conjugate Points, Math. Z., 155 (1977), 249-276.

[BE] Besse A.L, Manifolds all of whose Geodesics are Closed, Springer Verlag, Berlin Heidelberg New York, 1978.

[CC1] Clerc J.L, Sommes de Riesz et multiplicateurs sur un groupe de Lie compact, Ann. Inst. Fourier, 24-1 (1974), 149-172.

[CC2] Clerc J.L, Multipliers on Symmetric Spaces, Preprint, to appear.

[C-F] Colin de Verdière Y et Frisch M, Régularité Lipschitzienne et solutions de l'équation des ondes sur une variété riemannienne compacte, Ann. Scient. Ec. Norm. Sup., 9 (1976), 539-565.

[D-G] Duistermaat J.J and Guillemin V.W, The Spectrum of Positive Elliptic Operators and Periodic Bicharacteristics, Inventiones Math. 29 (1975), 39-79.

[DT] Duistermaat J.J, Fourier Integral Operators, Lecture Notes, Courant Institute, New York University, New York, 1973.

[G-S] Gelfand I.M and Shilov G, Generalized Functions I, Academic Press, New York and London, 1964.

[HR1] Hörmander L, On the Riesz Means of Spectral Functions and Eigenfunctions Expansions for Elliptic Differential Operators, Some Recent Advances in Basic Sciences, Yeshiva University Conference, Nov. 1966, 155-202.

[HR2] Hörmander L, On the Spectral Function of an Elliptic Operator, Acta Math., 121 (1968), 193-218.

[M-S] Magnus W, Oberhettinger F and Soni R.P, Functions of Mathematical Physics, 3rd Edition, Springer Verlag, Berlin Heidelberg New York, 1966.

[S-W] Stein E and Weiss G, Fourier Analysis on Euclidean Spaces, Princeton University Press, Princeton 1970.

[SZ] Schwartz L, Séminaire Schwartz 1969-1970, Exposé 14, Ecole Polytechnique, Paris 1970.

LABORATOIRE ASSOCIE AU C.N.R.S. n° 212
U.E.R. DE MATHEMATIQUES
UNIVERSITE PARIS 7
75221 PARIS CEDEX 05
FRANCE

Proceedings of Symposia in Pure Mathematics
Volume 36, 1980

SOME GEOMETRIC ASPECTS OF WAVE MOTION: WAVEFRONT DISLOCATIONS, DIFFRACTION CATASTROPHES, DIFFRACTALS

M.V. Berry

ABSTRACT. On the finest scales, waves $\psi = \rho e^{i\phi}$ are dominated by the singularities of their phase ϕ, called 'wavefront dislocations'. On the largest scales, when the wavelength is small in comparison with diffracting objects ('geometrical-optics limit'), waves are dominated by caustics (focal surfaces); stable caustics can be classified as catastrophes; caustics are decorated with characteristic interference patterns, called 'diffraction catastrophes'. When waves encounter fractals (objects with an infinite hierarchy of length scales) the short-wave limit is not geometrical optics; this is a new regime, and such waves are called 'diffractals'.

1. INTRODUCTION. In this article I describe wave motion unconventionally, in terms of the contrast and interplay between three morphologies. These are distinguished by the scale on which a wave is explored, and the scales of variation of diffracting objects or refracting media, in comparison with the wavelength λ.

The first morphology, to be discussed in section 2, occurs on the scale of λ, where the principal features are the wavefronts, or surfaces of constant phase. Wavefronts are dominated by their singularities, in the form of lines in space called 'wavefront dislocations' [1] by analogy with the similar structures that interrupt the regular stacking of atomic planes in crystals. It is typical for waves to be dislocated in this way.

The second morphology, to be discussed in section 3, occurs when λ is small in comparison with all scales of variation of the waves' environment. In the limit $\lambda \rightarrow 0$, geometrical optics gives a description of wave fields in terms of families of rays. These are dominated by their focal singularities, called caustics, which are envelopes of the ray families, typically in the form of surfaces in space. Stable caustics are classified by the catastrophe theory of Thom [2] and Arnol'd [3]. When λ is small but not zero, asymptotic approximations to wave functions can be obtained in terms of a hierarchy of interference patterns decorating the caustics; these are the 'diffraction catastrophes'.

1980 Mathematics subject classification 78.00.

The third morphology, to be discussed in section 4, occurs when waves propagate in environments characterised by a hierarchy of scales extending infinitely small. Mandelbrot [4] uses the term 'fractals' to describe such structures in terms of a Hausdorff-Besicovitch dimension that need not be an integer. A fractal never appears smooth, no matter how short are the waves used to explore it. Therefore the short-wave limit is not geometrical optics. For waves of this sort I introduced the term 'diffractals' [5].

Most of the material reviewed here has been published elsewhere, so I shall confine myself to making assertions for which justification can be found in the papers cited, and to emphasising mathematical problems raised by this 'morphological' approach to wave physics.

2. WAVEFRONT DISLOCATIONS. Let $\psi(X)$ be a complex scalar wave function, i.e. a smooth map from spacetime X = (x, y, z, t) to the complex plane whose polar coordinates are the wave amplitude ρ and the wave phase ϕ. Thus

$$\psi(X) = \rho(X)\, e^{i\phi(X)}, \qquad (1)$$

where ρ and ϕ are real. The wavefronts are the hypersurfaces $\phi =$ constant (mod 2π). Let C be a simple closed curve in X (fig. 1) and let C′ be the image of C in ψ. We define the dislocation strength S_C enclosed by C as the winding number of C′ about the origin $\psi = 0$, i.e.

$$S_C = \frac{1}{2\pi}\oint_C d\phi. \qquad (22)$$

The dislocations themselves are those regions in X through which C must pass in order for S_C to change. Obviously $\psi = 0$ on a dislocation. Therefore $\rho = 0$ and ϕ is undefined, so that dislocations are singularities of the wavefronts, as is clear from the example in fig. 1. Because the vanishing of ψ requires two conditions ($\mathrm{Re}\,\psi = \mathrm{Im}\,\psi = 0$), dislocations have codimension two, and so occur typically as lines in a plane or points in space (which move unless ψ is monochromatic).

For all waves except those in quantum mechanics, the quantity of direct physical significance is not ψ but $\mathrm{Re}\,\psi$. Dislocations correspond to births and deaths of crests, troughs and zeros of $\mathrm{Re}\,\psi$ (crests are loci of $\phi =$ mod 2π, troughs are loci of $\phi = \pi$ mod 2π). This can be seen by referring to fig. 1. Consider the waves passing points A, B, C. The graphs of $\mathrm{Re}\,\psi$ as a function of time are shown on fig. 2. Clearly the wave through C has one more crest and two more zeros than that through A, and these features are born at the point B through which the dislocation passes. In fact dislocations were discovered in this way [1] during an experiment in which ultrasonic waves were reflected from a rough surface in a laboratory simulation of the echo-sounding of the Antarctic glacier bed by radio waves (see also [37]).

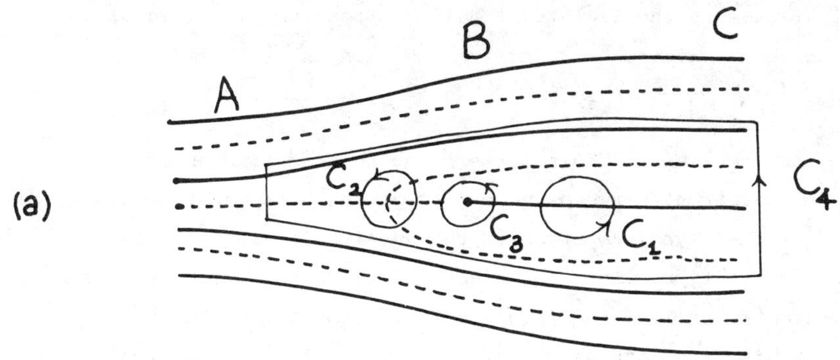

Fig. 1(a). Snapshot of a wave travelling up the page, showing crests (heavy lines) and troughs (dashed lines). The circuits C_1 and C_2 do not enclose any dislocations, while the circuits C_3 and C_4 each enclose a single dislocation.
(b) Images C_1' - C_4' in the ψ plane.

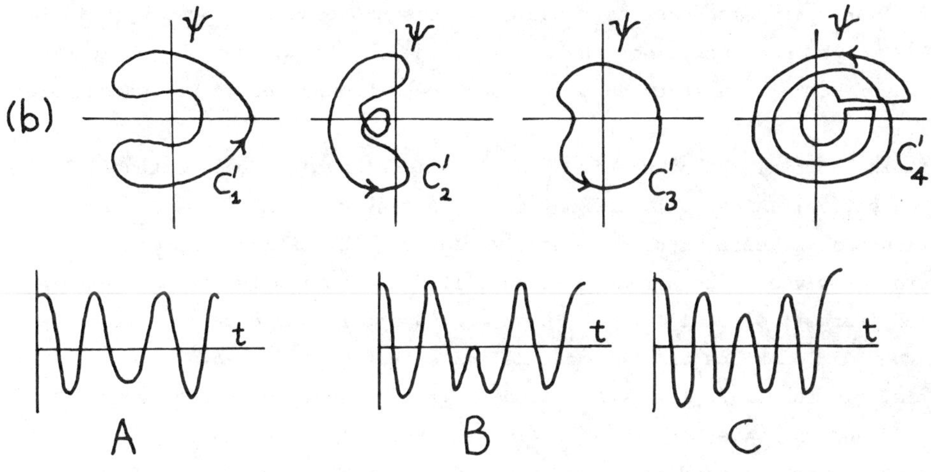

Fig. 2. Graphs of Re $\psi(t)$ at the points A, B, C on fig. 1, showing birth of a crest and two zeros as the dislocation passes B.

Of course ψ must satisfy a wave equation, but it is surprising that this seems to impose no restriction on the conformation, motion and interaction of dislocation lines. In the case of the simple wave equation

$$\Delta\psi = \frac{1}{c^2}\frac{\partial^2\psi}{\partial t^2}, \tag{3}$$

for example, a variety of dislocation morphologies can be exhibited in terms of explicit exact solutions. These solutions can be found in [1] , and take the form of dislocated versions of a 'host' monochromatic plane wave

$$\psi = e^{i\omega(z/c - t)} \tag{4}$$

The simplest case is the isolated 'edge' dislocation line (fig.1) moving rigidly with the host wavefronts; its wave function is

$$\psi = A\left(x + i\beta(z-ct)\right)e^{i\omega(z-ct)}, \qquad (5)$$

where A and β are constants. This corresponds to inserting an extra half-plane of 'crest' wavefront between the crests of the host wave. Dislocations can also be of 'screw' type, about which the wavefronts wind in helicoids, or of mixed 'edge-screw' type. They can be curved, and can form closed loops (holes in wavefronts, or isolated patches of wavefront). Dislocation lines can move rigidly with the host wavefronts, or parallel or perpendicular to them, or they may remain at rest. When dislocations encounter one another, they may annihilate, bounce off one another, or not interact at all.

The most important mathematical questions seem to be: for waves satisfying a given equation, which sorts and strengths of dislocation, and which dislocation interactions, occur stably in the plane, in space and in space-time? A promising approach to these questions has been initiated by Wright [6], who considered not the zeros of the complex ψ but the coalescing extrema of Re ψ as a function of t, using catastrophe theory to investigate stability.

It is surprising that wavefront dislocations have not been extensively studied before. After all, anybody since Huygens could have asked the question: Can wavefronts have edges? Moreover, dislocations are generic features of waves - the air in a sound-filled auditorium is threaded with a tangled 'forest' of dislocation lines. Perhaps this neglect stems from the fact that dislocations are the most delicate features of a wave. To discriminate them, ψ must be studied on length and times scales comparable with λ and the wave period respectively. Eyes and ears cannot do this - they are not 'phase-sensitive' but respond instead to $(\mathrm{Re}\,\psi)^2$ averaged over a number of cycles.

There have, however, been several partial anticipations of the dislocation concept. In 1836, Whewell [7] realised that the study of tides was greatly simplified by considering them as harmonic waves with a period of about 12 hours, whose wavefronts ('co-tidal lines') mark the loci of high water at different times. He discovered that the pattern of co-tidal lines has singularities ('amphidromic points') where the tide is always 'high' but has zero amplitude; one such singularity is shown on fig. 3. Of course these are wavefront dislocations, in this case points because the waves propagate in two dimensions. Then, in 1931, Dirac [8], while studying the effect of a magnetic field on the phase ϕ of an electron's wave function ψ, realised that there could exist circuits around which ϕ changed by a multiple of 2π , enclosing line zeros of ψ . However, he was more concerned with the ends of these lines, which are singularities not just of ψ but of the magnetic field, corresponding to magnetic monopoles. Next, in 1970, Riess [9] considered zeros ('nodal hypersurfaces') of quantum bound many-body systems such as molecules, and related the zeros to magnetic properties of the system. Finally, in work contemporary with ours, Hirschfelder and his collaborators [10] independently discovered dislocations in quantum mechanics, and called them 'quantised vortex lines', thus emphasising the vector field gradϕ rather than ϕ itself. All these authors confined themselves to monochromatic waves, and hence to stationary dislocations.

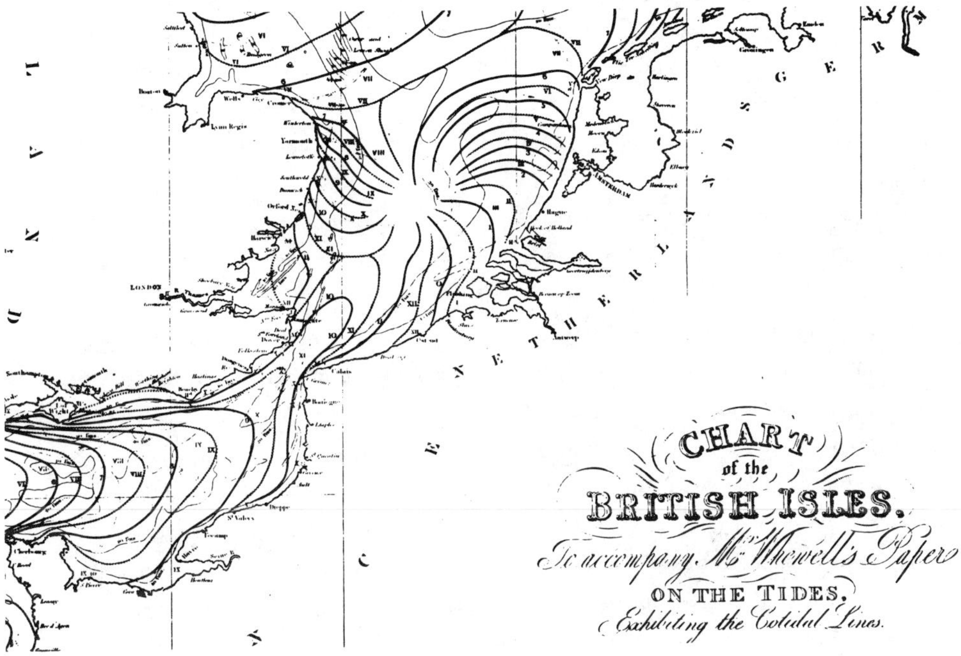

Fig. 3. Whewell's chart of co-tidal lines (wavefronts) in the North Sea [7] , showing a dislocation point.

It is important to distinguish wavefront dislocations from the singularities of 'geometrical-optics wavefronts' [3] . The latter are singularities of isosurfaces of solutions of the Hamilton-Jacobi equation and differ from wavefront dislocations in several ways. Firstly, the geometrical-optics wavefronts can cross (because several rays can pass through a point), whereas the true wavefronts ϕ = constant cannot cross (except on a dislocation). Secondly, the singularities of geometrical-optics wavefronts are ray caustics, which are typically surfaces in space (section 3), whereas wavefront dislocations are lines in space. And finally, while the solutions of the Hamilton-Jacobi equation are useful ingredients in short-wave approximations to ψ , the geometry of the geometrical-optics wavefronts does not itself possess direct physical significance; by contrast, the true wavefronts are directly detectable as crests and troughs of ψ.

3. DIFFRACTION CATASTROPHES. Question: which wave morphology is complementary (in the sense employed by Bohr in discussing the philosophy of quantum mechanics) to the wavefront dislocation? Answer: the caustic. Reason: at a dislocation, the wave's phase is singular and its amplitude vanishes. Observation of a dislocation requires discrimination of the wave's structure on wavelength scales - circumstances under which caustics are blurred by diffraction. At a caustic, on the other hand, the wave's geometrical-optics amplitude is infinite. The observation of a caustic takes place on scales large compared with the wavelength - circumstances under which phase details, and hence dislocations, are hard to discern. In other words, caustics are the singularities of ray theory, and dislocations are the singularities of wave theory.

Caustics are envelopes of ray families. Stable caustics are classified as catastrophes, that is as stable singularities of gradient maps. The reason for this ([11 - 14]), in outline, is that by Fermat's (or Hamilton's) principle rays are given by the extrema of optical distance (or action) and hence correspond to gradient maps, while caustics, on which neighbouring rays touch, are the loci of coalescence of the extrema and hence correspond to singularities of the maps.

Each catastrophe gives an equivalence class, consisting of caustics related by diffeomorphism (this is the sense in which the caustics are stable). The classification is in terms of codimension, at least for the simplest cases [3] . The catastrophe labelled i is defined by a generating polynomial $G_i\ (S_j;\ C_k)$ depending on 'state variables' S_j and 'control parameters' C_k. In ray theory, C_k could represent spacetime coordinates X, S_j could represent coordinates on a given initial wavefront, and G_i could represent optical path or action. The caustic inhabits control space, and is defined by the locus of C_k for which

$$\frac{\partial G_i}{\partial S_j} = 0 \quad \text{and} \quad \det\left\{\frac{\partial^2 G_i}{\partial S_j\,\partial S_{j'}}\right\} = 0 . \qquad (6)$$

In three-dimensional space, the stable caustics are smooth 'fold' surfaces (codimension one), 'cusp' lines (codimension two), and 'swallowtail', 'elliptic umbilic' and 'hyperbolic umbilic' points (codimension three). These forms can be seen in light focused by irregular glass (of the type used for bathroom windows) or by irregular water-droplet 'lenses' [11, 15, 16] .

Catastrophes play an important role in the physics of waves [27], analogous to that played by atoms in the physics of matter. It is helpful to think of them as 'elemental atomic forms', occupying a 'mesoscale' between 'macroscopic' caustic networks, formed by combinations of catastrophes, and 'microscopic' interference patterns decorating individual caustic singularities. On this analogy, wavefront dislocations are the elementary particles of wave physics.

'Macroscopic' caustic networks are a familiar sight as dancing lines of sunlight focused onto swimming-pool floors after refraction by the wavy water surface. The bright lines often meet in junctions, apparently violating catastrophe theory, according to which the only stable caustics in two dimensions are fold lines and cusp points. But the junctions are illusions, resulting from the fact that the patterns are imperfectly resolved. Figs. 4a and 4b show two commonly-observed caustic networks, and figs. 4c and 4d show their resolutions into detail suggested by catastrophe theory and confirmed by experiment [17, 18]. The study of caustic networks is still in its infancy; apart from some rules related to the indices of umbilic points [19, 14], no general results are known.

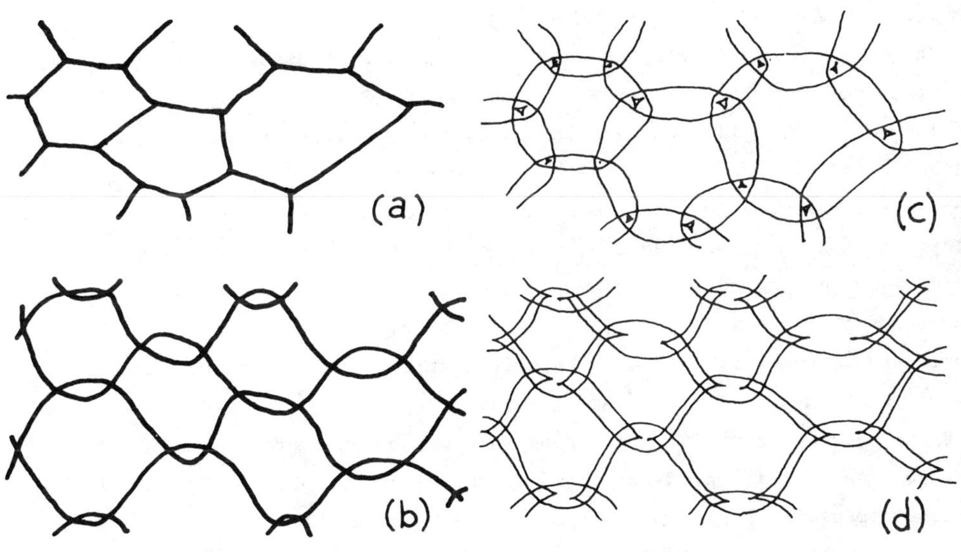

Fig. 4. Two sorts of 'swimming-pool' caustic network, seen with poor resolution (a,b) and good resolution (c,d).

On 'microscopic' scales, caustic singularities are softened by the effects of the finite wavelength λ. Asymptotic approximation of the solution of the wave equation [3, 13, 14, 20-22] yields the remarkable result that each catastrophe i gives rise to a canonical wave function $\psi_i(C_k;\lambda)$ describing diffraction near the caustic. The ψ_i take the form of oscillatory integrals, defined in terms of the generating polynomials G_i, as follows:

$$\psi_i(C_k;\lambda) = \frac{1}{\lambda^{n/2}}\int\cdots\int d^n S_i\, e^{\frac{2\pi i}{\lambda} G_j(S_j;C_k)} \quad . \tag{7}$$

(n is the corank of the catastrophe, that is the number of state variables S_j.) Thus catastrophe theory classifies stable short-wave diffraction patterns as well as stable caustics. The 'diffraction catastrophes' (7) constitute a new hierarchy of standard functions; our present state of knowledge concerning them can be summarised under three headings.

(i) Computations. In 1838, Airy [23] computed the fold diffraction catastrophe $\psi_{fold}(C_1;\lambda)$ (which in modern notation [24] is proportional to the Airy function Ai). In 1946, Pearcey [25] computed the cusp diffraction catastrophe $\psi_{cusp}(C_1, C_2;\lambda)$. The close agreement between theory and experiment is obvious from fig. 5. Notice in particular the wavefront dislocation points (figs. 5b, c), which appear on these intensity pictures as elongated dark regions where $|\psi_{cusp}| = 0$, occurring in pairs within the arms of the caustic and singly in two rows flanking the caustic. The ability of diffraction catastrophes to give an accurate description of these delicate features is remarkable.

The only higher diffraction catastrophe so far studied in detail is the elliptic umbilic [26]. Its wave function $\psi_E(C_1, C_2, C_3;\lambda)$ describes an elaborate diffraction pattern in space, whose maxima are stacked in a regular array like atoms in a distorted crystal, and whose dislocation lines form either hexagonally puckered rings or 'hairpins'. We found that, just as for the cusp, experiment and theory agree down to the finest details.

(ii) Projection identities. Because the G_i are polynomials, the diffraction catastrophes ψ_i satisfy partial differential equations; these are very complicated. However, they also satisfy a series of astonishing nonlinear identities [28]. These relate the intensity $|\psi_i|^2$ to an integral over the wave function ψ_i corresponding to the same catastrophe (or a particular hypersection of it) or to a less singular catastrophe. Here are three examples, written for simplicity for the case $\lambda = 2\pi$:

$$\left.\begin{aligned}
|\psi_{fold}(C_1)|^2 &= \sqrt{\tfrac{2}{\pi}}\, 2^{-1/3}\int_{-\infty}^{\infty} du\, \psi_{fold}\{2^{2/3}(u^2 + C_1)\} \\
|\psi_{cusp}(C_1, C_2)|^2 &= \sqrt{\tfrac{2}{\pi}}\int_{-\infty}^{\infty} du\, |6u|^{-1/3}\psi_{fold}\{2\,\mathrm{sgn}\,u\,(6|u|)^{-1/3}(u^3 + C_2 u + C_1)\} \\
|\psi_E(C_1, C_2, C_3)|^2 &= \frac{2^{1/3}}{\pi}\int_{-\infty}^{\infty} du_1 \int_{-\infty}^{\infty} du_2\, \psi_E\{2^{2/3}(C_1 + 2C_3 u_1 + 3(u_2^2 - u_1^2)), \\
&\qquad 2^{2/3}(C_2 + 2C_3 u_2 + 6u_1 u_2), 0\}
\end{aligned}\right\} \tag{8}$$

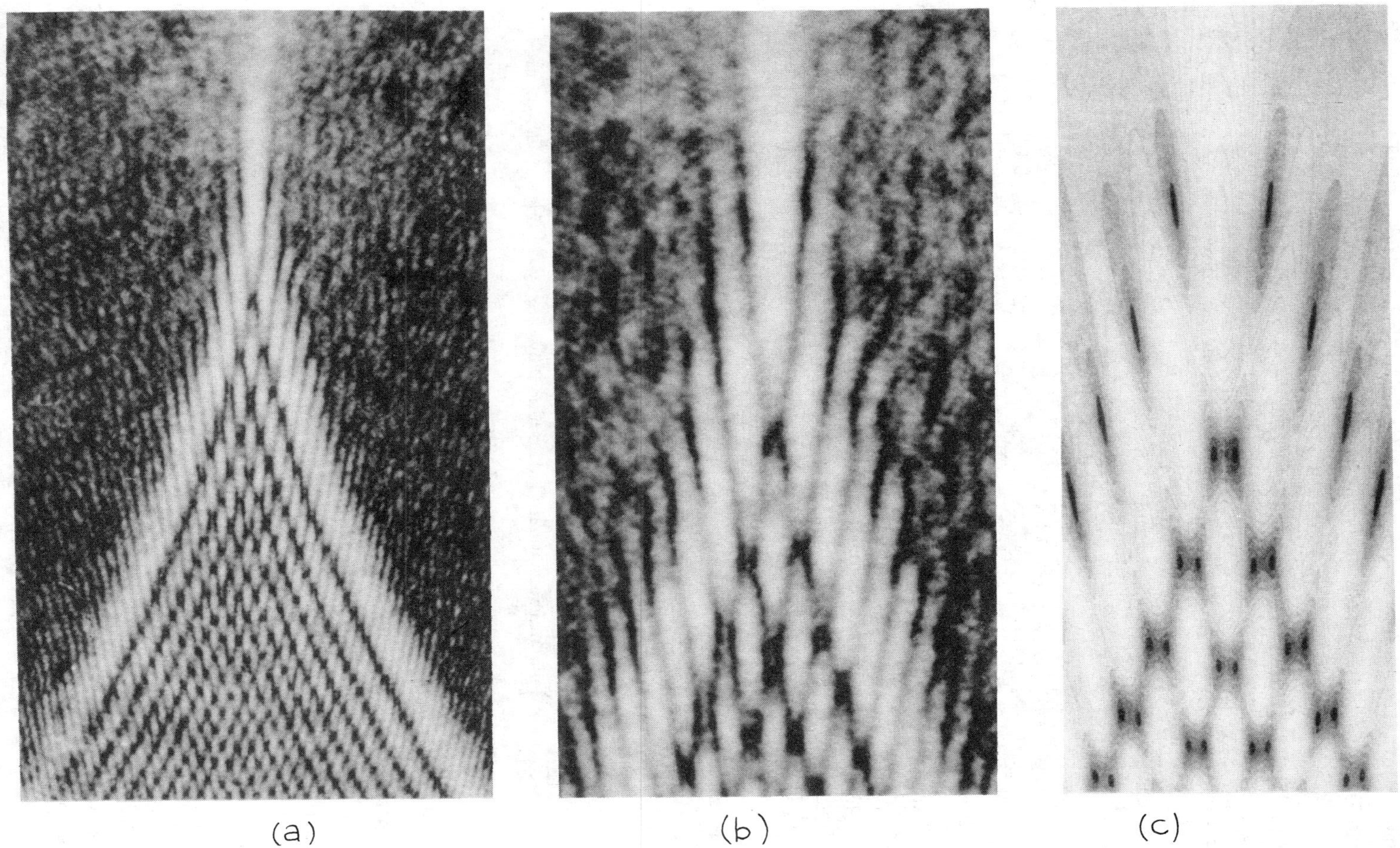

Fig. 5(a). Observation of diffraction near a cusp (courtesy of J.F. Nye); (b) enlargement of (a); (c) simulation of (b), obtained by shading a computed contour map of $\psi_{cusp}(C_1, C_2)$ (courtesy of F.J. Wright).

The first of these was already known [29, 30] as a relation between Airy functions.

These identities have a quantum-mechanical interpretation [28] in terms of the projections from phase space to control space of Wigner functions [30] associated with classical Lagrangian manifolds [21].

(iii) Wavelength scaling. As λ tends to zero, the wave intensity on a caustic becomes infinitely large and the scale of detail in interference fringes becomes infinitely small. These phenomena are governed by a scaling law giving the wave $\psi_i(C_k; \lambda)$ for wavelength λ in terms of the wave for some other wavelength λ_0. The law, derivable from (7), is

$$\psi_i\{C_k;\lambda\} = \left(\frac{\lambda_0}{\lambda}\right)^{\beta_i} \psi_i\left\{\left(\frac{\lambda_0}{\lambda}\right)^{\sigma_{ki}} C_k;\lambda_0\right\}, \tag{9}$$

where the exponents β_i and σ_{ki} are different for each catastrophe and depend on the precise form of the polynomial G_i. Therefore ψ_i need be computed only for a single wavelength. All scaling information about diffraction catastrophes is contained in the exponents, whose number is $K_i + 1$, where K_i is the codimension of the catastrophe i.

β_i governs the divergence of $|\psi_i|$ as $\lambda \to 0$, at the most singular point $C_k = 0$ in control space. It was introduced by Arnol'd [3] (who called it the 'singularity index') and studied in detail by Varchenko [31]. β_i increases with codimension K (and faster for umbilics than cuspoids). This is reasonable, since the larger K is, the greater is the number of rays coalescing to touch the caustic at its most singular point.

σ_{ki} govern the shrinking fringe spacings in different control directions C_k, as $\lambda \to 0$; these numbers were introduced by Berry [32]. The sum

$$\gamma_i \equiv \sum_{k=1}^{K} \sigma_{ki} \tag{10}$$

governs the behaviour of the K-dimensional hypervolume of the main diffraction maximum in control space; this scales as λ^{γ_i}. Romero [33] showed that γ_i is invariant under diffeomorphism. γ_i, which may be called the 'fringe index', also increases with K, showing that the intense region near the singularity, where all rays interfere constructively, shrinks very rapidly as $\lambda \to 0$ for the higher catastrophes. This shrinking is anisotropic in control space (as indicated by the fact that for a given catastrophe the σ_{ki} are not all equal), which explains why diffraction catastrophes formed by objects large in comparison with λ are observed [26] to be greatly elongated in certain directions.

Details of the computation of these exponents (and an account of difficulties encountered for higher catastrophes) are given by Berry [32] in connection with the statistics of twinkling starlight - a problem involving the whole hierarchy of catastrophes. Simplified treatments can be found in [14, 34].

4. DIFFRACTALS. Question: how is it possible to prevent the emergence of diffraction catastrophes as $\lambda \to 0$? Answer: by making waves encounter fractal objects. Reason: fractals [4] possess a hierarchy of structure which extends to infinitely small length scales. Therefore as $\lambda \to 0$ waves explore ever-finer details of the objects, which never appear smooth; this means that rays do not exist, so the geometrical-optics limit is never encountered. For this new regime in wave physics I employ the term 'diffractal' [5]. Examples of diffractals are radio waves reflected by landscapes or refracted by turbulence, light in fluids near the critical point, and modes of vibration of trees and sponges.

Mandelbrot [4] shows that the simplest characterisation of a fractal's geometry is its Hausdorff-Besicovitch dimension D, which need not be an integer. And for the three diffractal problems so far studied D indeed plays a central role in the (non-catastrophic) scaling laws governing the wavelength-dependence of a variety of wave properties. Here I shall give a very brief progress report on these problems, two of which involve scattering while the other involves confined modes.

No exact solution is yet available for any case where a wave strikes an impenetrable fractal object or propagates in a fractally inhomogeneous medium. But it seems a reasonable approximation to assume that such a structure will impose a fractal deformation on wavefronts encountering it. This suggests the following canonical mathematical problem. A plane wave with wavelength λ travels along the positive z axis. The wavefront $z = 0$ is deformed into a scale-free random fractal curve with dimension $D (1 \leq D < 2)$ and equation $z = h(x)$, and thereafter the wave $\psi(x,z)$ propagates freely towards $z = +\infty$, governed by the wave equation. The problem is to describe how the phase fluctuations at $z = 0$ develop as z increases into fluctuations of the intensity $|\psi|^2$. These intensity fluctuations are embodied in the 'second moment'

$$I_2 \equiv \frac{\langle |\psi(x,z)|^4 \rangle}{\langle |\psi(x,z)|^2 \rangle^2}, \tag{11}$$

where $\langle\ \rangle$ denotes averaging over the ensemble of $h(x)$.

A detailed discussion of this problem is given in [5] and summarised in [34]. It turns out that I_2 depends on D and the single parameter

$$\zeta \equiv \frac{z}{\lambda}\left(\frac{L}{\lambda}\right)^{\frac{D-1}{2-D}} . \tag{12}$$

The length L in this expression is the fractal curve's 'topothesy', i.e. the x separation between points on the curve whose connecting chord makes a root-mean-square slope of 1 radian with the x direction.

As ζ increases from 0 to ∞, I_2 varies from 1 (no intensity fluctuations) to 2 (Gaussian fluctuations). The form of this variation depends on D: for the 'rougher' fractals $1.5 \leqq D < 2$, I_2 increases monotonically with ζ and approaches 2 from below as ζ^{-1} ; for the 'less rough' fractals $1 < D < 1.5$, I_2 rises to a maximum of finite height at a finite value of ζ and approaches 2 from above as $\zeta^{-2(D-1)}$. This contrasts sharply with the behaviour of I_2 as $\lambda \to 0$ when the initial wavefront h(x) is smooth [32] ; then the maximum of I_2 is $O(\ln \lambda^{-1})$, giving rise to infinitely strong fluctuations in the geometrical-optics limit, which arise from the formation of caustics in the ray family (normals to the initial wavefront).

In the second scattering problem (unpublished), a quasi-monochromatic wave train is emitted isotropically from a source at distance z above a D-dimensional fractal landscape ($2 < D < 3$) with topothesy L; the echo $\psi(t,z)$ reflected from the landscape is received back at the source. For large z the echo has a long tail, arising from reflections by distant irregularities. The average intensity in this tail decays as

$$\langle|\psi(t,z)|^2\rangle \underset{\substack{t\to\infty \\ z\to\infty}}{\sim} \left(\frac{L}{\lambda}\right)^{2(D-2)} \frac{1}{z^{D-1}\, t^{4-D}} . \tag{13}$$

This algebraic dependence involving D contrasts strongly with reflection from a smooth surface, for which $\langle|\psi|^2\rangle \sim \exp(-O(t/z))$ when z and t are large.

The bound diffractal problem concerns the asymptotic mode distribution in fractal resonators [35]. Consider a D-dimensional region R with Hausdorff measure M_D, whose boundary ∂R is d-dimensional with Hausdorff measure m_d, where D and d need not be integers. Modes ψ_n are defined by

$$(\Delta + k_n^2)\psi_n = 0 \quad \text{in } R, \text{ and } \psi_n = 0 \text{ on } \partial R. \tag{14}$$

Let N(k) be the number of modes with eigenvalues $k_n < k$. Then I conjecture that

$$N(k) \underset{k\to\infty}{\sim} \frac{M_D k^D}{(D/2)!\,(4\pi)^{D/2}} - \frac{m_d k^d}{4(d/2)!\,(4\pi)^{d/2}} + \ldots \tag{15}$$

This formula generalises a well-known result [36] for the case where D is an integer and d = D - 1.

In [35] I list the far-reaching implications of the conjecture (15) and give a plausibility argument for its correctness. This is based on the scaling properties of fractals under magnification, together with the idea that modes with $k_n < k$ are insensitive to detail in R and ∂R on scales much smaller than k (the same argument suggests an interpretation of Δ in (14) when R is a fractal).

At this early stage of its development, diffractal theory poses a number of mathematical problems, which should be studied for their intrinsic interest and also because of the wide potential applicability of the subject. I conclude with a list:

(i) Study of the wavefront dislocations (section 1) in a scattered diffractal.

(ii) Search for exactly-soluble scattering diffractal problems. (Evolution of a wavefront in the form of a Weierstrass function [4] ? Scattering from a Cantor set of points on a line? Scattering from a triadic island in the form of Koch's snowflake curve [4] ?)

(iii) Rigorous examination of the conjecture (15).

(iv) Search for exactly-soluble fractal resonator problems. (Modes of a drum bounded by the Koch curve? Modes of a shivering Sierpinski sponge [4] ?)

5. CONCLUDING REMARKS. Space prevents more than reference to an important morphological problem in quantum mechanics that seems to cut across the categories of this paper. This is the nature of quantum states in the semiclassical limit (Planck's constant $\hbar \to 0$), for bound systems whose classical motion is nonintegrable (e.g. a nonsymmetric vibrating molecule); such classical motion exhibits stochasticity [38, 39] , i.e. a sensitive 'unpredictable' dependence on initial conditions. The problem was originally posed by Einstein [40] but neglected until recently [41] . Two promising approaches are to describe the quantum system using Wigner functions [30, 42] , thus retaining contact with the classical phase space, and to study a new class of model system constructed by quantising area-preserving maps of the plane [43] .

The morphologies described here are not restricted to waves satisfying a particular equation. Wavefront dislocations, diffraction catastrophes and diffractals can occur in any sort of wave, monochromatic or non-monochromatic, scalar or vector, in media that may be inhomogeneous, anisotropic or time-dependent (though for vector waves the map defining dislocations is more complicated). Nowadays much attention is concentrated on non-linear waves, but I hope I have shown by describing these three extreme regimes that linear wave theory still contains much to interest mathematicians.

BIBLIOGRAPHY

1. J.F. Nye and M.V. Berry, "Dislocations in wave trains", Proc. Roy. Soc. Lond., A336 (1974), 165-190.

2. R. Thom, Structural stability and morphogenesis, Benjamin, Reading, Mass., 1975. (First published in French 1972.)

3. V.I. Arnol'd, "Critical points of smooth functions and their normal forms", Russ. Math. Surv., 30: 5 (1975), 1-75.

4. B.B. Mandelbrot, Fractals: form, chance and dimension, W.H. Freeman, San Francisco, 1977. (First published in French 1975.)

5. M.V. Berry, "Diffractals", J. Phys. A, in press (1979).

6. F.J. Wright, "Wavefront dislocations and their analysis using catastrophe theory", to appear in Proceedings of 1978 symposium in Tübingen honouring Rene Thom, published by Springer 1979.

7. W. Whewell, "On the results of an extensive series of tide observations made on the coasts of Europe and America in June 1835", Phil. Trans. Roy. Soc. Lond., (1836), 289-307.

8. P.A.M. Dirac, "Quantised singularities in the electromagnetic field", Proc. Roy. Soc. Lond., A133 (1931), 60-72.

9. J. Riess, "Nodal structure of Scroedinger wave functions and its physical significance", Ann. Phys. New York, 57(1970), 301-321 (see also Phys. Rev. D, 2(1970) 647-633 and Phys. Rev. B, 13 (1976) 3862-3869.

10. J.O. Hirschfelder, A.C. Christoph and W.E. Palke, "Quantum mechanical streamlines. 1. Square-potential barrier". J. Chem. Phys., 61 (1974), 5435-5455 (see also 61 (1974) 5456-5459; 64 (1976) 760-785; 65 (1976), 470-486).

11. M.V. Berry, "Waves and Thom's theorem",Adv. in Phys., 25 (1976), 1-26.

12. T. Poston and I. Stewart, Catastrophe theory and its applications, Pitman, London 1978.

13. V. Guillemin and S. Sternberg, Geometric asymptotics, American mathematical society survey No. 14, Providence 1977.

14. M.V. Berry and C. Upstill, "Catastrophe optics: morphologies of caustics and their diffraction patterns", Progress in optics, in preparation.

15. J.F. Nye, "Optical caustics in the near field from liquid drops", Proc. Roy. Soc. Lond., A361 (1978), 21-41.

16. J.F. Nye, "Optical caustics from liquid drops under gravity: observations of the parabolic and symbolic umbilics", Phil. Trans. Roy. Soc. Lond., in press (1979).

17. M.V. Berry and J.F. Nye, "Fine structure in caustic junctions", Nature, 267 (1977), 34-36.

18. C. Upstill, "Light caustics from rippling water", Proc. Roy. Soc. Lond., A365 (1979), 95-104.

19. M.V. Berry and J.H. Hannay, "Umbilic points on Gaussian random surfaces", J. Phys. A., 11 (1977), 1809-1821.

20. J.J. Duistermaat, "Oscillatory integrals, Lagrange immersions and unfolding of singularities", Comm. Pure App. Math., 27 (1974), 207-281.

21. V.P. Maslov, Theorie des perturbations et methodes asymptotiques, Dunod, Paris, 1972 (originally published in Russian 1965).

22. D. Ludwig, "Uniform asymptotic approximations at a caustic", Comm. Pure App. Math., 19 (1966), 215-250.

23. G.B. Airy, "On the intensity of light in the neighbourhood of a caustic", Trans. Camb. Phil. Soc., 6 (1838), 379-403.

24. M. Abramowitz and I.A. Stegun (eds.), Handbook of mathematical functions, National Bureau of Standards, Washington, 1964.

25. T. Pearcey, "The structure of the electromagnetic field in the neighbourhood of a cusp of a caustic", Phil. Mag., 37 (1946), 311-317.

26. M.V. Berry, J.F. Nye and F.J. Wright, "The elliptic umbilic diffraction catastrophe", Phil. Trans. Roy. Soc. Lond., in press (1979).

27. M.V. Berry, "Les jeux de lumière dans l'eau". La Recherche, 92 (1978) 760-768.

28. M.V. Berry and F.J. Wright, "Phase-space projection identities for diffraction catastrophes', submitted to J. Phys. A (1979).

29. N.L. Balazs and G.G. Zipfel, "Quantum oscillations in the semi-classical fermion μ-space density" Ann. Phys. N.Y., 77 (1973), 139-156.

30. M.V. Berry, "Semi-classical mechanics in phase space", Phil. Trans. Roy. Soc. Lond., A287 (1977), 237-271.

31. A.N. Varchenko, "Newton polyhedra and estimation of oscillating integrals", Funct. Anal. Appl., 10 (1976), 13-38.

32. M.V. Berry, "Focusing and twinkling: critical exponents from catastrophes in non-Gaussian random short waves", J. Phys. A., 10 (1977) 2061-2081.

33. C. Romero, "On the Berry index", M.Sc. Thesis, University of Warwick, U.K., 1978.

34. M.V. Berry, "Catastrophe and fractal regimes in random waves", to appear in Proceedings of 1978 Tübingen symposium honouring Rene Thom, published by Springer 1979.

35. M.V. Berry, "Distribution of modes in fractal resonators", to appear in Proceedings of 1978 Tübingen symposium honouring Rene Thom, published by Springer 1979.

36. H.P. Baltes and E.R. Hilf, Spectra of finite systems, B.I. Wissenschaftsverlag, Mannheim, 1976.

37. M.E.R. Walford, P.C. Holdorf and R.G. Oakberg, "Phase-sensitive radio-echo sounding at the Devon Island ice cap, Canada", J. Glaciol., 18 (1977), 217-229.

38. V.I. Arnol'd and A. Avez, Ergodic problems of classical mechanics, Benjamin, New York, 1968.

39. M.V. Berry, "Regular and irregular motion", in Topics in nonlinear dynamics (Jorna, Ed.), Amer. Inst. Phys. Conf. Proc. No. 46, New York, 1978.

40. A. Einstein, "Zum Quantensatz von Sommerfeld und Epstein", Verh. Dt. Phys. Ges., 19 (1917), 82-92.

41. I.C. Percival, "Regular and irregular spectra", J. Phys. B., 6 (1973), L229-L332.

42. M.V. Berry, "Regular and irregular semiclassical wave functions", J. Phys. A., 10 (1977), 2083-2091.

43. M.V. Berry, N.L. Balazs, M. Tabor and A. Voros, "Quantum Maps", Ann. Phys. N.Y., in press 1979.

DEPARTMENT OF PHYSICS
UNIVERSITY OF BRISTOL
BRISTOL, U.K.

Current Address:
H.H. Wills Physics Laboratory,
Tyndall Avenue,
Bristol BS8 1TL,
U.K.

Proceedings of Symposia in Pure Mathematics
Volume 36, 1980

ON CHEEGER'S INEQUALITY $\lambda_1 \geqslant h^2/4$

Peter Buser

ABSTRACT. This is a report on Cheeger's lower bound for the first eigenvalue λ_1 of the Laplacian on a compact Riemannian manifold. After a short sketch of the proof of Cheeger's inequality, we show in the first chapter that on any compact differentiable manifold there exist Riemannian metrics for which the inequality becomes sharp. In the second chapter we show that at least in dimension two there is also an upper bound for λ_1 in terms of the isoperimetric constant. The third chapter contains applications. An earlier result on the number of eigenvalues in the interval $[0,1/4]$ for compact Riemann surfaces is extended to the noncompact case, as well as to compact hyperbolic manifolds af arbitrary dimensions. The treatment of the latter is based on the Margulis-lemma and on Gromov's volume estimate for manifolds of negative curvature, which for convenience will be exposed in chapter four. This last chapter has been worked out jointly with H. Karcher.

Since there is no further introduction, I use this place for my hearty thanks to R. Böhme und J. Elstrodt for their helpful discussions.

CHAPTER ONE. The inequality and examples.

In [10] J. Cheeger gave a lower bound for the first positive eigenvalue $\lambda_1 = \lambda_1(M)$ of the Laplace-Beltrami-Operator $\Delta = -\text{div grad}$ on a compact n-dimensional Riemannian manifold M, which is in terms of an isoperimetric constant $h = h(M)$ defined as follows:

$$h = h(M) = \inf_C J(C) \; ; \; J(C) = \frac{\mathrm{vol}_{n-1} C}{\min\{\mathrm{vol}_n M_1 , \mathrm{vol}_n M_2\}} , \tag{1.1}$$

where C runs through all compact codimension one submanifolds, which divide

---

1970 Mathematics Subject Classification 58 G 99, 53 C 20, 10 D 10.

Sonderforschungsbereich "Theoretische Mathematik" Bonn University, and partially supported by the Swiss National Science Foundation.

M into two disjoint connected open submanifolds M_1, M_2 with common boundary $C = \partial M_1 = \partial M_2$. Here for $1 \leq k \leq n$ the symbol $\mathrm{vol}_k(\ .\ )$ means the k-dimensional Riemann-Lebesgue-measure on M. If the dimension of the volume considered is clear from the context, we shall omit the index k. The one-dimensional measure however will for the most part be denoted by $\ell(\ .\ )$ instead of $\mathrm{vol}_1(\ .\ )$. We emphasize that in the above definition the submanifolds C need not be connected.

We also allow M to have piecewise differentiable boundary ∂M. Then the eigenvalue problem

$$\Delta u = \lambda u \quad \text{on} \quad M \tag{1.2}$$

is considered either subject to the Neumann condition

$$* du \mid \partial M = 0 \tag{1.3}$$

or to the Dirichlet condition

$$u \mid \partial M = 0\ . \tag{1.4}$$

For the former case the appropriate isoperimetric constant, which we then denote by $h_N(M)$, is defined by (1.1) also, where in addition C is allowed to have non-empty boundary, as long as the condition $\partial C = C \cap \partial M$ is satisfied. For the Dirichlet boundary condition (1.4) in contrast to [10], the isoperimetric constant, which we then denote by $h_D(M)$, has to be defined in a different way:

$$h_D = h_D(M) = \inf_A \frac{\mathrm{vol}_{n-1}(\partial A)}{\mathrm{vol}_n(A)}\ , \tag{1.5}$$

where A runs through all compact n-dimensional bordered submanifolds of M satisfying $A \cap \partial M = \emptyset$.

If we denote the first eigenvalue of the Laplacian by $\lambda_{1,N}(M)$ and $\lambda_{1,D}(M)$ according to whether we have Neumann- or Dirichlet condition (and by λ_1 if $\partial M = \emptyset$), then Cheeger's inequality states

1.6 THEOREM $\lambda_{1,*}(M) \geq \frac{1}{4} h_*^2(M)$, where either $* = N, D$ or $\emptyset$.

In special situations it may be useful to introduce variations of the isoperimetric constant for which one also has a Cheeger's inequality, (see e.g. Osserman [23]).

The proof of the theorem roughly runs as follows, for details we refer to [1] . Say M has non-empty boundary and consider a solution u of the Dirichlet eigenvalue problem with eigenvalue $\lambda_{1,D}$. It minimizes the Dirichlet-integral under the condition $\int_M u^2 = 1$, $u|\partial M = 0$. Using Cauchy's inequality one derives from this

(1.7) $$\lambda_{1,D} = \frac{\int_M \|\operatorname{grad} u\|^2}{\int_M u^2} \geq \frac{1}{4} H^2[u] \ ; \ H[u] := \frac{\int_M \|\operatorname{grad} u^2\|}{\int_M u^2} .$$

Define

$$L_t = \{p \in M | u^2(p) = t\} \ , \ A_t = \{p \in M | u^2(p) \geq t\} \ , \ 0 < t \leq m = \max_{x \in M} u^2(x)$$

$$h[u] = \inf_{0 < t \leq m} \frac{\operatorname{vol}_{n-1}(L_t)}{\operatorname{vol}_n(A_t)} ,$$

then for almost all t the set A_t satisfies the conditions for definition (1.5), and we compute, using the co-area formula

$$\int_M \|\operatorname{grad} u^2\| = \int_0^m d(\operatorname{vol}_{n-1}(L_t)) \geq h[u] \int_0^m \operatorname{vol}_n(A_t)dt = -h[u] \int_0^m t \cdot (\frac{d}{dt} \operatorname{vol}_n(A_t))dt$$

$$= h[u] \int_0^m t \operatorname{vol}_{n-1}(L_t)dt = h[u] \int_M u^2 .$$

Hence $H[u] \geq h[u]$. Now of course u^2 and hence all the level sets L_t are unknown. Therefore one is forced to consider everything which ever might be a level set. That is, one has to replace $h[u]$ by $h_D \leq h[u]$. Now the theorem follows from (1.7).

For the Neumann problem as well as for the case $\partial M = \emptyset$ (which may be regarded as a special case of the Neumann problem) we have $\int_M u = 0$, as seen from Green's formula. Hence the two subsets of M :

$$M^+ = \{p \in M | u(p) \geq 0\} \ , \ M^- = \{p \in M | u(p) \leq 0\}$$

are (using an appropriate approximation of u) without loss of generality submanifolds of M with piecewise differentiable boundary and with positive volume, say $\mathrm{vol}\, M^- \leq \mathrm{vol}\, M^+$. Again by Green's formula we have $\lambda_{1,N}(M) = \lambda_{1,*}(M^+) = \lambda_{1,*}(M^-)$ where $\lambda_{1,*}(M^{\pm})$ is the lowest eigenvalue for the mixed boundary condition $u = 0$ on $(\partial M^{\pm} \cap \mathrm{int}\, M)$ and $*du = 0$ on $(\partial M^{\pm} \cap \partial M)$. Now we reproduce the above proof on M^- and get the desired result with the exception that in the computation of $h_N(M)$ one also has to consider those submanifolds C of M, for which $M - C$ consists of more than only two open connected components. However, a straightforward argument as given in [33] or [4] shows that this amounts to the same, q.e.d.

It seems that until the papers of Cheng [12] and Yau [33] in 1975 and more recently in [23] , [24] , [25] , [16] no applications of Cheeger's inequality have been published. The reason may be twofold. First the handling with isoperimetric inequalities on compact manifolds is not always a pleasure and may look hopeless, and second from the proof of the inequality one would immediately conjecture that it can be improved. However there is a realm in Riemannian geometry where Cheeger's inequality plays an important rôle, namely hyperbolic geometry.

It is the following fact which governs all the applications and the examples below. Let H^n be hyperbolic n-space, i. e. the n-dimensional simply connected complete Riemannian manifold of constant sectional curvature $K = -1$. In polar coordinates $p = (\rho_p, \sigma_p)$ centered at some point $p_o \in H^n$, where $\rho_p = \mathrm{dist}(p, p_o)$ and $\sigma_p = \dot{\mu}_p(o) \in S^{n-1}$, μ_p denoting the geodesic of unit speed which connects p_o and p , the metric ds^2 of H^n is given by

$$(1.8) \qquad ds^2 = d\rho^2 + \sinh^2\rho \ d\sigma^2 ,$$

where $d\rho^2$ is the standard metric of the interval $(0,\infty)$ and $d\sigma^2$ is the standard metric of S^{n-1} . Now for any ball $B_\rho(p_o) = \{p \in H^n | \mathrm{dist}(p, p_o) \leq \rho\}$

of radius ρ we have from (1.8)

$$\frac{\mathrm{vol}_{n-1}(\partial B_\rho(p_o))}{\mathrm{vol}_n\ (B_\rho(p_o))} = \frac{\sinh^{n-1}\rho}{\int_0^\rho \sinh^{n-1}t\,dt} > n-1\ , \tag{1.9}$$

no matter how large ρ may be! The first consequence of this fact is [25] :

1.10 THEOREM. Cheeger's inequality for the Dirichlet condition is sharp in all dimensions.

<u>Proof.</u> By the classical isoperimetric inequality: $h_D(B_\rho(p_o)) > (n-1)$. On the other hand the first eigenfunction on $B_\rho(p_o)$ for the Dirichlet boundary condition is given in terms of an appropriate Legendre function and the eigenvalue $\lambda_{1,D}(B_\rho(p_o))$ is known to satisfy (see e.g. [13])

$$\lambda_{1,D}(B_\rho(p_o)) = \frac{(n-1)^2}{4} + O(\frac{1}{\rho^2})\ .$$

Hence $|\lambda_{1,D}(B_\rho(p_o)) - \frac{1}{4}h_D^2\ (B_\rho(p_o))| \to 0$ as $\rho \to \infty$, q.e.d.

We next give an example of a compact unbordered manifold of constant negative curvature with $\frac{1}{4}h^2$ as close to λ_1 as one whishes. We first state two lemmas which will be used again in chapter 3. Consider a simply connected compact domain $G \in H^2$ as shown in figure 1 ,

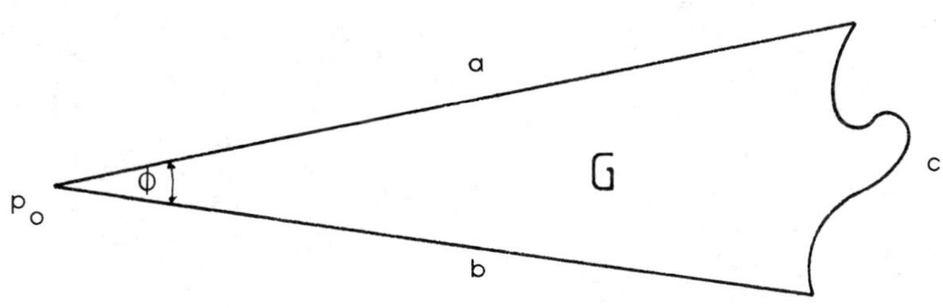

Fig. 1

whose boundary consists of two geodesic segment's a,b emanating from a point p_o under an angle $\phi \in (0,2\pi]$, and of a piecewise differentiable curve c , which connects the outer endpoints of a and b . The case that c is closed and ∂G consists of c only is also allowed.

1.11 LEMMA. Let G be as above, then

$$\frac{\ell(c)}{\mathrm{vol}\ G} > 1 .$$

Proof. We use polar coordinates $p \mapsto (\rho,\sigma)$, $\rho>0$, $\sigma\in S^1$ centered at p_0 . Passing from G to its convex hull if necessary, which increases the surface area of G and lowers $\ell(c)$, we may assume c to be parametrized as

$$t \mapsto c(t) = (\rho(t),t), \quad t\in[0,\phi].$$

From (1.8) $\ell(c) \geq \int_0^\phi \sinh\rho(t)\,dt > \int_0^\phi (\cosh\rho(t)-1)\,dt = \mathrm{vol}\ G$, q.e.d.

Next we consider a cylinder $Z = H^2/\Gamma$, where Γ is an infinite cyclic group without fixed points. The cylinder will be parametrized as

$$Z = \{p = (\rho,t) \mid \rho\in(-\infty,+\infty) , t\in \mathbb{R}/[t \mapsto t + \ell]\}$$

for some $\ell > 0$. The metric on Z is given by

$$(1.12) \qquad ds^2 = d\rho^2 + \cosh^2\rho\, dt^2 .$$

The curve $t \mapsto \gamma(t) = (0,t)$, $t\in[0,\ell]$ is the unique simple closed geodesic on Z and has length $\ell(\gamma) = \ell$. Now consider a doubly connected domain $A\subsetneq Z$ with one connected component of the boundary being γ , the other component of ∂A being a simple closed curve c, which is freely homotopic to γ , as drawn in figure 2.

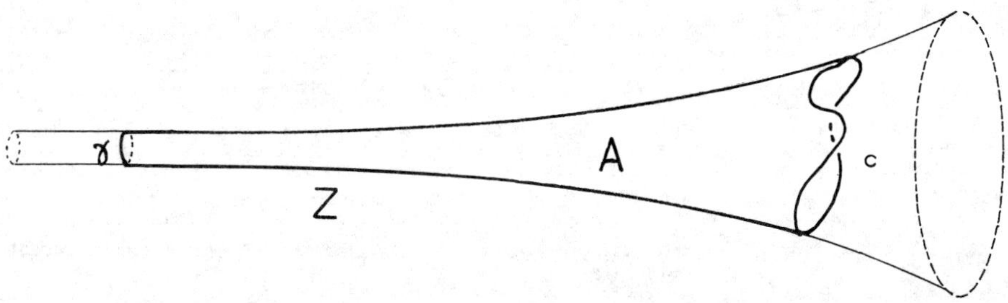

Fig. 2

1.13 LEMMA. Let A be as above, then

$$\frac{\ell(c)}{\mathrm{vol}\ A} > 1 .$$

Proof. As in the proof of Lemma 1.11 we may assume c is $t \mapsto c(t) = (\rho(t),t)$, $t \in [0,\ell]$, $\rho(t) > 0$. We then have from (1.12) :

$$\ell(c) \geq \int_0^\ell \cosh \rho(t)\, dt > \int_0^\ell \sinh \rho(t)\, dt = \mathrm{vol}\ A \ , \quad \text{q.e.d.}$$

The next example is from [6].

1.14 THEOREM. Cheeger's inequality is sharp for certain compact surfaces of constant negative curvature.

Proof. Take to samples of a rectangular hexagon D_δ in H^2 with three non-intersecting sides a,b,c of equal length L and three sides u,v,w of equal length δ between them, L large, δ small. To be precise: By hyperbolic trigonometry

$$\delta = 2 \operatorname{arcsinh} \frac{1}{2 \sinh L/2} . \tag{1.15}$$

Such hexagons D_δ exist for all $\delta > 0$. Glue the two samples together along the sides a,b,c, but not along u,v,w.

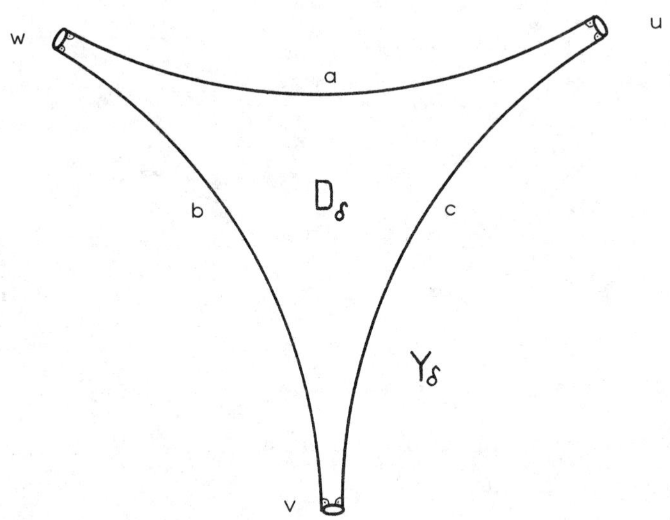

Fig. 3

What comes out is a bordered surface Y_δ with a Riemannian metric of constant curvature -1. The three boundary components $t \to \gamma_i(t)$, $\gamma_i(0) = \gamma_i(2\delta)$, $||\dot{\gamma}_i|| = 1$, $t \in [0,2\delta]$, $i = 1,2,3$ of ∂Y_δ are closed geodesics of length $\ell(\gamma_i) = 2\delta$. Thus Y_δ is a bordered Riemann surface of signature $(0,3)$. Finally in order to obtain a surface S_δ without boundary, in the same way as one passes from the southern hemisphere of S^2 to the projective plane, we use antipodal identification $\gamma_i(t) = \gamma_i(t+\delta)$, $t \in [0,2\delta]$ (mod 2δ), $i=1,2,3$. The surface S_δ thus obtained has Euler-characteristic $\chi(S_\delta) = -1$. Hence by Gauss-Bonnet

$$\text{vol}(S_\delta) = 2\pi \quad \text{for all} \quad \delta > 0 . \tag{1.16}$$

We estimate Cheeger's constant:

$$h(S_\delta) \geq 1 \quad \text{for all} \quad 0 < \delta \leq 2 \operatorname{arcsinh}\{1/(2 \sinh \pi)\} = 0.086.. \tag{1.17}$$

We prove this on Y_δ rather than on S_δ, which amounts to the same. Suppose c divides Y_δ into two open connected regions. If one of these regions looks as in figure 1 or 2, then $J(c) > 1$ by Lemma 1.11 and 1.13. In all other cases there exists a segment c' of c, which connects two points of ∂Y_δ and has length $\ell(c') \geq L \geq 2\pi$. Hence again $J(c) \geq 1$ by (1.15) and (1.16). This proves (1.17).

To finish the proof of theorem 1.14 we use <u>Cheng's inequality</u> [13]

$$\lambda_k(M) \leq \frac{(n-1)^2}{4} + \frac{a_n}{d^2} k^2 , \tag{1.18}$$

which holds for all compact n-dimensional Riemannian manifolds M with sectional curvature $K \geq -1$ (even for Riccivurvature $\geq -(n-1)$) and diameter $\text{diam}(M) \leq d$. Here the eigenvalues $0 < \lambda_1(M) \leq \lambda_2(M) \leq \dots$ of M are listed with their multiplicities. a_n is a constant depending only on the dimension n of M. Now as $\delta \to 0$, the diameter of S_δ grows to infinity, and by (1.17) and (1.18) we have

$$\frac{1}{4} h^2(S_\delta) \leq \lambda_k(S_\delta) \leq \frac{1}{4} h^2(S_\delta) + \frac{a_2 k^2}{\operatorname{diam}^2(S_\delta)} .$$

This proves the theorem. Moreover we have shown that Cheeger's inequality cannot be improved for higher eigenvalues either.

In higher dimension, the finding of constant curvature examples seems to be very difficult. However we have

1.19 THEOREM. Let M be an arbitrary compact unbordered differentiable manifold of dimension $n \geq 2$. Then for all integers $k \geq 1$ and for all $h > 0$, $\varepsilon > 0$ there exists a Riemannian metric on M such that $h(M) = h$ and

$$\lambda_k(M) \leq \frac{1}{4} h^2(M) + \varepsilon .$$

<u>Proof</u>. Since $\sqrt{\lambda_k}$ as well as h changes by a^{-1} if the metric is multiplied with a constant $a > 0$, it suffices to find a series of metrics such that $h \to (n-1)$ and $\lambda_k \to \frac{1}{4}(n-1)^2$. This will be first carried out on the bordered manifold

$$N = [0,b] \times S^{n-1} = \{(x,\sigma) \mid 0 \leq x \leq b , \sigma \in S^{n-1}\} , \; b \geq 3$$

with the Riemannian metric

$$ds^2 = dx^2 + \nu^2 f^2(x)\, d\sigma^2 ,$$

where we start with $f(x) = e^x$. The factor $\nu > 0$ will be chosen small in the sequel. The function u on N defined by

$$u(x) = \exp(-\tfrac{1}{2}(n-1)x) \cdot \sin(\tfrac{k\pi}{b} x + \xi) , \tag{1.20}$$

where ξ is such that $u'(0) = 0$, has zeroes α_1, $\alpha_2 = \alpha_1 + \frac{b}{k}, \ldots,$ $\alpha_k = \alpha_1 + \frac{k-1}{k} b\pi$. Furthermore $u'(b) = 0$. We let $\alpha_0 = 0$, $\alpha_{k+1} = 1$. The function u satisfies the differential equation

$$u''(x) + (n-1)u'(x) + \eta u(x) = 0 , \quad \eta = \frac{1}{4}(n-1)^2 + \frac{k^2\pi^2}{b^2} .$$

Thus the functions $u_i = u|N_i$ defined on the domains $N_i=[\alpha_i,\alpha_{i+1}]\times S^{n-1} \subsetneq N$, $i=0,\dots,k$ are eigenfunctions of the Laplacian for the boundary condition $*du_i \mid \partial N = 0$, $u_i \mid (\partial N_i - \partial N) = 0$. The minimax principle now says

$$\lambda_k(N) \le \frac{1}{4}(n-1)^2 + \frac{k^2\pi^2}{b^2} \tag{1.21}$$

independently of ν .

We show next that $h_N(N) \to (n-1)$ as $\nu \to 0$. For $n = 2$ the proof resembles the one of lemma (1.13), so let $n \ge 3$. For any subset $D \subsetneq N$ we define for $t \in [0,b]$:

$$D_t^- = \{(x,\sigma) \in N | x \le t\},\ D_t = \{(x,\sigma) \in N | x = t\},\ D_t^+ = \{(x,\sigma) \in N | x \ge t\}.$$

We let $c > 0$ be such that for all open subsets $U \subseteq S^{n-1}$ satisfying $m_{n-1}(U) \le (1 - \frac{1}{2}e^{-bn})m_{n-1}(S^{n-1})$ we have

$$m_{n-2}(\partial U)/m_{n-1}(U) \ge c \tag{1.22}$$

(here m_j is the j-dimensional Lebesque measure on the standard sphere S^{n-1}), and let $\nu \le \nu_0 := c/(ne^b)$. Consider a dissection $N - C = A \cup B$, with $A \cap B = \emptyset$, A,B open. <u>First case</u>: We assume $\text{vol}_{n-1}(A_t) \le (1 - \frac{1}{2}e^{-bn}) \cdot \text{vol}_{n-1}(N_t)$ for all $t \in [0,b]$. From (1.22) we conclude $\text{vol}_{n-2}(C_t) \ge n\,\text{vol}_{n-1}(A_t)$. Integration from 0 to b yields $J(C) \ge n$ for this case. <u>Second case</u>: We assume the condition of the first case is not fulfilled for A nor for B . Then eventually after exchanging the names of A and B we find $0 \le t_1 < t_2 \le b$ such that $\text{vol}_{n-1}(B_{t_1}) \le \frac{1}{2}e^{-bn}\,\text{vol}_{n-1}(N_{t_1})$, $\text{vol}_{n-1}(A_{t_2}) \le \frac{1}{2}e^{-bn}\,\text{vol}_{n-1}(N_{t_2})$ and $\text{vol}_{n-1}(A_t) \le (1 - \frac{1}{2}e^{-bn})\,\text{vol}_{n-1}(N_t)$ for all $t \in [t_2,b]$. We find $\text{vol}_{n-1}(C_{t_2}^+) \ge n\,\text{vol}_n A_{t_2}^+$ and $\text{vol}_{n-1}(C_{t_2}^-)\cdot\text{vol}_n(N_{t_2}^-) \ge (1 - e^{-bn})\,\text{vol}_{n-1}(N_{t_2})\cdot\text{vol}_n(A_{t_2}^-)$. Hence $J(C) \ge (n-1) + e^{-bn}$. Now we have proved together with (1.21)

$$\frac{1}{4}(n-1)^2 < \frac{1}{4}h_N^2(N) \le \lambda_k(N) < \frac{1}{4}(n-1)^2 + \frac{10k^2}{b^2} . \tag{1.23}$$

For the glueing below we alter the function f in a arbitrary small neighbourhood of 0 and b as indicated in figure 4

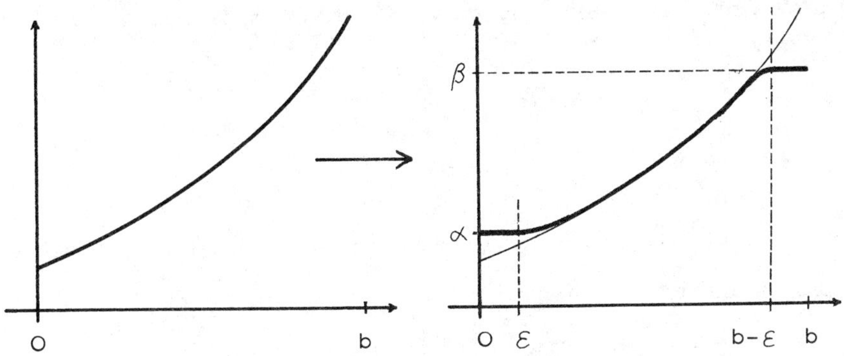

Fig. 4

such that $f|[0,\varepsilon] = \text{const} = \alpha$ and $f|[b-\varepsilon,b] = \text{const} = \beta$ for some small $\varepsilon > 0$ and such that (1.23) is still satisfied. Now take any compact bordered differentiable manifold P, whose boundary ∂P is diffeomorphic to S^{n-1}. Introduce a Riemannian metric on P such that a neighbourhood of ∂P becomes isometric to $S(\beta) \times [0,\varepsilon]$, $S(\beta)$ being the sphere of radius β in $\mathbb{R}^n$. Multiply this metric on P by ν. Glueing together N and P along their isometric boundary components we obtain a Riemannian manifold $M' = M'(\nu,\beta)$. The function u above may be continued as a testfunction on all M' by setting $u(p) = \text{const}$ for $p \in P$. So (1.21) is still valid for M', whereas $h_N(M')$ depends on $h_D(P)$. Since $h_D(P) = O(\frac{1}{\nu})$, $\text{vol}_n(N) = O(\nu^{n-1})$, $\text{vol}_n(P) = O(\nu^n)$ as $\nu \to 0$, we find $0<\nu_1 \leq \nu_0$ such that $\text{vol}_n(P) < e^{-2nb} \cdot \text{vol}_n(N)$, $h_D(P) > e^{2nb}$. Now define $D' = D \cap N$, $D'' = D \cap P$ for subsets $D \subsetneq M'$ and consider a dissection $M' - C = A \cup B$, $A \cap B = \emptyset$, A,B open, $\text{vol}_n A \leq \text{vol}_n B$. <u>First case:</u> $\text{vol}_n A' \geq 2e^{nb} \text{vol}_n A''$. Then $\text{vol}_n A' \leq (1 + 3e^{-2nb}) \text{vol}_n B'$ and $\text{vol}_{n-1}(C') \geq (n-1) \cdot (1 + e^{-2nb}) \text{vol}_n A$. <u>Second case:</u> $\text{vol}_n A' < 2e^{nb} \text{vol}_n A''$. Define $\tilde{C} = \partial P \cap \partial A''$, then $\text{vol}_{n-1}(C') \geq \frac{1}{4} \text{vol}_{n-1} \tilde{C}$ and therefore

$$\text{vol}_{n-1}(C) \geq \frac{1}{4} (\text{vol}_{n-1} \tilde{C} + \text{vol}_{n-1} C'') = \frac{1}{4} \text{vol}_{n-1}(\partial A'') > h_D(P) \cdot \text{vol}_n(A'') >$$
$$> e^{-n(b+2)} \cdot h_D(P) \cdot \text{vol}_n(A) > n \, \text{vol}_n(A)$$

. Thus we have shown that (1.23) is still valid, if N is replaced by M'.

Of course we can repeat the same procedure at the other end of N. Taking P at the right end of N and a spherical cap at the left end of N, we find a manifold $M = M(b)$, for which (1.23) is valid. This proves the theorem as $b \to \infty$.

CHAPTER TWO. An upper bound in terms of h.

The dealing with small eigenvalues sometimes turns into a severe problem. The following for instance, nicknamed as "does λ_1 go to zero?" is still unsolved (or at least seemed to be at the conference):

2.1 QUESTION. Does there exist a series $\{F_\nu\}_{\nu=1}^{\infty}$ of compact Riemann surfaces (curvature -1, no boundary) such that the genus of F_ν tends to infinity, whereas $\lambda_1(F_\nu)$ remains bounded below by a positive constant, possibly by $\frac{1}{4}$?

One looks therefore for necessary and sufficent conditions for λ_1 to go to zero. From the preceeding chapter one might conjecture that there is also an upper bound of the form $\lambda_1 \leq c \cdot h^2$, where c is a constant possibly depending on the dimension. However there are counterexamples.

For example take the torus $T = \mathbb{R}^2/[x \mapsto x + 2\pi,\ y \mapsto y + 2\pi]$ with the metric

$$ds^2 = f^2(x) \cdot (dx^2 + dy^2).$$

For $f \equiv 1$ we have a flat torus with $h(T) = 2/\pi$ and $\lambda_1(T) = 1$. Replacing f by the $\mathcal{L}^2$-function $\tilde{f}$:

$$\tilde{f}(x) = \begin{cases} 1, & x \neq \pi/2,\quad x \neq 3\pi/2 \\ \delta, & x = \pi/2,\quad x = 3\pi/2, \end{cases}$$

where $\delta > 0$ is arbitrarily small, we still get $\lambda_1 = 1$, since as regarded in $\mathcal{L}^2$, the Laplacian is still the same. But $h(T)$ changes drastically from $2/\pi$ to $\frac{2\delta}{\pi} \approx 0$. Of course the metric defined by $\tilde{f}$ is no longer differentiable, not even continuous, but with an appropriate smoothing of $\tilde{f}$ one can make this heuristic argument rigoros [7].

As one observes immediately, the curvature of this example becomes highly negative, and indeed at least in dimension two this is typical of all counter examples, i.e. we have the following

2.2 THEOREM. Let M be a compact unbordered two dimensional Riemannian manifold with curvature K bounded by $-1 \leq K < \infty$. Then

$$\lambda_1(M) \leq \min_{0<\tau<1} \frac{\tau}{1-\tau} \operatorname{arcsinh}^{-2}\left\{\frac{\tau}{2\pi\, h(M)}\right\} .$$

For sufficiently small values of $h(M)$ this upper bound is approximately $1.82 \cdot h(M)$. Using the fact that by Cheng's inequality (1.18) $\lambda_1(F) \leq 8$ for every compact Riemann surface of genus $g \geq 2$ (Huber has recently shown $\lambda_1(F) \leq \lambda_2(F) \leq 5.2$ for all F , see the article [18] in this book) and using the theorem of Bers [2], which states that any compact Riemann surface of genus $g \geq 2$ is dissectable into three holed spheres by $3g-3$ simple closed geodesics of length $\leq L_g$, where L_g is a number depending only on g, we get (see also the article [30] in this book):

2.3 THEOREM (Schoen, Wolpert, Yau [29]) For a compact Riemann surface F of genus $g \geq 2$ let $\ell(F)$ denote the length of the smallest separating chain of closed geodesics on F , then

$$a_g\, \ell^2(F) \leq \lambda_1(F) \leq b_g\, \ell(F) ,$$

where a_g and b_g are positive numbers depending only on g .

This theorem is used as a tricky argument in Wolpert's paper [32] on the length spectrum of Riemann surfaces.

Another application of the upper and lower bound of λ_1 in terms of h may be considered as a step towards the question 2.1 above. For a cubic graph G with $|G| = 2g-2$ vertices we define the partition number $h(G)$ as

$$h(G) = \min_{|H| \leq \frac{1}{2}|G|} \frac{(G,H)}{|H|} ,$$

where H runs through all subsets of the vertex set of G with at most $\frac{1}{2}|G|$ vertices, and where (G,H) is the number of edges of G with one endpoint in H the other endpoint in $G - H$. Among all cubic graphs with $2g-2$ vertices, there is one with the largest partition number, say $f(g) = \max_{|G|=2g-2} h(G)$. Now as carried out in [5], theorem 2.2 reduces the question above to a purely combinatorial problem:

2.4 THEOREM. "λ_1 goes to zero" if and only if $f(g)$ goes to zero as $g \to \infty$.

Proof of Theorem 2.2. We just give an outline, for details we refer to [7]. Consider a differentiable curve γ on M, $\gamma : [o,\ell] \to M$, $||\dot{\gamma}|| = 1$ of length $\ell(\gamma) = \ell$ and define

$$\gamma^r = \{p \in M | \text{ dist }(p,\gamma) < r\} ,$$

then

$$\text{(2.5)} \qquad \text{vol}(\gamma^r) \leq 2\pi(\cosh r - 1) + 2(\pi)\ell \sinh r .$$

This is obtained as follows (using Rauch's comparison theorem): The first term is an upper bound on the area of the circular disc of radius r centered at $\gamma(o)$ (that disc need not be homeomorphic to the unit circle $\{x \in \mathbb{R}^2 | \|x\| < 1\}$, it is just the set of points at a distance $< r$ from $\gamma(o)$). Now take a circle of radius r and let its center move along γ (i.e. consider the one parameter family $\{\eta_t\}$, $\eta_t = \{p \in M| \text{ dist}(p,\gamma(t)) = r\}$). As the center moves an infinitesimal dt, the circle sweeps out an annulus of area at most $(2\pi \sinh r)\, dt$. The circle runs over any point of γ^r except perhaps for the points in the disc first mentioned. Now (2.5) follows immediately.

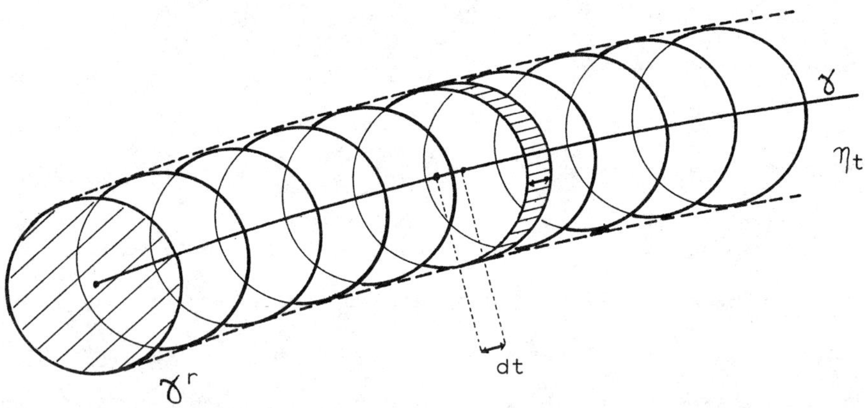

Fig. 5

A finer argument as illustrated in Fig. 5 (which up to now we can carry out only if the maximal rank radius of the exponential map $\exp|\gamma(t)$ is greater or equal r) shows that at least in the case $K \leq 0$ the factor π in the last term of (2.5) can be replaced by 1 . We have indicated this fact by putting π into brackets. The inequality then becomes sharp, equality holding for geodesic segments in H^2 for instance.

Now let C divide M into two connected open disjoint subsets A,B . The manifold C is the union $C = \gamma_1 \cup \ldots \cup \gamma_m$ of simple closed curves γ_i . Let r_o be the least upper bound of all r, for which $C^r := \gamma_1^r \cup \ldots \cup \gamma_m^r$ covers neither all of A nor all of B . Then one can show that for $r \leq r_o$ the circles η_t used above to prove (2.5) actually move over all points of γ_i^r . So the first term in (2.5) is superfluous in this situation i.e we have

$$\mathrm{vol}(C^r) \leq 2\pi\,\ell(C)\sinh r \quad \text{for } r \leq r_o$$

The testfunction f_A on A defined by

$$f_A(p) = \begin{cases} \mathrm{dist}(p,C), & \text{if } \mathrm{dist}(p,C) \leq r \\ r, & \text{if } \mathrm{dist}(p,C) \geq r \end{cases}, \quad p \in A$$

with $r = \operatorname{arcsinh} \frac{\tau}{2\pi\, J(C)}$, $0<\tau<1$ yields

$$\lambda_{1,D}(A) \le \operatorname{arcsinh}^{-2}\{\frac{\tau}{2\pi\, J(C)}\} ,$$

and the same holds for $\lambda_{1,D}(B)$. The theorem then follows from the minimax principle as $J(C) \to h(M)$.

CHAPTER THREE. Applications.

In this chapter we give new applications of Cheeger's inequality, concerning the low eigenvalues of hyperbolic manifolds. The chapter is divided into two parts. The first part deals with compact and also with non compact Riemann surfaces of finite volume. The second part deals with compact hyperbolic manifolds of higher dimension. It turns out that dimension three is the most difficult but also the most interesting case.

PART ONE. Riemann surfaces of finite volume.

The following theorem is a generalization of an earlier result [3] to Riemann surfaces which need not be compact:

3.1 THEOREM. Let F be a Riemann surface of finite type with signature $(g;0;m)$, where $2g+m \ge 3$. Then the number of eigenvalues of F in the interval $(0,1/4)$ is at most $4g + 2m - 3$.

Here as always in this chapter the eigenvalues of the Laplacian of F are counted with their multiplicities:

$$0 < \lambda_1(F) \le \lambda_2(F) \le \dots$$

In the language of Riemann surface theory, F is a Riemann surface of genus $g \ge 0$ with m cusps and without holes, endowed with the Poincaré metric. From the point of view of differential geometry, F is an m-times punctured surface of genus g carrying a complete metric of constant curvature $K = -1$

with finite volume. From Gauss-Bonnet:

$$\mathrm{vol}(F) = 2\pi(m+2g-2).$$

F has m points $\infty_1,\ldots,\infty_m$ at infinity, and each of these ∞_i ($\infty_i \notin F$ however) has a neighbourhood Z_i, which may be parametrized as

(3.2) $$Z_i = (\log \pi, \infty) \times S^1 = \{(x,\sigma) \mid x > \log \pi, \sigma \in S^1\}$$

with the metric tensor

(3.3) $$ds^2 = dx^2 + e^{-2x} d\sigma^2 ,$$

where dx^2 and $d\sigma^2$ are the line elements of $(0,\infty)$ and S^1 respectively. The fact just explained is nothing else than the collar theorem at infinity. The usual collar theorem states that for a closed geodesic γ of length $\ell(\gamma)$ on a Riemann surface ($K \equiv -1$, no boundary) the injectivity radius for the exponential map on the normal bundle of γ is at least $\operatorname{arcsinh}\{1/(\sinh \frac{1}{2}\ell(\gamma))\}$, [27], [9]. A straightforward argument leads to the collar theorem at infinity. Moreover it is easy to see that any two collars Z_i, Z_j at infinity are disjoint ($i \neq j$).

For the proof of theorem 3.1 we use the following theorem, which in the non compact case is no longer trivial:

3.4 THEOREM. There are only finitely many eigenvalues (counted with their multiplicities) of F in the interval $[0,\frac{1}{4})$.

For the proof of this theorem we refer to Roelcke [28] §§ 11, 12. It is possible to prove theorem 3.1 directly with the methods below, thereby giving a new proof of the fact that the continuous part of the spectrum is contained in $[\frac{1}{4},\infty)$. However this involves a lot of details in analysis and will be postponed to a future paper.

Assume now that there are more than $4g+2m-3$ eigenvalues of F in the interval $(0,\frac{1}{4})$. That is, there exist numbers $0<\eta_1<\eta_2<\ldots<\eta_p \leq \frac{1}{4}-\varepsilon$ for some $\varepsilon > 0$, such that η_i is an eigenvalue of multiplicity $\mu(\eta_i) \geq 1$ and such that

$$\sum_{i=1}^{p} \mu(\eta_i) \geq n := 4g+2m-2 .$$

Since the eigenvalues η_i have positive distance from the continuous part of the spectrum of F, to each η_i belongs a $\mu(\eta_i)$-dimensional subspace V_i of the Hilbert space $\mathcal{H}$ of those $\mathcal{L}^2$-functions f, which are C^∞ and for which Δf is again an $\mathcal{L}^2$-function. V_i is the space of eigenfunctions with eigenvalue η_i of the self adjoint extension $\tilde{\Delta}$ of Δ to $\mathcal{H}$. By the regularity theorem each function $u \in V_i$ is a C^∞-function and $||\text{grad } u||$ is in $\mathcal{L}^2(F)$ also. That is for any $u,v \in V := V_1 \ldots V_p$, now

$$(3.5) \qquad \int_F v\cdot\Delta u = \int_F \langle \text{grad } u, \text{grad } v\rangle = \int_F u\cdot\Delta v .$$

Choosing an orthonormal basis in each V_i, we deduce from (3.5) that the assumption above reads as follows: There exists a series $v_1,\ldots,v_n$ $(n=4g+2m-2)$ of functions satisfying $v_i \in C^\infty(F)\cap \mathcal{L}^2(F)$, $||\text{grad } v_i|| \in \mathcal{L}^2(F)$ such that

$$(3.6) \qquad \int_F v_i\cdot v_j = \delta_{ij} , \quad \int_F \langle \text{grad } v_i , \text{grad } v_j\rangle = \lambda_i(F)\cdot\delta_{ij} ,$$

where δ_{ij} is the Kronecker symbol and $\lambda_i(F)$ are eigenvalues of F with $0<\lambda_1(F)\leq\ldots\leq\lambda_n(F)\leq \frac{1}{4}-\varepsilon$. This has to be led to a contradiction.

In order to pass to grounds, where Cheeger's inequality lives, we cut off neighbourhoods

$$U_i = \{(x,\sigma) \in Z_i | x > T\} , \quad T >> \log \pi , \quad i=1,\ldots,m,$$

(see (3.2)) such that on the surface $F' = F - (U_1 \cup \ldots \cup U_m)$, the v_i still almost are eigenfunctions. To be precise: Let $\delta = \varepsilon/16n^2$. Then we can choose $T > \log \pi$ so large that the functions v_i, restricted to F' satisfy

$$(3.7)\qquad |\int_{F'} v_i v_j| \leq \delta\ , \qquad \int_{F'} v_i^2 \geq 1-\delta, \qquad (i \neq j)$$
$$|\int_{F'} \langle \operatorname{grad} v_i\ ,\ \operatorname{grad} v_j \rangle| \leq \delta\ ,\ \int_{F'} ||\operatorname{grad} v_i||^2 \leq \lambda_i\ .$$

Let $\phi_o,\ldots,\phi_{n-1}$ with $\phi_o \equiv 1$ be the first n eigenfunctions of F' with respect to the Neumann condition. There exists a linear combination

$$\Psi = \sum_{i=o}^{n} \alpha_i v_i\ ,\quad \sum_{i=o}^{n} \alpha_i^2 = 1$$

(v_o = const) such that

$$\int_{F'} \Psi\phi_\nu = o\ ,\quad \nu = o,\ldots,n-1\ .$$

Since $\int_F ||\operatorname{grad} v_j||^2 = \lambda_j$, from (3.5), (3.6) and (3.7)

$$\frac{\int_{F'} ||\operatorname{grad} \Psi||^2}{\int_{F'} \Psi^2} \leq \frac{\sum_{i=o}^{n} \alpha_i^2\lambda_i + n(n+1)\delta}{(1-\delta)\sum_{i=o}^{n} \alpha_i^2 - n(n+\delta)} \leq \lambda_n(F) + \frac{1}{2}\varepsilon < \frac{1}{4}\ .$$

Hence by the minimax principle

$$(3.8)\qquad \lambda_{n,N}(F') < \frac{1}{4}\ .$$

On the other hand, take a point $p \in F'$ and draw a system $\mathcal{E}$ of curves $a_1,b_1,a_1^{-1},b_1^{-1},a_2,b_2,\ldots,a_{2g},b_{2g},a_{2g}^{-1},b_{2g}^{-1},c_1,\ldots,c_m$ with endpoints at p, as indicated in figure 6 such that the connected components of $M - \mathcal{E}$ are m doubly connected regions $W_1,\ldots,W_m$ and one simply connected region D with boundary cycle $a_1b_1a_1^{-1}b_1^{-1}\cdot\ldots\cdot a_{2g}^{-1}b_{2g}^{-1}\cdot c_1\cdot\cdots\cdot c_m$, as indicated in figure 6. This is called a canonical dissection of F'. Since we have negative curvature on F', in each homotopy class of the curves a_i,b_i,c_j, $i=1,\ldots,2g$, $j=1,\ldots,m$, there is a unique goedesic loop $\bar{a}_i,\bar{b}_i,\bar{c}_j$. By a theorem of Epstein [14] there exists a topological deformation of F' into itself, leaving p and each point at the boundary $\partial F'$ fixed, transforming a_i into $\bar{a}_i$, b_i into $\bar{b}_i$ and c_j into $\bar{c}_j$. We may therefore assume that we have already $a_i = \bar{a}_i$, $b_i = \bar{b}_i$, $c_j = \bar{c}_j$. Thus D is a geodesic polygon in H^2, and each W_j is

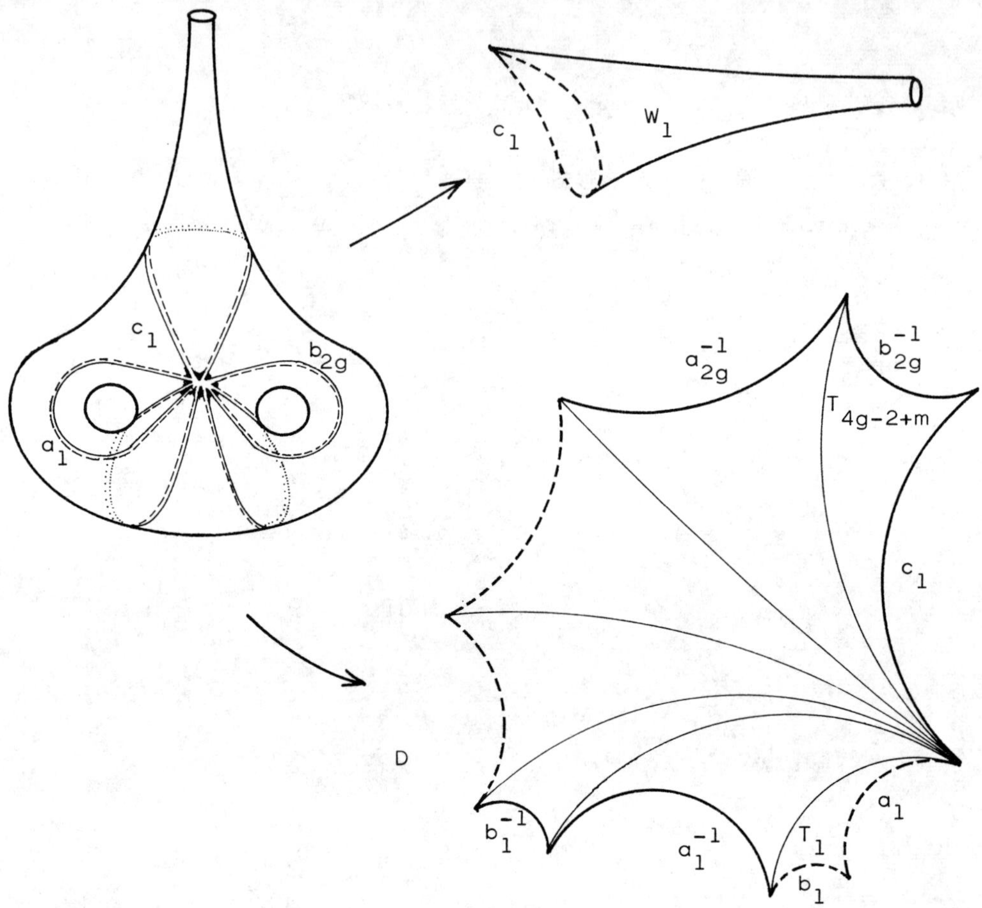

Fig. 6

a long "pipe", one boundary curve of W_j comes from a horocycle in H^2, the other boundary curve is a geodesic loop. It follows that W_j is locally convex. Finally we dissect the polygon D into 4g-2+m geodesic triangles as shown in figure 6. From lemma 1.11 and 1.13 in the same way as in the description of the example in chapter 1, we deduce that

$$(3.9) \qquad h_N(T_i) > 1, \quad h_N(W_j) > 1; \quad i = 1, \dots, 4g-2+m, \quad j = 1, \dots, m.$$

For the triangles this is so, because for any dissection at least one of the two components comes out as in figure 1. As for the W_j, consider a dissection $W_j - C = A \cup B$, $A \cap B = \emptyset$, and suppose neither A nor B looks like in figure 1. If C then is a closed curve homotopic to a boundary component, then one component, say A, looks almost like in figure 2 and one finds $J(C) > 1$ from (3.3) in the same way as in the proof of lemma 1.13. There remain two cases. First C has both endpoints on the horocycle-like boundary component, then since by (3.3) the pipe is very thin there, now $J(C) >> 1$. Second, if one endpoint of a connected component of C lies on the horocycle and the other endpoint on the geodesic loop, now $\ell(C)$ is large, while $\mathrm{vol}(W_j) < \pi/2$ by Gauss-Bonnet; hence again $J(C) >> 1$. This proves (3.9).

From (3.9), from Cheeger's inequality and from the minimax principle, we have $\lambda_{4g+2m-2,N}(F') > \frac{1}{4}$. This contradicts (3.8) as desired, q.e.d.

PART TWO. Compact hyperbolic n-manifolds.

For compact Riemann surfaces we have a striking phenomenon: According to theorem 3.1 we have $\lambda_{4g-2}(F) > \frac{1}{4}$ for all compact Riemann surfaces of genus $g \geq 2$.

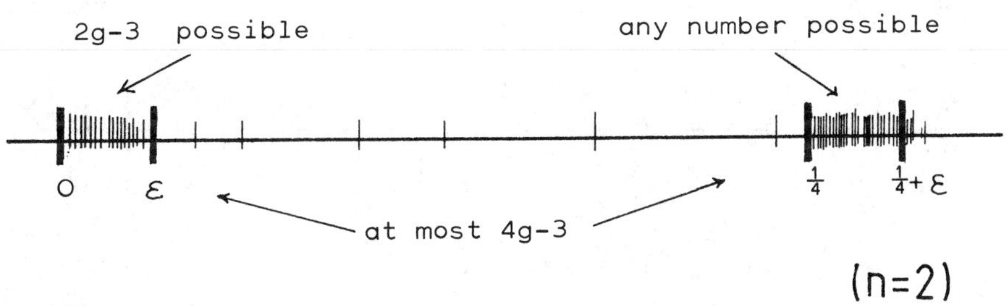

Fig. 7

On the other hand it is easy to give examples of compact Riemann surfaces of arbitrary large diameter for <u>any</u> genus $g \geq 2$. For example one may glue together $2g-2$ copies of the three holed sphere Y_δ, already used in chapter one. One gets a surface of genus g with diameter $\geq 2\log(1/\delta)$ if $\delta < 1$ (see (1.15)). Now by Cheng's inequality (1.18) as $\delta \to 0$ the number of eigenvalues in any interval $[\frac{1}{4}, \frac{1}{4} + \varepsilon]$, $\varepsilon > 0$, grows to infinity. Moreover since $\lambda_{1,D}(Y_\delta) \to 0$ as $\delta \to 0$, it follows that for any interval $[0,\varepsilon]$,

$\varepsilon > 0$, there exist compact Riemann surfaces with at least $2g-3$ eigenvalues in there. Schoen has recently shown that the latter result is sharp, i.e. for all $g \geq 2$ there exists a number c_g such that $\lambda_{2g-2}(F) \geq c_g$ for all compact Riemann surfaces of genus g. (See the article [30] in this book).

In this paragraph we investigate to what extent this phenomenon, as illustrated in figure 7, holds for compact hyperbolic (that is with constant curvature - 1) manifolds of higher dimensions. The critical number for dimension two is $\frac{1}{4}$. For arbitrary dimension n it is the number occurring in Cheng's inequality (1.18), as well as in the first example of chapter one, namely

$$\frac{(n-1)^2}{4} .$$

In higher dimensions the genus is replaced by the volume. It turns out that the phenomenon described above does not extend further than dimension three. There the diagram is

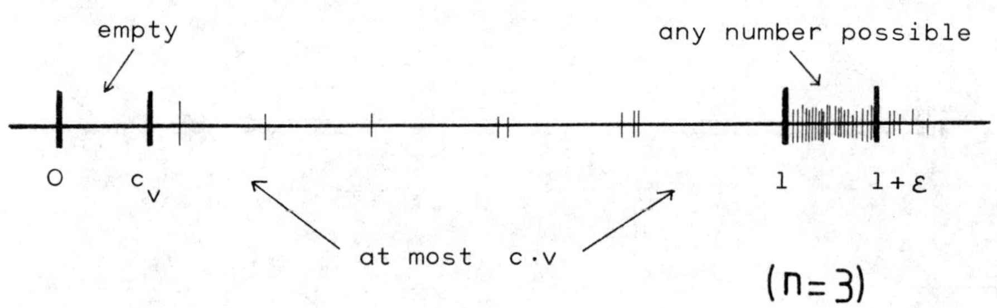

Fig. 8

For the sake of clearness we list the theorems together and give proofs afterwards. The results are qualitative rather than quantitative, since the true values of the constants are unknown. (As comes out in the proof, none smaller than e^{-64n^2}, but that seems to be far away from reality.)

The first result is in some sense easy and is listed for completeness:

3.10 THEOREM. For all $n \geq 3$ there exists a series $\{M_\nu\}_{\nu=1}^{\infty}$ of compact hyperbolic n-manifolds with $\mathrm{vol}(M_\nu) \to \infty$ such that for all $k = 1,2,...$

$$\lambda_k (M_\nu) \leq a \frac{k^2}{\mathrm{vol}^2(M_\nu)} \qquad (\nu \geq 4(k+1)) ,$$

where a is a constant depending on the series.

The next three theorems show the special situation in dimension three. The first result is in constrast to dimension two. For the proof see [30].

3.11 THEOREM (Schoen). For each $v > 0$ there exists $c_v > 0$ such that $\lambda_1(M) \geq c_v$ for all compact hyperbolic 3-manifolds M with $\mathrm{vol}(M) \leq v$.

3.12 THEOREM. There exists a constant $c > 0$ such that the number of eigenvalues of M in the interval $[0,1]$ is less than $c \cdot \mathrm{vol}(M)$ for all compact hyperbolic 3-manifolds M .

3.13 THEOREM. There exists a series $\{M_\nu\}_{\nu=1}^{\infty}$ of compact hyperbolic 3-manifolds with uniformly bounded volume such that for all $\varepsilon > 0$ the number of eigenvalues of M in the interval $[1,\varepsilon]$ tends to infinity as $\nu \to \infty$.

In dimension $n \geq 4$ the situation is again different. Schoen's result holds also: Combing Gromov's inequality [15]: $\mathrm{vol}(M) \geq \alpha_n(1+\mathrm{diam}(M))$ and Huber's inequality [17]: $\lambda_1(M) \geq \beta_n \exp(-\gamma_n \mathrm{diam}(M))$ (see also the general result of Peter Li [19]), one gets $\lambda_1(M) \geq \beta_n \exp(-\delta_n \mathrm{vol}(M))$. We prove here:

3.14 THEOREM. Let M be a compact hyperbolic n-manifold $n \geq 4$, then

$$\lambda_1(M) \geq a_n \ (\mathrm{vol}\ M)^{-2(n+8)} ,$$

where a_n is a constant depending only on the dimension.

Theorem 3.13 has no higher dimensional analogue:

3.15 THEOREM. There exist constants b_n, c_n depending only on $n \geq 4$ such that the number of eigenvalues of M in the interval $[0,x]$ is less than $b_n \mathrm{vol}(M)\cdot x^{n/2}$ for all compact hyperbolic n-manifolds M and for all $x \geq c_n(\mathrm{vol}\ M)^6$.

Proof of theorem 3.10. The examples are due to Millson [22]. By means of group theory for all $n \geq 3$ he gives examples of compact hyperbolic n-manifolds N with boundary ∂N consisting of two compact isometric totally geodesic (n-1)-dimensional submanifolds A, B. (For a particularly simple example of that kind in dimension three, confer also Löbell [20] and [21]). Take ν identical copies $N_1,\dots,N_\nu$ of N with boundary components A_ν, B_ν and make a necklace out of them by glueing together N_i and N_{i+1} along A_i, B_{i+1} , $i=1,\dots,\nu(\mathrm{mod}\,\nu)$. This gives the unborderd hyperbolic n-manifold M_ν with $\mathrm{vol}(M_\nu) = \nu\cdot\mathrm{vol}(N)$. Let ϕ be a testfunction on N with $\phi|A = 0$, $\phi|B = 1$ and define $D = \int_N ||\mathrm{grad}\ \phi||^2$. Consider ϕ as function on the $N_i, i=1,\dots,\nu$. For $\nu \geq 4(k+1)$ let $m = [\frac{\nu}{k+1}]$, $s = [\frac{m}{4}]$, where $[\ .\ ]$ are the Gauss brackets, e.g. $[7/2] = 3$ etc. and define $P_1 = N_1 \cup \dots \cup N_m$, $P_2 = N_{m+1} \cup \dots \cup N_{2m}, \dots, P_{k+1} = N_{km+1} \cup \dots \cup N_\nu$. The testfunction Ψ on P_1 defined by $\Psi(p) = (i-\phi(p))/s$, if $p \in N_i$ and $i=1,\dots,s$; $\Psi(p) = 1$ if $p \in N_j$, $j=s+1,\dots,m-s$; $\Psi(p) = 1 + (\phi - i)/s$ if $p \in N_{m-s+i}, i=1,\dots,s$, yields $\lambda_{1,D}(P_1) \leq \frac{200Dk^2}{\nu^2 \mathrm{vol}\, N}$, via the Rayleigh principle. The same holds for $P_2,\dots,P_{k+1}$. Hence the theorem by the minimax principle with $a = 200\ D\ \mathrm{vol}\ (N)$.

Proof of theorem 3.13. According to chapters IV and VI of Thurston's lecture notes [31], the following is true: For any number $a > 0$ there exists a number $b > a$ and an infinite sequence $\{M_\nu\}_{\nu=1}^{\infty}$ of mutually topologically distinct closed (= compact and without boundary) hyperbolic 3-manifolds M_ν with volume $\mathrm{vol}(M_\nu)$ satisfying

$$a \leq \mathrm{vol}(M_\nu) \leq b , \quad \nu = 1,2,\dots .$$

(Rather explicit examples are obtained by performing hyperbolic Dehn surgery on the complement of the figure eight knot, see § 4.6.) Take such a sequence.

We claim:

$$\operatorname{diam}(M_\nu) \to \infty \quad \text{as} \quad \nu \to \infty .$$

Proof. Let r_ν be the injectivity radius of M_ν . Suppose there is $\rho > 0$ and a subsequence $\{r_\mu\}$ of $\{r_\nu\}_{\nu=1}^{\infty}$ such that $r_\mu \geq \rho$ for all μ . Then Cheeger's finiteness theorem ([11] p. 133) says that among the manifolds M_μ there are only finitely many different diffeomorphism classes. (Roughly speaking this is so, because each M_μ admits an open covering with not more than $16b\rho^{-3}$ balls of radius $\rho/2$, and because there are only finitely many combinatorially different ways of arranging such balls). Thus each such subsequence $\{r_\mu\}$ is finite, i.e.

$$r_\nu \to 0 \quad \text{as} \quad \nu \to \infty .$$

A standard argument shows that under the condition $K = -1$ it follows from $\operatorname{vol}(M_\nu) > a$ and $r_\nu \to 0$ that

$$\operatorname{diam}(M_\nu) \to \infty \quad \text{as} \quad \nu \to \infty .$$

The theorem is now a consequence of Cheng's inequality, q.e.d.

Proof of theorem 3.12. It suffices to prove the theorem for orientable manifolds since for non orientable M the twofold orientable covering $\hat{M}$ of M has at least the same number of eigenvalues in $[0,1]$ and $\operatorname{vol} \hat{M} = 2 \operatorname{vol} M$. Thus M is orientable from now on.

1. Step. Let $\gamma_1, \ldots, \gamma_s$ be all simple closed geodesics on M of length

$$\ell(\gamma_i) \leq 2\varepsilon_0 := 4^{-7} , \; i=1,\ldots,s$$

(if there are any). The number $\varepsilon_0 = r_3/16$ comes from the Margulis lemma (4.1).

In (4.12) for each γ_i a number $d_{\gamma_i} > 0$ is defined with the following properties.

(3.16) The exponential map, restricted to vectors of length $< d_{\gamma_i}$ in the normal bundle of γ_i, is injective. In particular the set $T_{\gamma_i} = \{p \in M \mid \mathrm{dist}(p,\gamma_i) < d_{\gamma_i}\}$ is homeomorphic to $S^1 \times B^2$ (B^2 = open unit ball in $\mathbb{R}^2$). T_{γ_i} is called the small tube around γ_i.

(3.17) There exists a point p_o on the boundary ∂T_{γ_i}, where the injectivity radius of M equals $16\varepsilon_o$.

Let now γ be one of the γ_i. The tube T_γ is best visualized in the universal covering manifold H^3. With respect to a fixed covering

$$\Sigma : H^3 \to M$$

consider a geodesic $\tilde{\gamma} : \mathbb{R} \to H^3$, $\Sigma \circ \tilde{\gamma} = \gamma$. <u>Fermicoordinates</u> for points $q \in H^3$ with $\tilde{\gamma}$ as t-axis are defined as follows: There exists a unique perpendicular geodesic segment μ from γ to q, say $t \mapsto \mu(t)$, $t \in [0,\rho_q]$, $\|\dot{\mu}\| = 1$, $\mu(0) = \tilde{\gamma}(t_q)$ for some uniquely defined t_q, $\langle \dot{\mu}(0), \dot{\tilde{\gamma}}(t_q)\rangle = 0$, $\mu(\rho_q) = q$, $\rho_q = \mathrm{dist}(q,\tilde{\gamma})$. Identify the set of unit vectors at $\tilde{\gamma}(0)$ which are orthogonal to $\dot{\tilde{\gamma}}(0)$ with S^1, and let finally $\sigma_q \in S^1$ be the parallel translate of $\dot{\mu}(0)$ along $\tilde{\gamma}$ from $\tilde{\gamma}(t_q)$ to $\tilde{\gamma}(0)$. The assignment $q \mapsto (t_q,\rho_q,\sigma_q) = (t,\rho,\sigma)$ is the Fermi coordinate system. The hyperbolic line element ds^2 in these coordinates is

$$ds^2 = \cosh^2_\rho dt^2 + d\rho^2 + \sinh^2_\rho d\sigma^2 . \tag{3.18}$$

This formula looks the same in higher dimension as shall be used later. The decktransformation associated to γ (with respect to the covering $\Sigma : H^3 \to M$) is the mapping $L : H^3 \to H^3$ defined by

$$q = (t,\rho,\sigma) \mapsto L(q) = (t+\ell(\gamma),\rho,A\sigma) , \; A \in SO(2) .$$

Here we have used the fact that M is orientable, so A is a proper rotation.

Let q_o with $\mathrm{dist}(q_o,\tilde{\gamma}) = d_\gamma$ be a point in H^3 with $\Sigma(q_o) = p_o$. Without loss of generality: $q_o = (0,d_\gamma,\sigma_o)$. Since the injectivity radius of

M at p_o is $16\varepsilon_o$ (see (3.17)), now $\operatorname{dist}(q_o, L^k(q_o)) \geq 32\varepsilon_o$ for all $k \neq 0$. Hence a trivial calculation using (3.18) shows (recall $\ell(\gamma) \leq \varepsilon_o$)

$$d_\gamma > 32\varepsilon_o . \tag{3.19}$$

The set

$$D = \{(t,\rho,\sigma) \in H^3 \mid 0 \leq t < \ell(\gamma) \ , \ 0 \leq \rho < d_\gamma \ , \ \sigma \in S^1\}$$

is mapped one to one onto T_γ by Σ. Therefore by abuse of notation we use t,ρ,σ with $t \in [0,\ell(\gamma))$, $\rho \in [0,d_\gamma)$, $\sigma \in S^1$ as coordinates for points in T_γ in a natural way.

Since the mappings $(t,\rho,\sigma) \mapsto (t+a,\rho,\sigma)$, $a \geq 0$, and $(t,\rho,\sigma) \mapsto (t,\rho,V\sigma)$, $V \in SO(2)$ are isometries of H^3, we conclude that for some $k \neq 0$

$$\operatorname{dist}(q,L^k(q)) = 32\varepsilon_o \ , \quad \operatorname{dist}(q,L^m(q)) \geq 32\varepsilon_o \quad \text{for all other } m \neq 0, \tag{3.20}$$

is valid not only for the point q_o, but for <u>all</u> points $q \in H^3$ at a distance d_γ from $\tilde{\gamma}$. It follows from the Margulis lemma that the injectivity radius $r_p(M)$ of M at p satisfies $r_p(M) = 16\varepsilon_o$ for <u>all</u> $p \in \partial T_\gamma$. It is here, where we make use of the facts that M is orientable and has constant curvature.

$$r_p(M) > \varepsilon_o \qquad \text{for all} \qquad p \in M - (T_{\gamma_1} \cup \cdots \cup T_{\gamma_s}) . \tag{3.21}$$

<u>Proof</u>. If $r_p(M) \leq \varepsilon_o$, then we find a geodesic loop at p of length $\leq 2\varepsilon_o$. In the free homotopy class of that loop we find a closed geodesic, which is a power γ_i^m for some $1 \leq i \leq s$, $m \neq 0$. Draw the perpendicular μ from γ_i to p, $\mu : t \mapsto \mu(t)$, $t \in [0, \operatorname{dist}(p,\gamma_i)]$ and define $\gamma_{i,t}^m$ to be the unique geodesic loop at the point $\mu(t)$, which is homotopic at $\mu(t)$ to the curve $(\mu|[0,t]) \circ \gamma_i^m \circ (\mu|[0,t])^{-1}$. The function $t \mapsto \ell(\gamma_{i,t}^m)$ is monotonely increasing with value $\leq 2\varepsilon_o$ for $t = \operatorname{dist}(p,\gamma_i)$. But for $t = d_\gamma$ we have $\ell(\gamma_{i,t}^m) \geq 32\varepsilon_o$ from (3.20). Thus $p \in T_{\gamma_i}$ q.e.d.

From theorem 4.13 we have

(3.22) $$T_{\gamma_i} \cap T_{\gamma_j} = \emptyset\ ,\ i \neq j\ ,\quad i,j=1,\dots,s\ .$$

2. Step. We have to tile M by polyhedrons where Cheeger's constant is greater than 2 .

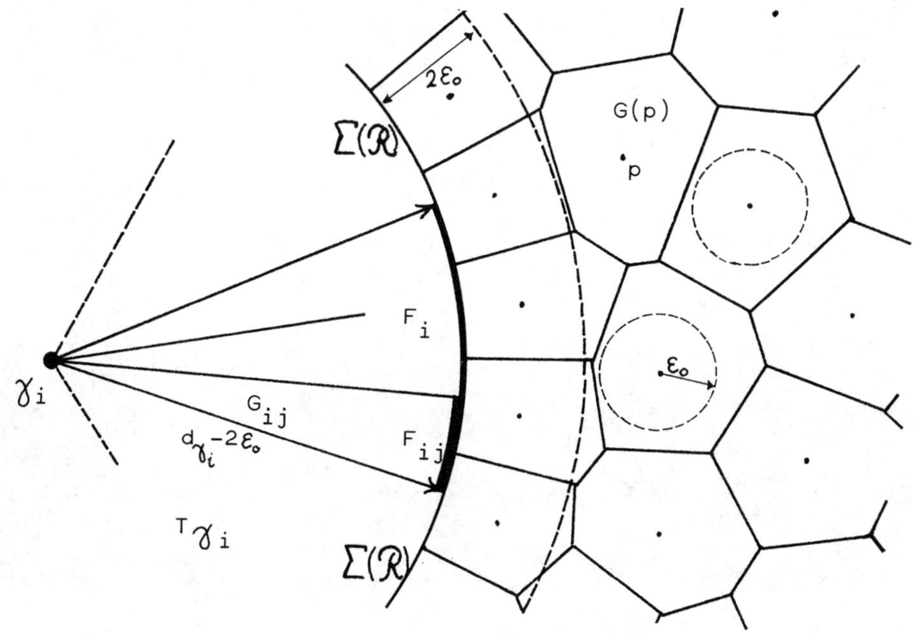

Fig. 9

We start inside T_{γ_i} . Let $\{p_1,\dots,p_m\} = \mathcal{P}_i$, $m=m(i)$, be a maximal set of points satisfying

$$\mathrm{dist}(p_j,\gamma_i) = d_{\gamma_i} - \varepsilon_o\ ,\ \mathrm{dist}(p_j,p_k) \geq 2\varepsilon_o\ ;\ 1 \leq j \neq k \leq m(i)\ ;\ i=1,\dots,s\ .$$

From (3.22) any two points in the union $\mathcal{P}_1 \cup \dots \cup \mathcal{P}_s$ have a distance greater or equal $2\varepsilon_o$ from each other. Therefore we find a maximal set $\mathcal{P}$ of points in M with the following property:

(3.23) $$\mathcal{P}_1 \cup \dots \cup \mathcal{P}_s \subseteq \mathcal{P},$$

(3.24) $$\mathrm{dist}(p,q) \geq 2\varepsilon_o \quad \text{for all} \quad p \neq q \in \mathcal{P},$$

(3.25) $\quad \operatorname{dist}(p,\gamma_i) \geq d_{\gamma_i} - \varepsilon_o$ for all $p \in \mathcal{P}$ and for all $i=1,\ldots,s$.

For each point $q \in \mathcal{P}$ we define

$$G(q) = \{p \in M \mid \operatorname{dist}(p,\gamma_i) \geq d_{\gamma_i} - 2\varepsilon_o \ , \ i=1,\ldots,s \ ;$$
$$\operatorname{dist}(p,q) \leq \operatorname{dist}(p,q') \quad \text{for all} \quad q' \in \mathcal{P} \ \}.$$

The tubes $T'_{\gamma_i} = \{p \in T_{\gamma_i} \mid \operatorname{dist}(p,\gamma_i) \leq d_{\gamma_i} - 2\varepsilon_o\}$, $i=1,\ldots,s$ have to be further dissected. First of all by an appropriate renumeration of the γ_i we have

$$d_{\gamma_1} \leq \cdots \leq d_{\gamma_r} \leq \frac{1}{4} < d_{\gamma_{r+1}} \leq \cdots \leq d_{\gamma_s} \ .$$

We leave $T'_{\gamma_1}, \ldots, T'_{\gamma_r}$ as they are. Let now $\gamma = \gamma_i$, $i > r$, and consider a covering curve $\tilde{\gamma}$ in H^3 : $\Sigma \circ \tilde{\gamma} = \gamma$, to which Fermicoordinates are adapted as above (3.18) . Define

$$\mathcal{R} = \{p \in H^3 \mid \operatorname{dist}(p,\tilde{\gamma}) = d_\gamma - 2\varepsilon_o \ \}.$$

Since the mappings $p = (t,\rho,\sigma) \mapsto (t+a,\rho,V\sigma)$, $a \in \mathbb{R}$ and $V \in O(2)$ are isometries of H^3 acting transitively on $\mathcal{R} \simeq S^1 \times \mathbb{R}$, now the metric inherited from H^3 on $\mathcal{R}$ has constant zero curvature. That is, $\mathcal{R}$ is a flat cylinder, and $\partial T'_\gamma$ is a flat torus, Σ restricted to $\mathcal{R}$ is a Riemannian covering map $\Sigma : \mathcal{R} \to \partial T'_\gamma$. Let $\Pi : \mathbb{R}^2 \to \mathcal{R}$ be the universal covering defined by

$$\mathbb{R}^2 \ni (t,s) \mapsto (t,\sigma) \in \mathcal{R} \ , \ \sigma := s \ (\mathrm{mod}\ 2\pi) \ .$$

Then $\Sigma \circ \Pi$ is the universal covering map from $\mathbb{R}^2$ to $\partial T'_\gamma$. From the classification theory of flat tori we find a fundamental domain $\tilde{F} = \tilde{F}_i$, which is a parallelogram with an acute angle ϕ satisfying $60^\circ \leq \phi \leq 90^\circ$ and two sides a, b . We let a be the one parallel to the s-axis of $\mathbb{R}^2$. From (3.21) and from the fact that $d_\gamma - 2\varepsilon_o \geq \frac{1}{4} - 2\varepsilon_o >> \varepsilon_o$, now $31\varepsilon_o < a,b < 33\varepsilon_o$. As indicated in figure 10, we replace $\tilde{F}$ by a less conventional fundamental domain, which consists of $n = n(i)$ congruent

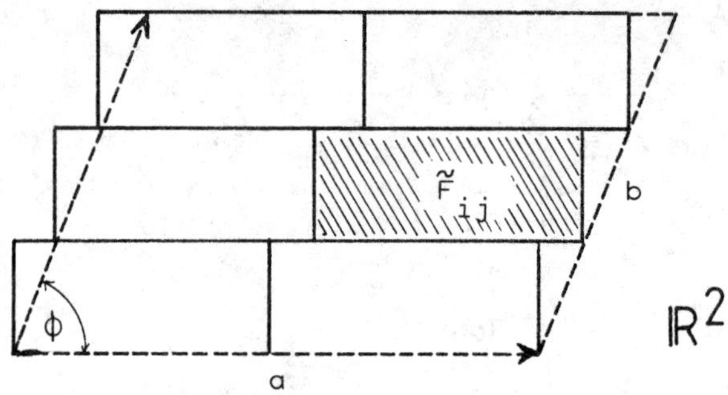

Fig. 10

rectangles $\tilde{F}_j = \tilde{F}_{ij}$, $j=1,\ldots,n(i)$, with sides parallel to the coordinate axes of lengths a_{ij} , b_{ij} satisfying

$$(3.26) \qquad 2\varepsilon_o \le a_{ij} = a_i \le 4\varepsilon_o \; ; \; 2\varepsilon_o \le b_{ij} = b_i \le 4\varepsilon_o \; , \; i=r+1,\ldots,s.$$

(The lengths in (3.26) and above are measured in $\mathbb{R}^2$.) The mapping $\Sigma \circ \Pi$ maps the rectangles $\tilde{F}_j$ bijectively onto rectangles $F_j := \Sigma \circ \Pi(\tilde{F}_j) \subseteq \partial T'_\gamma$, and the $F_1,\ldots,F_n, n=n(i)$ provide a tesselation of $\partial T'_\gamma = \partial T'_{\gamma_i}$.

Using the Fermicoordinates from above, we let now $G_j = G_{ij}$ be the subsets of T'_γ

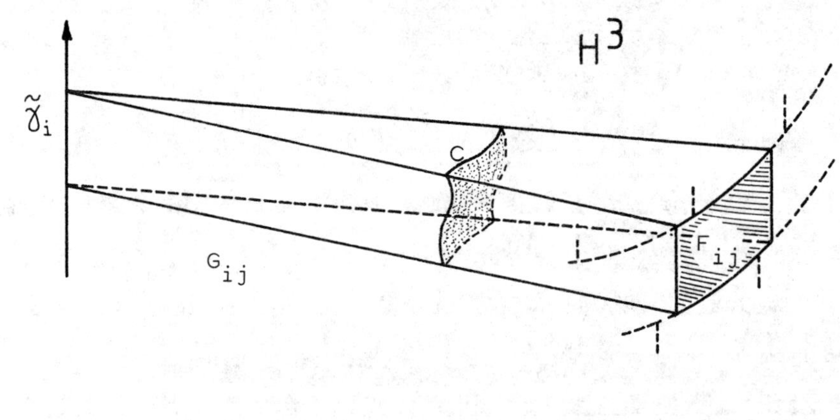

Fig. 11

defined by

$$G_{ij} = \{p = (t,\rho,\sigma) \in T'_\gamma \mid (t,d_\gamma - 2\varepsilon_o,\sigma) \in F_{ij}\} .$$

The set G_{ij} is again best visualized in the universal covering H^3. Figure 11 tries to show its shape. Since the F_j tile up $\partial T'_\gamma$, we have

(3.27) $\quad \text{int } G_{ij} \cap \text{int } G_{i\ell} = \emptyset \quad (1 \leq j \neq \ell \leq n(i)),\ G_{i1} \cup \ldots \cup G_{in} = T'_{\gamma_i} \quad (i=r+1,\ldots,s).$

We have now the sets $G(p)$, $p \in \mathcal{P}$, the tubes T'_{γ_i} $i=1,\ldots,r$ and the hyperbolic pieces of cheese G_{ij}, $j=1,\ldots,n(i)$, $i=r+1,.,s$. From (3.27), (3.25), (3.24), (3.23) and (3.22) if follows that the interiors of any two of these sets have empty intersection, while the union of these sets is M. Thus we have a dissection of M into polyhedrons. If $p \in \mathcal{P}$, then as follows from (3.24) and (3.25), each point q at a distance $\text{dist}(q,p) \leq \varepsilon_o$ belongs to $G(p)$. Since from (3.21) and (3.20) we know that the injectivity radius of M at p is at least ε_o, we see that $G(p)$ contains a ball of radius ε_o. Hence

$$\text{vol}(G(p)) \geq \frac{4\pi}{3}\varepsilon_o^3 := \frac{2}{c} ,$$

and the same inequality holds for the tubes T'_{γ_i} $i=1,\ldots,r$ and the G_{ij} $j=1,\ldots,n(i)$; $i=r+1,\ldots,s$, also, as seen from (3.17), (3.18), (3.19) and (3.26). Hence

(3.28) The number of the above polyhedrons which tile M, is at most $\frac{c}{2}\,\text{vol}(M)$.

<u>3. Step.</u> We evaluate Cheeger's constant h_N for the above tiles. The $G(p)$, $p \in \mathcal{P}$ are the easiest. A very rough elementary consideration with a glance at figure 9 and (3.19) shows that

(3.29) $\quad \min\{\text{dist}(q,p) \mid q \in \partial G(p)\} \geq \varepsilon_o , \ \max\{\text{dist}(q,p) \mid q \in \partial G(p)\} \leq 3\varepsilon_o$

$$\max\{\alpha_q \mid q \in \partial G(p)\} \leq 75^{\circ} .$$

In the third inequality, α_q is the angle at q between the geodesic from p to $q \in \partial G(p)$ and the outer normal of the (piecewise differentiable) boundary $\partial G(p)$ at q. The third inequality is a consequence of the second one and of the fact that $\mathrm{dist}(p, \partial T'_{\gamma_i}) \geq \varepsilon_o$ for all $i=1,\ldots,s$. The main argument for the second inequality is of course that the set $\mathcal{P}$ is maximal, thus if $q \in M - (T'_{\gamma_1} \cup \ldots \cup T'_{\gamma_s})$ and $q \in \partial G(p)$, then $\mathrm{dist}(q,p) \leq 2\varepsilon_o$, for otherwise we could add q to $\mathcal{P}$ without violating the properties (3.24) and (3.25).

We introduce polar coordinates $q \mapsto (\rho,\sigma)$, $\sigma \in S^2$, $\rho = \mathrm{dist}(q,p)$ with center p. For each $q = (\rho,\sigma) \in G(p)$ there is $\bar{q} = (\bar{\rho},\sigma)$ on $\partial G(p)$. The set $B(p) = \{q \in G(p) \mid \mathrm{dist}(q,p) \leq \varepsilon_o\}$ is contained in $G(p)$ and is a hyperbolic ball. The mapping $\Phi : G(p) \to B(p)$ defined by $(\rho,\sigma) \mapsto \Phi(\rho,\sigma) = (\varepsilon_o \frac{\rho}{\bar{\rho}}, \sigma)$ is a piecewise diffeomorphism from $G(p)$ onto $B(p)$. If we replace the hyperbolic metric $ds^2 = d\rho^2 + \sinh^2_\rho \, d\sigma^2$ on $B(p)$ by the euclidean metric $d\bar{s}^2 = d\rho^2 + \rho^2 \, d\sigma^2$, then Φ becomes a piecewise diffeomorphism between $G(p)$ and a euclidean ball $U_{\varepsilon_o} = \{x \in \mathbb{R}^3 \mid \|x\| \leq \varepsilon_o\}$. From (3.29) one derives

$$h_N(G(p)) \geq \frac{1}{60} \cdot h_N(U_{\varepsilon_o}) . \tag{3.30}$$

Now as a consequence of the classical isoperimetric inequality in $\mathbb{R}^n$ we have for the unit ball $E^n = \{x \in \mathbb{R}^n \mid \|x\| \leq 1\}$ (m_k is the euclidean k-measure)

$$h_N(E^n) = \frac{m_{n-1}(E^{n-1})}{\frac{1}{2} m_n (E^n)} = \frac{2\Gamma(\frac{n}{2}+1)}{\sqrt{\pi}\,\Gamma(\frac{n+1}{2})} \geqq (\frac{2n}{\pi})^{\frac{1}{2}} . \tag{3.31}$$

From (3.30) and (3.31), since $h_N(U_{\varepsilon_o}) = \frac{1}{\varepsilon_o} h_N(E^3)$:

$$h_N(G(p)) \geq 4^4 \quad \text{for all } p \in \mathcal{P} . \tag{3.32}$$

Now the T'_{γ_i} for $i \leq i \leq r$: We replace the hyperbolic metric

$$ds^2 = \cosh \rho \, dt^2 + d\rho^2 + \sinh^2 \rho \, d\sigma^2$$

in the above described Fermicoordinates by the euclidean metric

$$ds^2 = dt^2 + d\rho^2 + \rho^2\, d\sigma^2$$

and identify T'_{γ_i} supplied with this new metric with the subset $T = \{(x,y,z) \in \mathbb{R}^3 \mid x^2+y^2 \leq d_{\gamma_i} - 2\varepsilon_o \ ; \ o \leq z \leq \ell(\gamma_i)\}$ of $\mathbb{R}^3$. Since $d_{\gamma_i} < \frac{1}{4}$, now

$$h_N(T'_{\gamma_i}) \geq \frac{1/4}{\sinh \frac{1}{4} \cosh \frac{1}{4}} h_N(T) \geq 0.95\ h_N(T) .$$

Define the projection $(x,y,z) \mapsto \Psi(x,y,z) = (x,y,0)$ on T. Consider a codimension one submanifold C dividing T into the open subsets $T - C = A \cup B$, $A \cap B = \emptyset$. If $m_2(\Psi(C)) \geq 10^{-4} m_2(\Psi(T))$, then $J(C) \geq 2m_2(\Psi(C))/m_3(T) \geq 10^{-4}/\varepsilon_o > 3.2$. If $m_2(\Psi(C)) < 10^{-4}$, then $J(C) \geq 0.99\ h_N(\{(x,y) \in \mathbb{R}^2 \mid x^2+y^2 \leq d_\gamma - 2\varepsilon_o\}) \geq 3.9 \cdot h_N(E^2) > 4$. This yields

$$h_N(T'_{\gamma_i}) \geq 3 \quad , \ i=1,\dots,r . \tag{3.33}$$

The interesting case is provided by the sets G_{ij}. Let G be one of them, belonging to $T'_\gamma = T'_{\gamma_i}$. The striking fact is that, even if d_γ is large, i.e. even if G is long and thin, $h_N(G)$ does not go below 2. This is again so because we have curvature -1 on M (it would be true also if we had variable curvature $K \leq -1$). We first give a short heuristic proof of this fact and then a correct one. We use again the Fermicoordinates with metric (3.18). The claim is

$$h_N(G) > 2 \quad \text{for all} \quad G = G_{ij} \ , \ j=1,\dots,n(i);\ i=r+1,\dots,s. \tag{3.34}$$

Heuristic proof. The only good way to dissect G by a codimension one submanifold C is as drawn in figure 11. It follows at once from (3.18) that

$$J(C) \geq \frac{\cosh d_\gamma \cdot \sinh d_\gamma}{\int_0^{d_\gamma} \cosh \rho \sinh \rho \, d\rho} = 2 \coth d_\gamma \geq 2 + e^{-2d_\gamma} := 2 + \delta. \tag{3.35}$$

Pedantic proof. We first consider the case that $\rho_m = \max\{\mathrm{dist}(p,\gamma)\ p \in C\} < d_\gamma - 2\varepsilon_o$, that is, C does not touch the boundary $\partial T'_\gamma$. Then one of the two sets A, B, into which G is dissected, say A, satisfies $\partial A \cap \partial T'_\gamma = \emptyset$.

For each $p \in A$, $p = (t,\rho,\sigma)$, there is a point $p' = (t,\rho',\sigma) \in C$ with $\rho' > \rho$. It follows as in (3.35) that $\mathrm{vol}_2(C)/\mathrm{vol}_3(A) \geq 2 \coth \rho_o \geq 2 + \delta$. In this case we are through.

Next we make the assumption that for some C we have

$$J(C) < 2 + \delta$$

and lead this to a contradiction. We introduce the following notation: For any subset $U \subseteq G$ we define

$$U_\rho^- = \{p \in U | \rho_p \leq \rho\} \quad ; \quad U_\rho^+ = \{p \in U | \rho_p \geq \rho\}$$

$$\rho \in (0, d_\gamma - 2\varepsilon_o]$$

$$U_\rho = \{p = (t_p, \rho_p, \sigma_p) \in U | \rho_p = \rho\} .$$

By an appropriate approximation we may without loss of generality suppose that the C_ρ are one dimensional submanifolds of the G_ρ , with the exception of a finite number of ρ , where some of the connected components (in the relative topology of G_ρ) of C_ρ reduce to a point, while the other connected components still are one dimensional submanifolds. In particular we have $\mathrm{vol}_2(C_\rho) = 0$ for all $\rho \in (0, d_\gamma - 2\varepsilon_o)$.

Now first of all we find a point $\rho_1 \in [d_\gamma - \frac{1}{32}, d_\gamma - \frac{1}{32} + 96\varepsilon_o]$ such that possibly after exchanging the names of A and B

$$\text{(3.36)} \qquad \mathrm{vol}_1(C_{\rho_1}) \leq \frac{1}{8}\varepsilon_o \quad , \quad \mathrm{vol}_2(A_{\rho_1}) \leq \frac{1}{4}\varepsilon_o^2 .$$

<u>Proof.</u> Suppose the first inequality did not hold. Then $\mathrm{vol}_2 C \geq \mathrm{vol}_2(C_{\rho_1}^+) \geq 12\varepsilon_o^2$. But from (3.26) and (3.18) we know that (putting $d = d_\gamma - 2\varepsilon_o$)

$$\mathrm{vol}_3(G) \leq 16\varepsilon_o^2 \int_0^d \cosh\rho \cdot \sinh\rho \, d\rho / \cosh d \cdot \sinh d \leq 8\varepsilon_o^2 .$$

It would follow that $J(C) \geq 3$, which contradicts the assumption on $J(C)$. Thus the first inequality in (3.36) is proved. Now since all the G_ρ are flat rectangles with sides of length $\leq 4\varepsilon_o$, a simple argument shows that

(3.37) $$h_N(G_\rho) \geq \frac{1}{2\varepsilon_o} \quad \text{for all} \quad \rho \in (0, d_\gamma - 2\varepsilon_o) .$$

Without loss of generality we have $\mathrm{vol}_2(A_{\rho_1}) \leq \frac{1}{2}\mathrm{vol}_2(G_{\rho_1})$. Since $\mathrm{vol}_1(C_{\rho_1}) \leq \frac{1}{8}\varepsilon_o$, from (3.37) now $\mathrm{vol}_2(A_{\rho_1}) \leq \frac{1}{4}\varepsilon_o^2$ as claimed.

(3.38) $$\mathrm{vol}_2(A_\rho) \leq \frac{1}{2}\mathrm{vol}_2(G_\rho) \quad \text{for all} \quad \rho \in [\rho_1, d_\gamma - 2\varepsilon_o] .$$

<u>Proof</u>. If we should find $\rho' \in [\rho_1, d_\gamma - 2\varepsilon_o]$ such that (3.38) does not hold for ρ' , then we define

$$\tilde{C} = C^-_{\rho_1} \cup A_{\rho_1} .$$

As seen in the first case, $\tilde{C}$ satisfies

$$\mathrm{vol}_2(\tilde{C})/\mathrm{vol}_3(A^-_{\rho_1}) = (\mathrm{vol}_2(C^-_{\rho_1}) + \mathrm{vol}_2 A_{\rho_1})/\mathrm{vol}_3(A^-_{\rho_1}) \geq 2 + \delta.$$

Moreover

$$\mathrm{vol}_2 C^+_{\rho_1}/\mathrm{vol}_3(A^+_{\rho_1}) \geq 6 ,$$

(as follows from the fact that $\mathrm{vol}_2(G_{\rho_1}) \geq 3\varepsilon_o^2$, hence $\mathrm{vol}_2(C^+_{\rho_1}) \geq \frac{1}{2}\mathrm{vol}_2(G_{\rho_1}) - \frac{1}{4}\varepsilon_o^2 \geq \frac{5}{12}\mathrm{vol}_2(G_{\rho_1})$, but $\mathrm{vol}_3(A^+_{\rho_1}) \leq \mathrm{vol}_3(G^+_{\rho_1}) < \frac{1}{16}\cdot\mathrm{vol}_2 G_{\rho_1})$.
Since $\mathrm{vol}_2(C^+_{\rho_1}) \geq \frac{3}{2}\varepsilon_o^2 - \frac{1}{4}\varepsilon_o^2 > 2\,\mathrm{vol}_2(A_{\rho_1})$, (see (3.36)) we calculate

$$\mathrm{vol}_2(C) \geq \mathrm{vol}_2(C^-_{\rho_1}) + \mathrm{vol}_2(A_{\rho_1}) + \frac{1}{2}\mathrm{vol}(C^+_{\rho_1}) \geq$$

$$\geq (2 + \delta)(\mathrm{vol}_3(A^-_{\rho_1}) + \mathrm{vol}_3(A^+_{\rho_1})) ,$$

which contradicts the assumption $J(C) < 2 + \delta$. Now (3.38) is proved.

(3.39) $$\text{If} \quad \mathrm{vol}_1(C_\rho) \leq \frac{1}{8}\varepsilon_o , \quad \text{then} \quad \mathrm{vol}_2(A_\rho) \leq \frac{1}{\pi}(\mathrm{vol}_1 C_\rho)^2 .$$

<u>Proof</u>. Since C_ρ is shorter than the distance of two non meeting sides of the flat rectangle G_ρ , each connected component of C_ρ is either a closed curve, or a segment with both endpoints on the same side of G_ρ , or a segment

with the endpoints on either of two sides which have a common vertex. It follows from the classical isoperimetric inequality in the plane, that in this situation the volume of A_ρ (which by (3.38) is smaller or equal $\mathrm{vol}_2(B_\rho)$) would be maximized by A_ρ being a quadrant of a circle of radius $r = 2\,\mathrm{vol}_1(C_\rho)/\pi$. This proves (3.39).

Now we start an induction. There exists $\rho_2 \in [\rho, \rho_1 + 8\varepsilon_o]$ such that

(3.40) $$\mathrm{vol}_1(C_{\rho_2}) \leq \frac{1}{16}\varepsilon_o .$$

<u>Proof</u>. Otherwise $\mathrm{vol}_2(C^+_{\rho_1}) \geq \frac{1}{16}\varepsilon_o \cdot 8\varepsilon_o \geq 2\,\mathrm{vol}_2(A_{\rho_1})$. As before $\mathrm{vol}_2(C^-_{\rho_1}) + \mathrm{vol}_2(A_{\rho_1}) \geq (2+\delta)\,\mathrm{vol}_3(A^-_{\rho_1})$. But now we use (3.38) and (3.37) to conclude that $\mathrm{vol}_2(C^+_{\rho_1}) > \frac{1}{\varepsilon_o}\,\mathrm{vol}(A^+_{\rho_1})$. This again gives $J(C) \geq 2 + \delta$.

Now from (3.40) and (3.39) we already have $\mathrm{vol}_2(A_{\rho_2}) < (1/16)^2\varepsilon_o$; therefore we can this time use the same proof as before, to show that there is $\rho_3 < \rho_2 + 4\varepsilon_o$ such that $\mathrm{vol}_1(C_{\rho_3}) \leq \frac{1}{32}\varepsilon_o$. Then we find $\rho_4 < \rho_3 + 2\varepsilon_o$ with $\mathrm{vol}_1(C_{\rho_4}) \leq \frac{1}{64}\varepsilon_o$ and so on. Thus we find a convergent series $\{\rho_i\}_{i=1}^{\infty}$ with limit $\rho_o \leq d_\gamma - \frac{1}{32} + 112\varepsilon_o < d_\gamma - 2\varepsilon_o$, such that C_{ρ_o} is a finite set of points. Since C is a submanifold, and since A (as well as B) is a connected open set, it follows that A is located left of the membrane G_{ρ_o} , hence C must be of the form considered in the first case, which involves $J(C) \geq 2 + \delta$. This is the desired contradiction proving (3.34).

<u>4. Step</u>. From (3.32), (3.33), (3.34) by Cheeger's inequality we have $\lambda_{1,N}(G) > 1$ for any set G among the G_{ij} , $G(p)$, T'_{γ_i} , which tile up M . Since by (3.28) there are at most $2c\,\mathrm{vol}(M)$ such G , we find $\lambda_{c'} > 1$, $c' = [\frac{c}{2}\,\mathrm{vol}\,M]$, by the minimax principle for orientable M, and $\lambda_{2c'} > 1$ in general, q.e.d.

3.41 REMARK. The constant given in this proof is $c = 4^{22}$. This is far away from reality. In fact with a lot of effort one can refine the above method to get down to $c \sim 10^6$. However it seems likely that with different methods one should get much better estimates for c .

Since the results of theorems 3.14 and 3.15 are of qualitative nature, we avoid anoying calculations by introducing the following notation:

3.42 DEFINITION. Let P, Q be geometric quantities of an n-dimensional manifold, then the term $P <_* Q$ has the following meaning: There exists a constant $c > 0$, depending only on the dimension n of the manifold, such that $P \leq cQ$.

<u>Proof</u> of theorems 3.14 and 3.15. Since $n \geq 4$, we know from corollary 4.15 that the injectivity radius $r(M)$ of M satisfies

(3.43) $$r = r(M) >_* (\mathrm{vol}\ M)^{-\sigma n}, \quad \sigma n = \left[\frac{n+1}{2}\right] / \left[\frac{n-2}{2}\right] .$$

Let

$$\delta = \frac{1}{4} \min\{\frac{1}{n}, r\} ,$$

and fill in a <u>maximal</u> set $\mathcal{P}$ of points satisfying $\mathrm{dist}(p,q) \geq \delta$ for all $p,q \in \mathcal{P}$ $p \neq q$. As in the proof of theorem 3.12 above, we define the Dirichlet regions

$$G(p) = \{p' \in M \mid \mathrm{dist}(p',p) \leq \mathrm{dist}(p',q) \quad \text{for all} \quad q \in \mathcal{P}\} .$$

The boundary of $G(p)$ lies within a geodesic ball of radius δ , for otherwise we should have a point p' with $\mathrm{dist}(p',p) > \delta$ and $\mathrm{dist}(p',q) \geq \mathrm{dist}(p',p) > \delta$ for all other $q \in \mathcal{P}$ as well. This contradicts the condition that $\mathcal{P}$ is maximal. On the other hand by the condition on the distance of the points in $\mathcal{P}$, $G(p)$ contains a geodesic ball of radius $\delta/2$. From the smallness $\delta < 1/2n$ we see that we may calculate within $G(p)$ as if $G(p)$ were a polyhedron in $\mathbb{R}^n$, the errors in the volume estimates are at most by a factor $(1+1/n)^n$. Thus we have $\mathrm{vol}(G(p)) >_* \delta^n$, and from this

(3.44) The number $\#(\mathcal{P})$ of Dirichlet regions $G(p)$, $p \in \mathcal{P}$ satisfies

$$\#(\mathcal{P}) <_* \frac{\mathrm{vol}(M)}{\delta^n} .$$

Consider one of the sides D of the polyhedron $G(p)$, it is part of a totally geodesic $(n-1)$-dimensional submanifold $\tilde{D}$, defined in a neighbourhood of p which is at least so large that it contains a perpendicular μ from p to $\tilde{D}$. The length of μ is at least $\delta/2$, the distance from p to any point q on D at most δ. Since $\delta < \frac{1}{2n}$, the angle between the geodesic from p to q and the outer normal of D at q is at most 61°. Hence as in the proof of theorem 3.12 there exists a piecewise diffeomorphism from $G(p)$ to a euclidean ball $E_{\delta/2}$ of radius $\delta/2$ showing that (confer (3.31))

$$(3.45) \qquad h_N(G(p)) \geq \frac{1}{2^{n+2}} \cdot h_N(E_{\delta/2}) \geq \frac{1}{\delta\, 2^{n+1}} \frac{b_{n-1}}{b_n},$$

where b_k is the volume of the unit ball in $\mathbb{R}^k$. Hence

$$h_N(G(p)) >_* \frac{1}{\delta}, \quad \lambda_{1,N}(G(p)) >_* \frac{1}{\delta^2}, \quad \lambda_{\#}(\mathcal{P}) >_* \frac{1}{\delta^2}.$$

Combining this result with (3.43) and (3.44), we have $<_*(\mathrm{vol}\, M)\, x^{n/2}$ eigenvalues of M in any interval $[0,x]$, for which $x >_* (\mathrm{vol}\, M)^{2\sigma n}$. This proves theorem 3.15.

In order to prove theorem 3.14 we estimate $h(M)$. Let C be a codimension 1 submanifold, which divides M into $M - C = A \cup B$, $\mathrm{vol}\, A \leq \mathrm{vol}\, B$. Two cases are possible: Either $\mathrm{vol}\, C \geq b_{n-1}(\delta/64)^{n-1}$ or $\mathrm{vol}\, C < b_{n-1}(\delta/64)^{n-1}$. We first consider the latter case and claim:

$$(3.46) \qquad \mathrm{vol}(G(p) \cap A) \leq \frac{1}{2}\, \mathrm{vol}\, G(p) \quad \text{for all } p \in \mathcal{P}.$$

Suppose not. From the fact that $\mathrm{vol}\, A \leq \frac{1}{2}\, \mathrm{vol}\, M$ and because the Dirichlet regions $G(p)$, $p \in \mathcal{P}$ tile all M, we find $G_1 = G(p_1)$ and $G_2 = G(p_2)$ such that

$$\mathrm{vol}(G_1 \cap A) > \frac{1}{2}\, \mathrm{vol}\, G_1, \quad \mathrm{vol}(G_2 \cap A) < \frac{1}{2}\, \mathrm{vol}\, G_2.$$

From (3.45) and from $\mathrm{vol}_{n-1}(G_i \cap C) \leq b_{n-1}(\delta/64)^{n-1}$ we conclude that even

$$(3.47) \qquad \mathrm{vol}(G_1 \cap B) < 2b_n (\tfrac{\delta}{2})^n\, 16^{(2-n)}, \quad \mathrm{vol}(G_2 \cap A) < 2b_n (\tfrac{\delta}{2})^n\, 16^{(2-n)}.$$

Recall that each G_i contains a ball of radius $\delta/2$ centered at p_i, $i=1,2$. Join p_1 and p_2 with a shortest geodesic segment γ :

$$t \mapsto \gamma(t) , \quad t \in [-d, d] , \quad \gamma(-d) = p_1 , \quad \gamma(d) = p_2 , \quad \|\dot{\gamma}\| = 1 ,$$

where $2d = \mathrm{dist}(p_1,p_2)$. Extend γ to the interval $[-d-\delta, d+\delta]$ and define W to be the set of points $p \in M$, for which a geodesic segment μ_p exists, satisfying

$$s \mapsto \mu(s) , \quad s \in [0,\rho_p] , \quad \mu(0) = \gamma(t_p) \quad \text{for some} \quad t_p \in [-d-\delta, d+\delta] ,$$

$$\mu(\rho_p) = p , \quad \langle\dot{\mu}(0) , \dot{\gamma}(t_p)\rangle = 0 , \quad \rho_p = \mathrm{dist}(p,\gamma) \leq \delta .$$

Since the injectivity radius of M is $\geq 4\delta$, the t_p is uniquely determined, and we can introduce Fermicoordinates $p \mapsto (t_p,\rho_p,\sigma_p)$, $\sigma_p \in S^1$ for the points $p \in W$, where S^1 is interpreted as the set of unit vectors at $\gamma(0)$ perpendicular to γ . In these coordinates W is the set

$$W = \{p = (t,\rho,\sigma) \mid t \in [-d-\delta,d+\delta] , \quad r \leq \delta \}.$$

Of course $G_1 , G_2 \subset W$. We define

$$D = \{p = (t,\rho,\sigma) \in W \mid t = 0\} .$$

From (3.18), D is a totally geodesic $(n-1)$-ball perpendicular to γ centered at $\gamma(0)$.

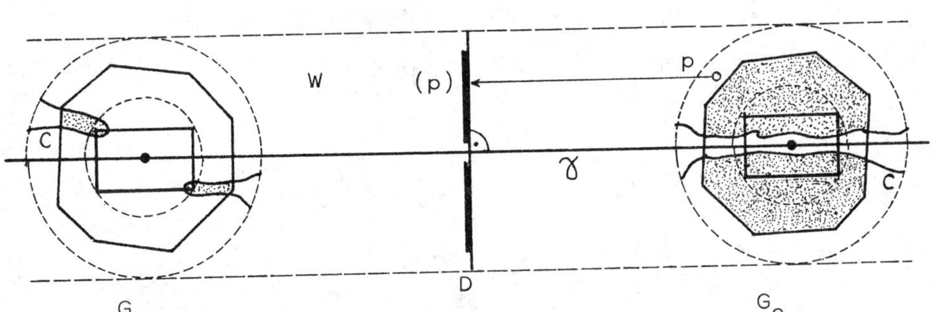

Fig. 12

The projection Ψ defined by $\Psi(t,\rho,\sigma) = (0,\rho,\sigma)$, $(t,\rho,\sigma) \in W$, satisfies

$$\mathrm{vol}_{n-1}(C \cap W) \geq \mathrm{vol}_{n-1}(\Psi(C \cap W)),$$

as seen from (3.18). Therefore we have

$$\mathrm{vol}_{n-1}(\Psi(C \cap W)) \leq b_{n-1}(\delta/64)^{n-1}. \tag{3.48}$$

On the other hand consider the set

$$A_1 = \{p = (t,\rho,\sigma) \in G_1 \cap A \mid -d-\delta/4 \leq t \leq -d+\delta/4;\ 0 \leq \rho \leq \delta/4;$$
$$\Psi(p) \notin \Psi(W \cap C)\}.$$

As seen from (3.18), (3.47) and (3.48), it satisfies

$$\mathrm{vol}_{n-1}(\Psi(A_1)) \geq b_{n-1}\left(\frac{\delta}{4}\right)^{n-1}\left[1 - \frac{4\,b_n}{8^{n-2}\,b_{n-1}} - 16^{1-n}\right] > \frac{1}{2}\,b_{n-1}\left(\frac{\delta}{4}\right)^{n-1}.$$

Now if $p \in G_2$ satisfies $\Psi(p) \in \Psi(A_1)$, then clearly $p \in A$. Thus the set

$$A_2 = \{p = (t,\rho,\sigma) \in G_2 \mid \Psi(p) \in \Psi(A_1),\ d-\delta/4 \leq t \leq d+\delta/4\}$$

is contained in $G_2 \cap A$ and has volume

$$\mathrm{vol}_n(A_2) \geq \frac{\delta}{2}\,\mathrm{vol}_{n-1}\Psi(A_1) > b_{n-1}\left(\frac{\delta}{4}\right)^n > 2b_n\left(\frac{\delta}{2}\right)^n \frac{1}{2^{n+2}}.$$

Since $n \geq 4$, this contradicts the second inequality in (3.47) and proves (3.46).

From (3.46) and from the fact that the Dirichlet regions $G(p)$, $p \in \mathcal{P}$ tile all M, we find $\mathrm{vol}_{n-1}(C)/\mathrm{vol}_n A \geq \min_{p \in \mathcal{P}} h_N(G(p))$. Hence from (3.45)

$$\frac{\mathrm{vol}_{n-1}(C)}{\mathrm{vol}_n(A)} \geq \frac{1}{2^{n+2}\,\delta} \mathrel{>_*} 1$$

Now we consider the case $\mathrm{vol}(C) \geq b_{n-1}(\delta/64)^{n-1}$. Here from (3.43):

$$\frac{\mathrm{vol}_{n-1}(C)}{\mathrm{vol}_n(A)} \geq \frac{2\,b_{n-1}\,\delta^{n-1}}{64^{n-1}\,\mathrm{vol}(M)} \mathrel{>_*} (\mathrm{vol}\,M)^{-1-(n-1)\sigma_n}.$$

This shows that $h(M) \gtrsim_* (\mathrm{vol}\ M)^{-(1+(n-1)\sigma n)}$, and theorem 3.14 follows from Cheeger's inequality. q.e.d.

CHAPTER FOUR. The size of hyperbolic manifolds.

We use Gromov's method [15],[8] to estimate the volume of tubes on compact hyperbolic manifolds. The results hold for variable curvature, but for constant curvature the treatment is particularly simple. Throughout the chapter M is a compact hyperbolic manifold of dimension $n \geq 3$.

For $p \in M$ consider the set Π_p of all geodesic loops at p. For $a,b \in \Pi_p$, ab is defined to be the unique geodesic loop in the homotopy class of the composed curve. $a \circ b$ (first b then a). Π_p with this product is canonically isomorphic to the fundamental group of M.

4.1 MARGULIS LEMMA. If $a,b \in \Pi_p$ satisfy $\ell(a), \ell(b) \leq 4^{-(n+2)} := 2r_n$, then the subgroup of Π_p spanned by a and b is cyclic.

Proof. (For details and more generality see [15], [8]).

If α is any curve with endpoints at p, then for tangent vectors $X \in T_pM$ let $A(X) \in T_pM$ be the parallel translate of X along α. The holonomy map A belongs to $O(n)$. We work with the biinvariant Finsler metric on $O(n)$ defined by the maximal rotation angle norm:

$$|A| = \max_{X \in T_pM} \sphericalangle(X, A(X)), \qquad A \in O(n).$$

If A is the holonomy of α, we write $r(\alpha) := |A|$. Finally for $\alpha \in \Pi_p$ we define

$$||\alpha|| = \max \{r(\alpha); 3\ell(\alpha)\}.$$

1. Step: Let $\alpha, \beta \in \Pi_p$ and denote by $[\alpha,\beta] = \alpha\beta\alpha^{-1}\beta^{-1}$ their commutator. If $||\alpha||, ||\beta|| \leq 0.44$, then

$$||[\alpha,\beta]|| \leq 0.99 \min \{||\alpha||, ||\beta||\}. \tag{4.2}$$

Proof. Consider the universal covering $\Sigma : H^n \longrightarrow M$. Take $p_1 \in H^n$ with $\Sigma(p_1) = p$ and consider geodesic segments $\tilde{\alpha}$ from p_1 to p_2 , $\tilde{\tilde{\beta}}$ from p_2 to p_3 , $\tilde{\beta}$ from p_1 to p_4 , $\tilde{\tilde{\alpha}}$ from p_4 to p_5 , as drawn in figure 13, such that $\Sigma(p_i) = p$, $i = 1,\dots,5$, $\Sigma(\tilde{\alpha}) = \Sigma(\tilde{\tilde{\alpha}}) = \alpha$, $\Sigma(\tilde{\beta}) = \Sigma(\tilde{\tilde{\beta}}) = \beta$. The geodesic segment from p_3 to p_5 is then a lift of $[\alpha,\beta]$.

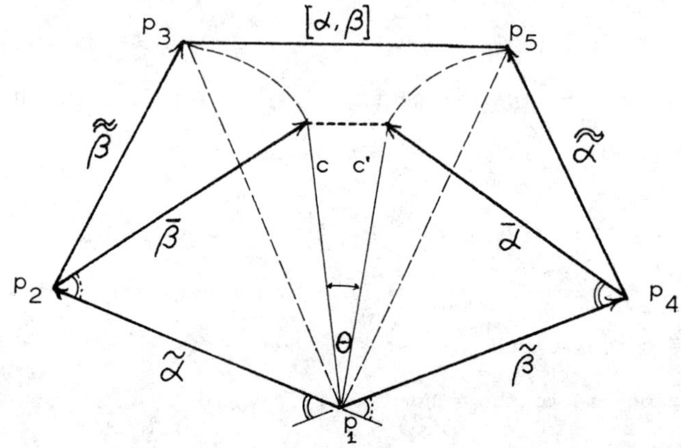

Fig. 13

Let $\overline{\beta}$ be the parallel translate of $\tilde{\beta}$ along $\tilde{\alpha}$ and let $\overline{\alpha}$ be the parallel translate of $\tilde{\alpha}$ along $\tilde{\beta}$. The segments $\tilde{\alpha},\overline{\beta}$ $\tilde{\beta},\overline{\alpha}$ lie in a two dimensional totally geodesic submanifold, which we identify with H^2 . The triangles $\tilde{\alpha},\overline{\beta},c$ and $\tilde{\beta},\overline{\alpha},c'$ are isometric. Hence the angle θ between c and c' equals the difference between π and the sum of the inner angles of the triangle $\tilde{\alpha}\overline{\beta}c$, i.e.

$$\theta = \mathrm{vol}_2(\tilde{\alpha},\overline{\beta},c) \le \frac{1}{2}\,\ell(\alpha)\ell(\beta) \quad . \tag{4.3}$$

From $\measuredangle(\overline{\beta},\tilde{\tilde{\beta}}) \le r(\alpha)$, $\measuredangle(\overline{\alpha},\tilde{\tilde{\alpha}}) \le r(\beta)$ and from (4.3):

$$\ell([\alpha,\beta]) \le r(\alpha)\cdot\sinh \ell(\beta) + r(\beta)\cdot\sinh \ell(\alpha) + \frac{1}{2}\,\ell(\alpha)\ell(\beta)\cdot\sinh\,(\ell(\alpha)+\ell(\beta)) . \tag{4.4}$$

If A,B are the holonomies along α,β respectively, then the holonomy along $\alpha\beta\alpha^{-1}\beta^{-1}$ is almost $ABA^{-1}B^{-1}$, since there is a homotopy of surface area F between $\alpha\beta\alpha^{-1}\beta^{-1}$ and $\alpha\circ\beta\circ\alpha^{-1}\circ\beta^{-1}$ satisfying

$$F \le \ell(\alpha)\ell(\beta) + \frac{1}{2}\,\ell([\alpha,\beta])\cdot(\ell(\alpha) + \ell(\beta)) \quad , \tag{4.5}$$

and generally, if there is a homotopy of area F between two curves c_1 and

c_2 , then their holonomies A_1, A_2 satisfy the following property (see [8]):

4.6 LEMMA . $$|A_1 A_2^{-1}| \leq ||R|| \cdot F ,$$

where $||R||$ ist the norm of the curvature tensor of M , which in our case is equal to 1 . From (4.5) and (4.6), since $|[A,B]| \leq 2|A||B|$ for $A,B \in O(n)$, now

$$(4.7) \quad r([\alpha,\beta]) \leq 2r(\alpha)r(\beta) + \ell(\alpha)\ell(\beta) + \frac{1}{2}\ell([\alpha,\beta])(\ell(\alpha)+\ell(\beta)) .$$

(4.4) together with (4.7) proves (4.2).

<u>2. Step:</u> Powers of a and b satisfy the hypothesis of (4.2):

4.8 LEMMA. Let $A \in O(n)$, $\phi \in (0,\pi)$ then $|A^k| \leq \phi$ for some $1 \leq k \leq 2 + 2(\frac{2\pi}{\phi})^{[\frac{n}{2}]}$.

For the proof see [8] . Now take $\phi = 2\pi/15$, $m = 2(\frac{2\pi}{\phi})^{[\frac{n}{2}]} + 2$ and $1 \leq k \leq m$ such that $|A^k| \leq \phi$, A the holonomy of a . There is a homotopy of surface area

$$F \leq \frac{1}{2} \ell^2(a)(1+2+\ldots+k-1) \leq 2^{-8}$$

between the curve $a \circ \ldots \circ a$ (k times) and the geodesic loop a^k . From lemma 4.6 now

$$(4.9) \qquad ||a^k||, ||b^j|| \leq 0.44 \text{ for some } j,k \geq 1 .$$

<u>3. Step:</u> Let $\alpha = a^k$, $\beta = b^j$, and define $c_1 = [\alpha,\beta]$, $c_{i+1} = [c_i,\beta]$. Because of (4.9) and (4.2), now $||c_\nu|| \to 0$ as $\nu \to \infty$. Since Π_p is discrete: $||c_\mu|| = 0$, i.e. c_μ is trivial loop for some smallest integer μ. Since $[c_{\mu-1},\beta] = 1$, the decktransformations associated to $c_{\mu-1}$ and β (relative to the covering $\Sigma : H^n \longrightarrow M$) have the same invariant geodesic. Since $c_{\mu-1} = [c_{\mu-2},\beta]$, this holds for $c_{\mu-2}$ and β also. By induction, the decktransformations associated to α and β have the same invariant geodesic, and the same holds for a and b. So by the theorem of Preissman [26] , a and b

belong to the same maximal cyclic subgroup of Π_p, q.e.d.

4.10 DEFINITION. For a simple closed geodesic γ on M we define d'_γ to be the injectivity radius for the exponential map on the normal bundle of γ, and let

$$T'_\gamma = \{p \in M \mid \text{dist}(p,\gamma) < d'_\gamma\} .$$

T'_γ is called the tube around γ. It is the largest set swept out by geodesics of equal length emanating perpendicularly from γ, which is topologically equivalent to $S^1 \times B^{n-1}$, B^{n-1} an open $(n-1)$-ball.

If γ had selfintersections, T'_γ would be empty. This shows that for dimension $n \geq 3$ there is no lower bound for $\text{vol}(T'_\gamma)$, in terms of $\ell(\gamma)$, hence no general collar theorem (confer below (3.3)). However we have

4.11 THEOREM. If $n \geq 3$ and $\ell(\gamma) \leq 2r_n := 4^{-(n+2)}$, then T'_γ contains a point p, where the injectivity radius (of M at p) is $r_p(M) = r_n$.

Proof: Take a geodesic $t \mapsto \mu(t)$, $t \in [0,\infty)$ emanating perpendicularly at some point from γ, say at $\gamma(0)$. For $\nu \in \mathbb{Z}$, $t \geq 0$, define $\gamma_t^\nu \in \Pi_{\mu(t)}$ to be the geodesic loop in the homotopy class of the curve $(\mu|[0,t]) \circ \gamma^\nu \circ (\mu|[0,t])^{-1}$. For each $\nu \neq 0$, the length $\ell(\gamma_t^\nu)$ tends monotonely to infinity as $t \to \infty$. (This is best seen in the iniversal covering of M). Hence we find t_o such that for some $k \neq 0$ we have

$$\ell(\gamma_{t_o}^k) = 2r_n , \qquad \ell(\gamma_{t_o}^\nu) > 2r_n \qquad \text{for all other } \nu \neq 0 .$$

The injectivity radius of M at $\mu(t_o)$ is $r_{\mu(t)}(M) = r_n$, for otherwise we find $b \in \Pi_{\mu(t_o)}$ with $\ell(b) < 2r_n$, thus b is different from all the $\gamma_{t_o}^\nu$, which contradicts the Margulis lemma.

Now take μ to be the shortest curve with endpoints on γ, which is not homotopic to zero, when homotopies with endpoints moving freely on γ are considered. μ is a geodesic of length $2d'_\gamma$, which meets γ perpendicularly at both endpoints. Extend μ to a geodesic on $[0,\infty)$. If the above t_o would satisfy $t_o \geq t_1 := 2d'_\gamma$ (recall $||\dot{\mu}|| = 1$), then for some $\nu \neq 0$, now $\ell(\gamma^\nu_{t_1}) < 2r_n$. On the other hand, γ interpreted as element of $\Pi_{\mu(t_1)}$ is different from all the $\gamma^\nu_{t_1}$ (since $\mu|[0,t_1]$ is not homotopic to zero) and has length $\ell(\gamma) < 2r_n$ also; which contradicts the Margulis lemma. Hence $t_o < t_1$ and $\mu(t_o)$ T'_γ, q.e.d.

4.12 DEFINITION. Let γ be a simple closed geodesic on M with length $\ell(\gamma) < 2r_n$. According to theorem 4.11 the number $d_\gamma = \min\{\mathrm{dist}(p,\gamma) \mid p \in T'_\gamma,\ r_p(M) = r_n\}$ is well defined and satisfies $0 < d_\gamma \leq d'_\gamma$. The set $T_\gamma = \{p \in M \mid \mathrm{dist}(p,\gamma) < d_\gamma\}$ is topologically equivalent to $S^1 \times B^{n-1}$ and is called <u>the small tube around</u> γ.

4.13. THEOREM. Let $\gamma \neq \delta$ be simple closed geodesics on M with $\ell(\gamma), \ell(\delta) < 2r_n$, then $T_\gamma \cap T_\delta = \emptyset$.

Proof. If $p \in T_\gamma \cap T_\delta$, draw a perpendicular μ from γ to p : $t \mapsto \mu(t)$, $t \in [0, \mathrm{dist}(p,\gamma)]$, w.l.o.g. $\mu(0) = \gamma(0)$, and a similar perpendicular $\bar{\mu}$ from δ to p. For $k \in \mathbb{Z}$ define $\gamma^k_p \in \Pi_p$ to be the geodesic loop homotopic at p to the curve $\mu \circ \gamma^k \circ \mu^{-1}$. Similarly δ^k_p is the geodesic loop homotopic to $\bar{\mu} \circ \delta^k \circ \bar{\mu}^{-1}$. Since $p \in T_\gamma \cap T_\delta$, now $\ell(\gamma^k_p) < 2r_n$ and $\ell(\delta^m_p) < 2r_n$ for some $m,k \neq 0$. Hence by the Margulis lemma γ^m_p and δ^k_p generate a cyclic subgroup of Π_p. This is impossible, q.e.d.

Theorem 4.11 yields a lower bound on the volume of the small tube T_γ, which does not go to zero when the length of γ goes to zero. For $n \geq 4$ a much stronger result holds.

4.14 THEOREM If $\dim M = n \geq 4$ and if γ is a closed geodesic on M with length $\ell(\gamma) \leq 2r_n$, then $\mathrm{vol}(T_\gamma) \gtrsim_* \ell(\gamma)^{-\left[\frac{n-2}{2}\right]/\left[\frac{n+1}{2}\right]}$.

Here we have used the notation introduced in definition 3.42.

4.15 COROLLARY. The injectivity radius $r(M) = \min\{r_p(M) \mid p \in M\}$ of a compact hyperbolic manifold M of dimension $n \geq 4$ satisfies
$r(M) >_* (\mathrm{vol}\ M)^{-\left[\frac{n+1}{2}\right]/\left[\frac{n-2}{2}\right]}$.

4.16 REMARK. The proof given here yields an order of magnitude for the constants involved, which is e^{-5n^2} for 4.14 and e^{-12n^2} for 4.15. The true values are unknown.

Proof of theorem 4.14. Let again $\Sigma : H^n \longrightarrow M$ be the Riemanian covering. Consider a covering curve $\tilde{\gamma} : \mathbb{R} \longrightarrow H^n$, $\Sigma \circ \tilde{\gamma} = \gamma$, and introduce Fermicoordinates $p = (t,\rho,\sigma)$, $t \in \mathbb{R}$, $\rho = \mathrm{dist}\,(p,\gamma)$, $\sigma \in S^{n-2} := \{$set of all unit vectors at $\tilde{\gamma}(0)$ orthogonal to $\dot{\tilde{\gamma}}(0)\}$. The metric is

$$ds^2 = \cosh^2\rho dt^2 + d\rho^2 + \sinh^2\rho\ d\sigma^2 \quad . \tag{4.17}$$

The decktransformation associated to γ is the mapping

$$p = (t,\rho,\sigma) \longmapsto L(p) = (t + \ell(\gamma),\rho,A\sigma), \quad \text{with } A \in O(n-1) .$$

By abuse of notation, A is the holonomy of γ.

By the Margulis lemma we have $2d_\gamma + \frac{1}{2}\ell(\gamma) > 2r_n$. Since $\ell(\gamma) \leq 2r_n$, now $d_\gamma \geq \frac{1}{2}r_n$ and

$$\cosh d_\gamma \leq \frac{3}{r_n} \sinh d_\gamma . \tag{4.18}$$

Let $\phi = r_n/\sinh d_\gamma$. By lemma 4.8 there exists an integer $m <_* (\sinh d_\gamma)^{\left[\frac{n-1}{2}\right]}$ such that

$$|A^k| > \phi , \quad \text{for } 1 \leq k \leq m-1 , \quad |A^m| \leq \phi \quad . \tag{4.19}$$

By definition there is a point $p_o \in \partial T_\gamma$, where $r_{p_o}(M) = 2r_n$. Let q_o with $\mathrm{dist}(q_o,\tilde{\gamma}) = d_\gamma$ and $\Sigma(q_o) = p_o$ be a point above p_o. Without loss of generality $q_o = (0,d_\gamma,\sigma_o)$ for some $\sigma_o \in S^{n-2}$. Since the injectivity of M at p_o is $2r_n$, now $\mathrm{dist}(q_o,L^m(q_o)) \geq 2r_n$. From (4.17) and from (4.19) we have $\mathrm{dist}\,((0,d_\gamma,\sigma_o),(0,d_\gamma,A^m\sigma_o)) \leq \phi\cdot\sinh d_\gamma = r_n$, and again from (4.17):

$\mathrm{dist}((0,d_\gamma,A^m\sigma_o),\ (m\cdot\ell(\gamma),d_\gamma,A^m\sigma_o)) \leq m\cdot\ell(\gamma)\cdot\cosh d_\gamma$. Hence together with (4.18)

$$\ell(\gamma)\cdot m\cdot\sinh d_\gamma \geq \frac{1}{3} r_n^2 \quad . \tag{4.20}$$

Consider the set $U_o \subset S^{n-2}$ defined by $U_o = \{\sigma \in S^{n-2} \mid \sphericalangle(\sigma,\sigma_o) < \phi/2\}$. From the first inequality (4.19) we deduce that

$$A^j(U_o) \cap A^k(U_o) = \emptyset \qquad \text{for all} \quad 1 \leq k \neq j \leq m-1 \quad .$$

Hence the volume of the set

$$D_m = \{(t,\rho,\sigma) \in H^n \mid 0 \leq t \leq m\cdot\ell(\gamma),\ 0 \leq \rho < d_\gamma\ ,\ \sigma \in U_o\}$$

is m times the volume of

$$D_1 = \{(t,\rho,\sigma) \in H^n \mid 0 \leq t \leq \ell(\gamma),\ 0 \leq \rho < d_\gamma\ ,\ \sigma \in U_o\} .$$

(Here we use the fact that H^n has constant curvature to make sure that the map $(t,\rho,\sigma) \mapsto (t+\ell(\gamma),\rho,\sigma)$ is volume preserving). Since the set

$$\hat{D}_1 = \{(t,\rho,\sigma) \in H^n \mid 0 \leq t < \ell(\gamma),\ 0 \leq \rho < d_\gamma\ ,\ \sigma \in S^{n-2}\}$$

is a fundamental domain for the small tube T_γ , now

$$\mathrm{vol}(T_\gamma) = \mathrm{vol}(\hat{D}_1) = \frac{\mu(S^{n-2})}{\mu(U_o)}\ \mathrm{vol}\ D_1 > \phi^{2-n}\ \mathrm{vol}\ D_1 = \phi^{2-n}\ \frac{\mathrm{vol}\ D_m}{m} . \tag{4.21}$$

Here μ is the Lebesgue measure on the standard S^{n-2} . Since there is a ball of radius $r_n/5$ contained in D_m , we have vol $D_m > 1$.

From (4.20), now $\ell >_* 1/(m\cdot\sinh d_\gamma) >_* (\sinh d_\gamma)^{-[\frac{n+1}{2}]}$. From (4.21) vol $T_\gamma >_* \phi^{2-n}\ (\sinh d_\gamma)^{-[\frac{n-1}{2}]} > (\sinh d_\gamma)^{[\frac{n-2}{2}]}$. This yields

$$\mathrm{vol}\ T_\gamma >_* \ell(\gamma)^{-[\frac{n-2}{2}]/[\frac{n+1}{2}]} \quad , \qquad \text{q.e.d.}$$

REFERENCES

1. Berger, M., Grauduchon, P., Mazed, E., Le spectre d'une variété riemannienne, Lect. Notes in Mathematics 194, Berlin-Heidelberg-New York: Springer 1971.

2. Bers, L., Spaces of degenerating Riemann surfaces, in Discontinuous Groups and Surfaces (College Park), Annals of Math. Studies 79, Princeton University Press (1974), 43-55.

3. Buser, P., Riemannsche Flächen mit Eigenwerten in $(0,\frac{1}{4})$, Comment. Math. Helv. 52 (1977), 25-34.

4. Buser, P., Ueber eine Ungleichung von Cheeger, Math. Z. 158 (1978), 245-252.

5. Buser, P., Cubic graphs and the first eigenvalue of a compact Riemann surface, Math. Z. 162 (1978), 87-99.

6. Buser, P., Beispiele für λ_1 auf kompakten Mannigfaltigkeiten, Math. Z. 165 (1979), 107-133.

7. Buser, P., Ueber den ersten Eigenwert des Laplace-Operators auf kompakten Riemannschen Flächen, to appear in Comment. Math. Helv. 54 (1979).

8. Buser, P., Karcher, H., Diskrete Gruppen und kleine Krümmung nach Gromov, preprint.

9. Chavel, I., Feldman, E., Cylinders on surfaces, Comment. Math. Helv. 53 (1978), 439-447.

10. Cheeger, J., A lower bound for the smallest eigenvalue of the Laplacian, Problems in Analysis, A symposium in honour of Bochner, Princeton University Press, Princeton, N.J. (1970), 195-199.

11. Cheeger, J., Ebin, D., Comparison theorems in Riemannian Geometry, North-Holland Publishing Company, Amsterdam - Oxford (1975).

12. Cheng, S.-Y., Eigenfunctions and eigenvalues of the Laplacian, Differential Geometry, Proc. Sympos. Pure Math. 27, Amer. Math. Soc., Providence R.I. (1975), 185-193.

13. Cheng, S.-Y., Eigenvalue comparison theorems and its geometric applications, Math. Z. 143 (1975), 289-297.

14. Epstein, D.B.A., Curves on 2-manifolds and isotopies, Acta Math. 115 (1966), 83-107.

15. Gromov, M., Manifolds of negative curvature, preprint.

16. Hoffman, D., Lower bounds on the first eigenvalue of the Laplacian on Riemannian manifolds (to appear).

17. Huber, H., Ueber den ersten Eigenwert des Laplace-Operators auf kompakten Mannigfaltigkeiten konstanter negativer Krümmung. Archiv. Math. 26 (Basel 1975), 178-182.

18. Huber,H., On the spectrum of the Laplace-Operator on compact Riemann surfaces, these Proceedings.

19. Li, Peter, Ph. d. Thesis, UC Berkeley, 1979.

20. Löbell, F., Beispiele geschlossener dreidimensionaler Clifford-Kleinscher Räume negativer Krümmung, Ber. d. Math. phys. Kl. d. Sächs. Akad. d. Wiss. zu Leipzig 83 (1931), 167-174.

21. Löbell, F., Zur Konstruktion geschlossener Clifford-Kleinscher Räume negativer Krümmung, Sitz. Ber. d. Bayer Akad. d. Wiss. Math. Natw. Kl. (1955), München 1956.

22. Millson, J , On the first Betti number of a constant negatively curved manifold, Annals of Math. 104 (1976), 235-247.

23. Osserman, R., A note on Hayman's theorem on the bass note of a drum, Comment. Math. Helv. 52 (1977), 545-555.

24. Osserman, R., The isoperimetric inequality, Bull. Amer. Math. Soc. 84 (1978), 1182-1238.

25. Osserman, R., Bonnesen-style isoperimetric inequalities, Amer. Math. Monthly 86 (1979), 1-29.

26. Preissmann, A., Quelques propriêtés globales des espaces de Riemann. Comment. Math. Helv. 15 (1942), 175-216.

27. Randol, W., Cylinders in Riemann surfaces, Comment. Math. Helv. 54 (1979), 1-5.

28. Roelcke, W., Das Eigenwertproblem der automorphen Formen in der hyperbolischen Ebene, I: Math. Annalen 167 (1966), 292-337, II: Math. Annalen 168 (1967), 261-324.

29. Schoen, R., Wolpert, S., Yau, S.T., Estimates for the first eigenvalue of a compact Riemann surface, preprint.

30. Schoen, R., Geometric bounds on the low eigenvalues, these Proceedings.

31. Thurston, W., The geometry and topology of 3-manifolds, lecture notes, Princeton 1977/78.

32. Wolpert, S., The length spectra as moduli for compact Riemann surfaces, Annals of Math. 109 (1979), 323-351.

33. Yau, S.T., Isoperimetric constants and the first eigenvalue of a compact Riemannian manifold, Ann. Sci Ecole Norm. Sup., IV. Sér. 8 (1975), 487-507.

34. Yau, S.T., On the estimate of the first eigenvalue of a compact Riemannian manifold, these Proceedings.

DEPARTEMENT OF MATHEMATICS
BONN UNIVERSITY
BONN, GERMANY

Current Adress:
Sonderforschungsbereich 40
"Theoretische Mathematik"
Beringstrasse 4

D-5300 Bonn (Germany)

Proceedings of Symposia in Pure Mathematics
Volume 36, 1980

LOWEST-EIGENVALUE INEQUALITIES

Isaac Chavel[1)]

ABSTRACT. We present here some 1-dimensional eigen value inequalities motivated by geometric considerations.

In [6] E.T. Kornhauser and I. Stakgold proposed that if Ω is a simply connected domain in $\mathbb{R}^2$ with compact closure and sufficiently smooth boundary, $\lambda(\Omega)$ the lowest eigenvalue of the Laplacian acting on functions on Ω with Dirichlet (i.e., zero) boundary data, $\mu(\Omega)$ the lowest positive eigenvalue with Neumann (i.e., free) boundary data, and D is the disk with area equal to that of Ω, then

(1) $$\mu(\Omega) \leq \mu(D) ,$$

(2) $$\mu(D) < \lambda(D) ,$$

(3) $$\lambda(D) \leq \lambda(\Omega) .$$

Thus,

(4) $$\mu(\Omega) < \lambda(\Omega)$$

for all relatively compact Ω with sufficiently smooth boundary.

We first note that (4) was derived in [12] independent of the inequalities (1)-(3); and that in [11] it was remarked that (4) implies that the closure of any nodal domain of $\mu(\Omega)$ cannot be contained in Ω.

Next we note that (2) is known to be valid in virtue of explicit approximations to $\lambda(D)$, $\mu(D)$. In §1 below we consider the inequality (2) for a geodesic disk in a standard space form of constant sectional curvature of arbitrary dimension. The inequality is not always valid.

The inequality (3) is the celebrated Faber-Krahn inequality, and is

---

1980 Mathematics Subject Classification 58G20, 35P15

1) Partially supported by NSF Grant MSC 77-02757

valid when Ω and D (now a geodesic disk with same volume as Ω) are both contained in any standard space form of constant sectional curvature of arbitrary dimension. In fact (3) is a consequence of the isoperimetric inequality claiming that the $(\dim \Omega - 1)$-dimensional volume of the boundary of Ω is not less than the corresponding volume of the boundary of D. Similarly, (3) is also valid when Ω is contained in a complete simply connected 2-dimensional Riemannian manifold whose Gauss curvatures are bounded above by a constant $\varkappa$, and D is a geodesic disk, with the same area as Ω, contained in the standard space form of constant Gauss curvature $\varkappa$, cf. [4; 8; 9; 10] for results and references.

As to inequality (1), it was first established for Ω, D in $\mathbb{R}^2$ by G. Szego [13], and extended to Ω, D in $\mathbb{R}^n$, for arbitrary n, by H. Weinberger [14]. Furthermore it is no longer required that Ω be simply connected. It is quite straightforward to adjust Weinberger's argument to be valid when Ω, D are both contained in the standard space form of constant negative sectional curvature. If, on the other hand, the space form in question has constant positive sectional curvature $\varkappa$ then Weinberger's argument requires that Ω be contained in a sufficiently small geodesic disk, e.g., if $\dim \Omega = 2$ then Ω must be contained in a geodesic disk of radius $\pi/4\sqrt{\varkappa}$. To my knowledge this is the current state of the art.

In §3, we present a 1-dimensional isoperimetric inequality of the type (1), where Ω, D are both geodesic 2-disks of revolution about their center, with the Gauss curvatures of Ω bounded above by a constant equal to the constant Gauss curvature of D, and with the same area.

The argument there requires an inequality showing that the length of the boundary of Ω is not less than that of D. Such an inequality was considered in [1], and we present here an elementary Sturmian argument (also valid in higher dimensions via R.L. Bishop's extension of

Rauch's comparison theorem [2]) as an alternative proof.

§1. The eigenvalues $\lambda(\Omega)$, $\mu(\Omega)$ for geodesic disks in space forms

We fix $\varkappa \in \mathbb{R}$, n a positive integer ≥ 2, and let M be the simply connected space form of constant sectional curvature $\varkappa$ and dimension n. We also let S (resp. C) be the solution of the scalar equation

$$y'' + \varkappa y = 0 \tag{5}$$

with initial data: $y(0) = 0$, $y'(0) = 1$ (resp. $y(0) = 1$, $y'(0) = 1$). Then S,C are a basis of solutions of (5) and satisfy

$$C = S', \quad C' = -\varkappa S$$

$$C^2 + \varkappa S^2 = 1,$$

$$(C/S)' = (S'/S)' = -S^{-2}.$$

If we let $\mathbb{S}^{n-1}$ denote the unit sphere of $\mathbb{R}^n$ with $\Box$ denoting the Laplacian acting on functions on $\mathbb{S}^{n-1}$, then the distinct eigenvalues of $\Box$ are given by

$$\nu = \ell(\ell+n-2), \quad \ell = 0,1,2,\ldots \tag{6}$$

For $\ell = 0$ the eigenspace consists of the constant functions, and for $\ell = 1$ the eigenspace is spanned by the orthogonal (with respect to $L^2(\mathbb{S}^{n-1})$) functions

$$x^j \,|\, \mathbb{S}^{n-1}, \quad j = 1,\ldots,n$$

where $\{x^j\}$ are the coordinate functions on $\mathbb{R}^n$. For $\ell = 1$, $\nu = n-1$.

Now let $p \in M$, exp: $M_p \to M$ be the exponential map from M_p, the tangent space of M at p, onto M, and let Δ be the Laplacian acting on functions on M. $\mathbb{S}_p$ will denote the unit sphere in M_p; and for $\varkappa \leq 0$, t will range over $[0,\infty)$, whereas for $\varkappa > 0$, $t \in [0,\pi/\sqrt{\varkappa}]$. For a function $F: M \to \mathbb{R}$ of the form

$$F(\exp t\xi) = T(t)G(\xi), \tag{7}$$

where $\xi \in \mathbb{S}_p$, one has by the standard calculation that F is an eigenfunction of Δ with eigenvalue λ if and only if (i) G is an eigenfunction of $\Box$ on $\mathbb{S}_p$ and ii) for the eigenvalue ν of G given by (6), T must satisfy

(8) $$(S^{n-1}T')' + \{\lambda - \ell(\ell+n-2)S^{-2}\}S^{n-1}T = 0 \quad .$$

For F to be differentiable at p we must have

(9) $$T'(0) = 0 \qquad \text{when } \ell = 0 ,$$
$$T(0) = 0 \qquad \text{when } \ell = 1,2,\ldots \quad .$$

For $\beta > 0$ we let $B(\beta)$ denote the geodesic disk centered at p with radius β. Then the Dirichlet (resp. Neumann) boundary data for $B(\beta)$ are given by

(10) $$T(\beta) = 0 \qquad (\text{resp. } T'(\beta) = 0) \ .$$

When given a fixed ℓ, the set of λ, for which a non-trivial solution of (8) satisfying (9) exists with respective boundary condition (10), is an infinite discrete set of non-negative numbers - when $\ell > 1$ they are all positive. The set of eigenfunctions of each of the boundary value problems of the form (7) are known to determine a complete orthogonal basis of $L^2(B(\beta))$.

Next we note that if for each $\ell = 1,2,\ldots$ we set $\overline{\lambda_\ell}$ (resp. $\overline{\mu_\ell}$) be the lowest eigenvalue of (8) with (9) and respective boundary data (10) then Strum's theorem implies: if $\ell < k$ then $\overline{\lambda_\ell} < \overline{\lambda_k}$, $\overline{\mu_\ell} < \overline{\mu_k}$. So the search for the lowest eigenvalue of both boundary value problems is restricted to considering $\ell = 0,1$. The lowest eigenvalue for the Neumann problem is 0, with constant functions for eigenfunctions, so our interest is in the lowest positive eigenvalue. Direct calculation verifies

<u>Lemma 1</u>: For any solution T of $(8\colon \ell = 0)$ the function $V = T'$ satisfies $(8\colon \ell = 1)$.

<u>Theorem 2</u>: The lowest eigenvalue for the Dirichlet boundary data $\equiv\colon \lambda(\beta)$ is the lowest eigenvalue of $(8\colon \ell = 0)$ -- in particular $\lambda(\beta) > 0$, and the lowest positive eigenvalue for the Neumann boundary data $\equiv\colon \mu(\beta)$ is the lowest eigenvalue of $(8\colon \ell = 1)$.

Thus if $x: M \to \mathbb{R}^n$ is a Riemann normal coordinate system determined by an orthonormal basis of M_p, then a basis of the eigenfunctions of $\mu(\beta)$ is given by the functions

$$(x^j/\|x\|)T(\|x\|) \quad , \quad j = 1,\dots,n$$

where $\|x\|^2 = \Sigma_j (x^j)^2$ and T satisfies

$$(S^{n-1}T') + \{\mu(\beta) - (n-1)S^{-2}\}S^{n-1}T = 0 ,$$

$$T(0) = T'(\beta) = 0 , \quad T'|[0,\beta) > 0 .$$

Finally, if $\varkappa \le 0$ then $\mu(\beta) < \lambda(\beta)$ for all $\beta > 0$. If, on the other hand, $\varkappa > 0$, then $\mu(\beta) < \lambda(\beta)$ for $\beta < \pi/2\sqrt{\varkappa}$, $\mu(\pi/2\sqrt{\varkappa}) = \lambda(\pi/2\sqrt{\varkappa}) = n$, and $\mu(\beta) > \lambda(\beta)$ for $\beta \in (\pi/2\sqrt{\varkappa}, \pi/\sqrt{\varkappa})$.

<u>Theorem 3</u>: If $\varkappa < 0$ then $\mu(\beta)/\lambda(\beta) \to 0$ as $\beta \to \infty$. If $\varkappa = 0$, then $\mu(\beta)/\lambda(\beta) = \text{const.}$, and if $\varkappa > 0$ then $\mu(\beta)/\lambda(\beta) \to \infty$ as $\beta \to \pi/\sqrt{\varkappa}$.

We first remark that Theorem 2 follows from Lemma 1 used in conjunction with standard Sturmian arguments; we only add that for $\beta = \pi/2\sqrt{\varkappa}$ we have $T(t) = \cos\sqrt{\varkappa}\, t$ provides the solution to the Dirichlet problem and $V(t) = T'(t) = \sqrt{\varkappa}\sin\sqrt{\varkappa}\, t$ for the Neumann problem.

For Theorem 3 just note that when $\varkappa < 0$ we have $\lambda(\beta) \ge (n-1)^2\varkappa/4$ by [7]. To estimate $\mu(\beta)$ from above, set $\varkappa = -1$, and use the Poincare model of hyperbolic space: i.e., on $\mathbb{B}^n$, the unit ball in $\mathbb{R}^n$, the metric is given by

$$ds^2 = \frac{4\|dx\|^2}{\{1 - \|x\|^2\}^2} .$$

A geodesic disk about $x = 0$ of radius β is given by the Euclidean disk $\mathbb{B}^n(\alpha)$ about $x = 0$ of radius $\alpha = \tanh \beta/2$. For a test function in the Rayleigh quotient pick the coordinate function $F(x) = x^1$. Let $d\mu$ denote the Lebesgue measure on $\mathbb{R}^n$. Then

$$2^n \int_{\mathbb{B}^n(\alpha)} \frac{F(x)\,d\mu}{\{1 - \|x\|^2\}^n} = 0 ,$$

so F is an admissible test function. The Dirichlet integral of F is given by

$$2^n \int_{\mathbb{B}^n(\alpha)} \frac{d\mu}{\{1-\|x\|^2\}^{n-2}} \le \begin{array}{ll} \text{const.} & n = 2 \\ \text{const.}\,|\log(1-\alpha)| & n = 3 \\ \text{const.}\,(1-\alpha)^{-n+3} & n > 3 \end{array}$$

The L^2-integral of F is estimated below, for $\alpha > \frac{1}{2}$, by the L^2-integral over the domain $\{x: x^1 > ((x^2)^2 + \ldots + (x^n)^2)^{\frac{1}{2}}, \frac{1}{2} < \|x\| < \alpha\}$. Thus one easily obtains

$$2^n \int_{\mathbb{B}^n(\alpha)} \frac{(x^1)^2}{\{1-\|x\|^2\}^n}\, d\mu \ge \text{const.}\ (1-\alpha)^{-n+1} ,$$

from which one has $\mu(\beta) \to 0$ as $\beta \to \infty$ (my thanks to B. Randol for this argument).

For $\varkappa = 0$ one simply has $\lambda(\beta)$, $\mu(\beta)$ are both given by const./β^2.

For $\varkappa > 0$ it is a result of MacDonald [5, p. 408; 3] that $\lambda(\beta) \to 0$ as $\beta \to \pi/\sqrt{\varkappa}$. To estimate $\mu(\beta)$ from below fix $\varkappa = 1$ and consider the stereographic projection of S^n to $\mathbb{R}^n$. Then the metric is given by

$$ds^2 = \frac{4\|dx\|^2}{\{1+\|x\|^2\}^2}$$

and a Euclidean disk $\mathbb{B}^n(\alpha)$ of radius α about $x = 0$ corresponds to a geodesic disk in S^n of radius β where $\alpha = \tan \beta/2$. Set $r = \|x\|$, and let grad denote the Euclidean gradient. Then the eigenfunctions of $\mu(\beta)$ are of the form $F^j(x) = x^j f(r)/r$, $j = 1,\ldots,n$. An explicit calculation shows that

$$\Sigma_j \|\text{grad } F^j\|^2 = f'^2 + (n-1) f^2/r^2 .$$

Thus

$$\mu(\beta) = \frac{\int_0^\alpha \{f'^2 + (n-1)f^2/r^2\}\{1+r^2\}^{2-n}\, r^{n-1}\, dr}{\int_0^\alpha f^2\{1+r^2\}^{-n}\, r^{n-1}\, dr} \ge n-1$$

§2. An isoperimetric inequality for geodesic disks

Theorem 4: Let $\beta > 0$, $\varphi, \psi : [0,\beta] \to [0,\infty) \in C^k$, $k \geq 2$ such that on $(0,\beta)$ we have

(11) $$\varphi, \psi > 0 \ , \ (\varphi/\psi)' > 0 \ ,$$

(12) $$\varphi(x) \sim \psi(x) \quad \text{as} \quad x \to 0 \ .$$

Furthermore assume that

(13) $$(\psi'/\psi)' < 0 \ , \ \psi' > 0$$

on $(0,\beta)$. Set

(14) $$\Phi(x) = \int_0^x \varphi \ , \quad \Psi(x) = \int_0^x \psi$$

and define $y: [0,\beta] \to [0,\infty)$ by

(15) $$\Phi(y(x)) = \Psi(x) \ .$$

Then y is well-defined, $y|(0,\beta] \in C^{k+1}$ and

(16) $$0 \leq y \leq x \ , \quad x \in [0,\beta] \ ,$$

(17) $$0 < y' \leq 1 \ , \quad x \in (0,\beta) \ ,$$

(18) $$\psi \leq \varphi \circ y \ , \quad x \in (0,\beta] \ .$$

If for some $x_0 \in (0,\beta]$ we have equality on the right side of any of (16), (17), (18) at $x = x_0$, then $\varphi = \psi$ on $[0,x_0]$.

Proof: Of course (11), (12) imply that

(19) $$\psi \leq \varphi \ , \quad x \in [0,\beta]$$

with equality at $x_0 \in (0,\beta]$ if and only if $\varphi = \psi$ on $[0,x_0]$. Thus y is well-defined and (16) is immediately valid on $[0,\beta]$ with equality at $x_0 \in (0,\beta]$ if and only if $\varphi = \psi$ on $[0,x_0]$.

For the differential properties of y and inequalities (17), (18) we consider the differential equation

(20) $$(\varphi \circ z)z' = \psi$$

on $(0,\beta)$, fix $x_1 \in (0,\beta)$, $y_1 \in (0,x_1]$ for which $\Phi(y_1) = \Psi(x_1)$, and let $y_1(x)$ be the solution of (20) on $(0,\beta]$ satisfying $y_1(x_1) = y_1$. Then $y_1(x)$ satisfies (15); thus $y(x) = y_1(x) \in C^{k+1}$, and

(21) $$(\varphi \circ y)y' = \psi \ ,$$

(22) $$(\varphi' \circ y)y'^2 + (\varphi \circ y)y'' = \psi' \ .$$

From (11)-(13), (22), and the fact that $y' > 0$, we have

(23) $$y'' \le y'(1-y')(\psi' \circ y)/(\psi \circ y) \ .$$

An immediate consequence of (23) is that at any $x \in (0,\beta)$ where we have $y'(x) > 1$ we must have $y''(x) < 0$.

Assume there exists $x_2 \in (0,\beta)$ such that $y'(x_2) > 1$. The previous remark implies now that $y'(x) > y'(x_2)$ on $(0,x_2)$ which in turn implies for any ϵ in $(0,x_2)$

$$y(x_2) - y(\epsilon) > y'(x_2)(x_2 - \epsilon) \ .$$

Let $\epsilon \to 0$. Then we have

$$y(x_2) \ge y'(x_2)\, x_2 > x_2$$

a contradiction to (16). So (17) is proven, and (18) follows from (17), (21).

For the case of equality in (17) or (18) we first note that the argument leading to (23) implies that for any x satisfying $y(x) < x$ we also have the strict inequality

(24) $$y'' < y'(1 - y')(\psi' \circ y)/(\psi \circ y) \ .$$

Assume $y'(x_0) = 1$ for a given $x_0 \in (0,\beta)$. We claim $y(x_0) = x_0$. If not then $y(x_0) < x_0$ and (24) implies $y''(x_0) < 0$. Thus $y'(x) > 1$ for sufficiently small $x - x_0 < 0$, which contradicts (17).

<u>Remark 5</u>: It is easy to show that Theorem 4 remains valid if we replace the second inequality in (13) by the hypothesis that there exists $x_3 \in (0,\beta)$ such that $\psi'|(0,x_3) > 0$, $\psi'|(x_3,\beta) < 0$.

<u>Remark 6</u>: If for some $\ell \ge 1$ we also have

$$\varphi \sim \psi \sim x^{\ell} \quad \text{as} \quad x \to 0$$

then

(25) $$\lim_{x\to 0} \frac{y(x)}{x} = \lim_{x\to 0} y'(x) = 1 \ ,$$

(26) $$\lim_{x\to 0} y(x)y''(x) = 0 .$$

Remark 7: If $\varphi_\varkappa(x) = \varkappa^{-n+1} \sinh^{n-1} \varkappa x$, $\psi(x) = \sinh^{n-1} x$ and $y_\varkappa(x)$ is given by (15) for each $\varkappa > 1$ one easily shows that for each $x > 0$,

$$y_\varkappa(x) \sim \frac{(n-1)}{n}\frac{\log \varkappa}{\varkappa} \quad \text{as} \quad \varkappa \to \infty .$$

§3. A 1-dimensional Neumann isoperimetric inequality

Theorem 8: Let β, φ, ψ be as in Theorem 4 with

(27) $$\varphi \sim \psi \sim x + O(x^3) \quad \text{as} \quad x \to 0 .$$

Consider the eigenvalue problems

(28) $$(\varphi u')' + \{\mu_\varphi - \varphi^{-2}\}\varphi u = 0$$

(29) $$u(0) = u'(y(\beta)) = 0 \;,\; u'|[0,y(\beta)) > 0 \;,$$

and

(30) $$(\psi v')' + \{\mu_\psi - \psi^{-2}\}\psi v = 0$$

(31) $$v(0) = v'(\beta) = 0 \;,\; v'|[0,\beta) > 0$$

then

(32) $$\mu_\varphi \le \mu_\psi$$

with equality if and only if $\varphi = \psi$ on $[0,\beta]$.

Proof: We use Prufer's method. For convenience assume $u'(0) = v'(0) = 1$ and define $\tau\colon [0,\beta] \to [0,\pi/2]$ by

$$\tau = \arctan (v/\psi v') \quad .$$

Then $\tau \in C^k$, $\tau(0) = \pi/4$, $\tau(\beta) = \pi/2$, and $\tau\colon (0,\beta) \to (0,\pi/2)$. The differential equation for τ is given by

(33) $$\tau' = \psi^{-1} \cos 2\tau + \mu_\psi \sin^2 \tau .$$

One now explicitly checks that the Taylor expansion for τ about $x = 0$ is given by

(34) $$\tau(x) = \pi/4 + \mu_\psi x^2/8 + o(x^2) \quad .$$

Next one notes that since $\mu_\psi > 0$ we have $\tau(x) > \pi/4$ for sufficiently small $x > 0$. If for any $x_0 \in (0,\beta)$ we have $\tau(x_0) = \pi/4$ then (33) implies that $\tau'(x_0) > 0$. One has, therefore, that τ: $(0,\beta) \to (\pi/4,\pi/2)$.

In a similar manner we define θ: $[0,y(\beta)] \to [0,\pi/2]$ by

$$\theta = \arctan(u/\varphi u') \quad ;$$

however, here we also define Θ: $[0,\beta] \to [0,\pi/2]$ by

$$\Theta = \theta \circ y \quad .$$

Then for Θ we have $\Theta(0) = \pi/4$, $\Theta(\beta) = \pi/2$, Θ: $(0,\beta) \to (\pi/4,\pi/2)$, $\Theta'(0) = 0$, and $\Theta''(x) \to \mu_\varphi/4$ as $x \to 0$. The differential equation for Θ is

$$\Theta' = y'^2 \psi^{-1} \cos 2\Theta + \mu_\varphi \psi \sin^2 \Theta . \tag{35}$$

We now wish to show that $y(\beta) < \beta$ implies $\mu_\varphi < \mu_\psi$. So assume $y(\beta) < \beta$ and $\mu_\varphi \geq \mu_\psi$. Then $y'(\beta) < 1$, and

$$(\Theta - \tau)' = (y'^2 \cos 2\Theta - \cos 2\tau)\psi^{-1} + (\mu_\varphi \sin^2\Theta - \mu_\psi \sin^2 \tau)\psi$$

$$\geq (\cos 2\Theta - \cos 2\tau)\psi^{-1} + \mu_\psi\left\{\frac{\sin^2\Theta - \sin^2\tau}{\Theta - \tau}\right\}(\Theta - \tau)\psi \quad .$$

For any $x_1, x_2 \in [0,\beta]$, $x_1 < x_2$ such that $\Theta - \tau < 0$ on (x_1,x_2) we therefore have

$$(\Theta - \tau)' > \mu_\psi(\Theta - \tau)\psi \quad ,$$

which implies for Ψ given by (14)

$$\{e^{-\mu_\psi \Psi}(\Theta - \tau)\}' > 0 \quad .$$

Thus

$$(\Theta - \tau)(x_1) < e^{-\mu_\psi(\Psi(x_2)-\Psi(x_1))}(\Theta - \tau)(x_2) \quad .$$

In particular if we are given any $x_3 \in (0,\beta)$ for which $(\Theta - \tau)(x_3) < 0$ then $\Theta - \tau < 0$ on $[0,x_3]$. But for $x = \beta$ we have $\Theta(\beta) = \tau(\beta) = \pi/2$, $y'(\beta) < 1$ and

$$(\Theta - \tau)'(\beta) \geq (1 - y'^2(\beta))\psi^{-1}(\beta) > 0$$

Thus $(\Theta - \tau)(x) < 0$ for sufficiently small $x - \beta < 0$, which implies $\Theta - \tau < 0$ on $[0,\beta)$ -- a contradiction, since $\Theta(0) = \tau(0) = \pi/4$.

REFERENCES

[1] T. Aubin, Problemes isoperimetriques et espaces de Sololev, J. Diff. Geometry, 11(1976), 573-598.

[2] R. Bishop and R. Crittenden, Geometry of Manifolds, Academic Press, New York, 1964.

[3] I. Chavel and E. A. Feldman, Spectra of domains in compact manifolds, J. Funcl. Anal. , 30(1978, 198-222.

[4] ____________________, Isoperimetric inequalities on curved surfaces, to appear in Adv. in Math.

[5] E. W. Hobson, The Theory of Spherical and Ellipsoidal Harmonics, Cambridge University Press, London, 1931.

[6] E. T. Kornhauser and I. Stakgold, A variational theorem for $Su + \lambda u = 0$ and its application, J. of Math. and Physics, 31(1952), 45-54.

[7] H. P. McKean, Jr., An upper bound to the spectrum of Δ on a manifold of negative curvature, J. Diff. Geometry, 4(1970), 359-366.

[8] R. Osserman, The isoperimetric inequality, Bull. A.M.S., 84(1978), 1182-1238.

[9] __________, Bonneson-style isoperimetric inequalities, Amer. Math. Monthly, 86(1979, 1-29.

[10] J. Peetre, A generalization of Courant's modal line theorem, Math. Scand., 5(1957), 15-20.

[11] A. Pleijel, Remarks on Courant's modal line theorem, Comm. Pure and Appl. Math., 9(1956), 543-550.

[12] G. Polya, Remarks on the foregoing paper, J. Math. and Physics, 31(1952), 55-57.

[13] G. Szego, Inequalities for certain eigenvalues of a membrane of given area, J. Rat. Mech. and Anal., 3(1954), 343-356.

[14] H. F. Weinberger, On the isoperimetric inequality for the u-dimensional free membrane problem, J. Rat. Mech. and Anal., 5(1956), 633-636.

DEPARTMENT OF MATHEMATICS
CITY COLLEGE OF C.U.N.Y.
New York, N.Y. 10031

Proceedings of Symposia in Pure Mathematics
Volume 36, 1980

ON THE HODGE THEORY OF RIEMANNIAN PSEUDOMANIFOLDS

Jeff Cheeger

ABSTRACT. We give an extension of Hodge Theory to a certain class of metrics on pseudomanifolds and briefly describe some applications to the topology of singular algebraic varieties which admit a Kahler metric in this class.

0. INTRODUCTION

In [5], we described an extension of the spectral theory of the Laplace operator $\Delta = d\delta + \delta d$ to riemannian manifolds with cone-like singularities and indicated some consequences of a further extension to the piecewise smooth case. Two striking features were the role played by the new homology theory of Goresky-MacPherson [18], and the necessity in certain instances, to introduce "ideal boundary conditions" - even when the space in question had no boundary in the topological sense. The connection with the work of Goresky-MacPherson (and also Morgan [24]) was pointed out by Dennis Sullivan who observed that our computations of L^2-cohomology and harmonic forms, agreed with the answers obtained by Goresky-MacPherson in the conical case.

Details of the approach of [5] will appear in [7]; see also [9]. Here, our main purpose is to give a more direct route to that part of the theory which is concerned with the Hodge Theorem. This approach yields only the boundedness and not the complete continuity of the Green's operator in the compact

1980 Mathematics Subject Classification 58A14
Partially supported by NSF Grant MCS7802679-02

case. For the most part, we also avoid the issue of "ideal boundary conditions". However, the present discussion does provide the version of Stokes' Theorem which is required in [7], and in that sense it is a necessary preliminary. Moreover, the methods we are going to describe are "softer" than those [7], in that they depend only on <u>local</u> <u>regularity</u> for Δ. Thus, they are more easily suited to generalization. We illustrate this point by simulatiously treating both cones and "horns" (in the sense of Section 2). There is strong reason to think that metrics which are considerably more general than these can also be treated by closely related techniques.

The most interesting application of the Hodge Theorem proved here concerns the topology of singular algebraic varieties which admit a Kahler metric of the particular type under consideration. It is a formal consequence of our theory that the Goresky-MacPherson cohomology of these varieties has many of the special properties of the ordinary cohomology of nonsingular Kahler manifolds. This was conjectured jointly with MacPherson. It will be discussed further in [8].

Certain applications of the more elaborate theory of [7], go beyond the methods of the present work. Two of these, which were briefly indicated in [5], are described in more detail in Section 7.

The remainder of this paper is divided into seven sections as follows:

1. Preliminaries, Piecewise Smooth Equivalence
2. Collars and Horns, Stokes' Theorem
3. Collars and Horns, the Poincaré Lemma and Bounded d^{-1}
4. Some Standard Lemmas
5. Handlebodies and the Hodge Theorem

6. The Pairing With Goresky-MacPherson
7. Applications and Further Results.
8. Appendix

We are greatful to Mark Goresky, Bob MacPherson, Dennis Sullivan and Michael Taylor, for inspiring conversations. We refer to [3], [11], [20], [28], for some previous works on L^2-cohomology and to the article of Singer [26], for discussion of possible extensions of Hodge Theory to singular spaces.

1. PRELIMINARIES

Let Y be <u>any</u> (in general incomplete) riemannian manifold with metric g and $\partial Y = \phi$. Let Λ^i, Λ^i_0 denote the spaces of C^∞ i-forms and C^∞ i-forms of compact support, and let L^2 denote the space of square integrable i-forms with measurable coefficients. Let $\overline{d\Lambda_0^{i-1}}$, $\overline{\delta\Lambda_0^{i-1}}$ denote the closures in L^2 of the images of Λ_0^{i-1}, Λ_0^{i+1} under the operators d, δ. Finally, let $\mathcal{H}^i$ denote the space of L^2 harmonic forms θ satisfying $d\theta = \delta\theta = 0$.

Recall that $d\alpha = \eta$ in the <u>weak</u> sense means that

$$(1.1)\quad \langle \eta, \beta\rangle = \langle \alpha, \delta\beta\rangle$$

holds for all $\beta \in \Lambda_0^{i+1}$. If we define the domain of d by

$$(1.2)\quad \operatorname{dom} d = \{\alpha \in \Lambda^i \cap L^2 \mid d\alpha \in L^2\},$$

then (1.1) holds for all $\alpha \in \operatorname{dom} d$, with $\eta = d\alpha$. Since $\{\beta \in \Lambda_0^{i+1}\}$ is dense in L^2, d has a well defined <u>strong closure</u>, $\bar{d}$. $\alpha \in \operatorname{dom} \bar{d}$, $d\alpha = \eta$, if there exists $\eta \in L^2$, and $\alpha_j \to \alpha$, $d\alpha_j \to \eta$. By results of [10], [15], <u>$d\alpha = \eta$ in the weak sense if and only if $\alpha \in \operatorname{dom} \bar{d}$ and $\bar{d}\alpha = \eta$.</u> The proof depends on the existence of regularization operators R_ϵ; (see the Appendix, Section 8).

Let

(1.3) $H^i_{(2)}(Y) = \ker d_i/\text{range } d_{i-1}$,

(1.4) $H^i_{(2),\#}(Y) = \ker \bar{d}_i/\text{range } \bar{d}_{i-1}$,

denote the i-th L^2-cohomology groups of Y, for $d, \bar{d}$. Note that there are natural maps $\iota_{\mathcal{H},\#} = \iota_{(2)} \circ \iota_{\mathcal{H}}$, as follows.

(1.5)
$$\begin{array}{ccc} \mathcal{H}^i & \xrightarrow{\iota_{\mathcal{H}}} & H^i_{(2)}(Y) \\ & \underset{\iota_{\mathcal{H},\#}}{\searrow} & \downarrow \iota_{(2)} \\ & & H^i_{(2),\#}(Y) \end{array}$$

In fact, the map $\iota_{(2)}$ is always an isomorphism. This follows from the existence of bounded operators A_ϵ, which preserve smoothness and satisfy $\bar{d}A_\epsilon + A_\epsilon \bar{d} = I - R_\epsilon$; see Section 8. Thus we can use either $H_{(2)}(Y)$ or $H_{(2),\#}(Y)$ as convenience dictates. If $\iota_{\mathcal{H},\#} : \mathcal{H}^i \to H_{(2),\#}(Y)$ is an isomorphism, we will say that the Strong Hodge Theorem holds. There is a natural pseudo norm on $H^i_{(2),\#}(Y) \simeq H^i_{(2)}(Y)$, given by

(1.6) $\|u\| = \inf_{\alpha \in u} \|\alpha\|$.

It follows trivially that (1.6) is a norm if and only if $\bar{d}_{i-1}$ has closed range. As a direct consequence of the Hodge Theorem 1.1, this is equivalent to the surjectivity of $\iota_{\mathcal{H},\#}$. It also follows directly from Theorem 1.1 that in this case, $\iota_{\mathcal{H},\#}|(\text{range } d_{i-1})^\perp$ is an isometric **injection.** Moreover, since $\bar{d}_i$ is a closed operator, $\ker \bar{d}_i$ is a closed subspace. Since $\bar{d}_{i-1}$ is also closed, as a consequence of the Open Mapping Theorem, if $H_{(2),\#}(Y)$ is finite dimensional, then range $\bar{d}$ is closed; (in case range $\bar{d}_{i-1}$ is not closed, one commonly replaces range $\bar{d}_{i-1}$ by its closure in (1.4) and studies the resulting groups).

The <u>injectivity</u> of $\iota_{\mathcal{H},\#}$ follows in particular if <u>Stokes' Theorem holds in the L^2 sense</u>. This means that

$$(1.7)\quad \langle \bar{d}\alpha, \beta\rangle = \langle \alpha, \bar{\delta}\beta\rangle,$$

for all $\alpha \in \operatorname{dom} \bar{d}, \beta \in \operatorname{dom} \bar{\delta}$, or equivalently, by the discussion above, for all $\alpha \in \operatorname{dom} d, \beta \in \operatorname{dom} \delta$. Then with a suitable, with a choice of domain, Δ is essentially self-adjoint, (see [17]). By a fundamental result of Gaffney [16], (1.7) holds for Y <u>complete</u>; but here we are primarily interested in certain <u>incomplete</u> cases.

We now state the general Hodge Theorem of Kodaira, (see [10], [23]). Y is still arbitrary.

THEOREM 1.1 (Kodaira). There is an orthogonal direct sum decomposition

$$L^2 = \overline{\delta\Lambda_0^{i+1}} \oplus \overline{d\Lambda_0^{i-1}} \oplus \mathcal{H}^i,$$

which leaves invariant the space $\Lambda^i \cap L^2$.

Since $\alpha \in \ker \bar{d}$ clearly implies, $\alpha \in \overline{d\Lambda_0^{i-1}} \oplus \mathcal{H}^i$, it follows from Theorem 1.1, that the Strong Hodge Theorem is equivalent to $\overline{d\Lambda_0^{i-1}} = \operatorname{range} \bar{d}_{i-1}$. By use of the Hodge $*$ operator, if necessary on a 2-fold cover, this is equivalent to $\overline{\delta\Lambda_0^{n-i+1}} = \operatorname{range} \bar{\delta}_{n-i+1}$. Theorem 1.1 is of necessity, much weaker than the corresponding statement for closed manifolds. In that case we have,

A) $H^i_{(2),\#}(Y) = H^i_{(2)}(Y)$ is finite dimensional,

B) Stokes' Theorem holds in the L^2 sense,

and thus the Strong Hodge Theorem holds. Moreover, we can define d^{-1}, $\bar{d}^{-1}$ by setting $d^{-1}, \bar{d}^{-1} | (\operatorname{range} d)^{\perp} \equiv 0$, and

$$(1.8)\quad d^{-1}(\eta), \bar{d}^{-1}(\eta) = \alpha_{ce}$$

for $\eta \in$ range $d, \bar{d}$. Here $d\alpha = \eta$, $\bar{d}\alpha = \eta$, respectively and α_{ce}, α_e, $\alpha_{\mathcal{H}}$ are the components of α corresponding to the decomposition in Theorem 1.1. Note that

$$(1.9)\quad d\alpha_{\mathcal{H}} = \bar{d}\alpha_e = \bar{\delta}\alpha_{ce} = 0.$$

Thus for such η, $d(d^{-1}\eta)$, $\bar{d}(\bar{d}^{-1}\eta) = \eta$. δ^{-1}, $\bar{\delta}^{-1}$ are defined similarly. The identity

$$(1.10)\quad \bar{d}A_\epsilon\bar{d}\alpha = \bar{d}\alpha - dR_\epsilon\alpha$$

implies

$$(1.11)\quad \bar{d}A_\epsilon\eta = \eta - dR_\epsilon\alpha.$$

Since A_ϵ preserves smoothness, if $\eta \in \Lambda^i \cap L^2$, then $A_\epsilon\eta + R_\epsilon\alpha \in \Lambda^i \cap L^2$. Thus range $\bar{d} \cap \Lambda^i$ = range d. Then, since $\pi_{ce} : \Lambda^i \cap L^2 \to \Lambda^i \cap L^2$, in fact, $\bar{d}^{-1}$: range $\bar{d} \cap \Lambda^i \to \Lambda^i \cap L^2$. Finally, by the Open Mapping Theorem, <u>if range $\bar{d}$ is closed, then $\bar{d}^{-1}$ is a bounded operator</u>. If Y is a closed manifold, then moreover,

C) $\bar{d}^{-1}$, $\bar{\delta}^{-1}$ are compact operators.

It is important to note that <u>every property</u> we have mentioned so far depends only on the <u>equivalence class</u> of the metric g, on Y. We say that g' is <u>equivalent</u> to g if for some $c > 0$,

$$(1.12)\quad \frac{1}{c}\, g \le g' \le cg.$$

For example, the spaces L^2, $\overline{d\Lambda_0^{i-1}}$, range $\bar{d}$, depend only on the equivalence class of g and the Strong Hodge Theorem is just the statement $\overline{d\Lambda_0^{i-1}}$ = range $\bar{d}$. We emphasize once and for all that in the sequel, although it will be convenient to carry out certain computations for <u>special metrics</u>, the results will automatically hold for <u>equivalent metrics</u>.

There is a more general notion of equivalence which will be

important, namely, piecewise smooth equivalence. Let Y, Y' be riemannian manifolds with metrics g, g' and smooth triangulations T, T'; (if Y, Y' are open, T, T' are infinite). A homeomorphism $f : Y' \to Y$ will be called a piecewise smooth equivalence if there exist smooth subdivisions S of T and S' of T' for which the restriction of f to each closed top dimensional simplex $(\overline{\sigma'})^n \in S'$ is a diffeomorphism onto some $\overline{\sigma}^n \in S$. f will be called a piecewise smooth riemannian equivalence if for some constant c and all simplices $(\overline{\sigma'})^n$,

(1.13) $\frac{1}{c} g' \leq f^*(g) \leq cg'$.

Let Σ^i denote the i-skeleton of some S as above, and let $\alpha \in \operatorname{dom} d_{Y-\Sigma^{n-1}}$. α will be called a piecewise smooth form on Y (relative to T), if for each $(n-1)$-simplex $\sigma^{n-1} \in S$, the tangential component of α (relative to a piecewise smooth local coordinate system with $\sigma^{n-1} = \{x \mid x_n = 0\}$) is continuous across σ^{n-1}. We will temporarily denote the form $d\alpha \mid (Y - \Sigma^{n-1})$ by $\hat{d}\alpha$, and regard it as a form in $L^2(Y)$. Clearly, if $f : Y' \to Y$ is a piecewise smooth equivalence, then f induces a norm bounded isomorphism such that $\hat{d}f^* = f^*\hat{d}$. The following Lemma 1.1 which will be proved by an argument that is somewhat in the spirit of the arguments of Section 2, is in essence, well known; compare [16], [21]. It is considerably weaker than the Stokes' Theorem of Sections 2 and 5, which is required in order to treat the Strong Hodge Theorem for the singular spaces considered in this paper. In particular, the existence of the phenomenon of "ideal boundary conditions" is not revealed; compare [16]. However, the special case of the Strong Hodge Theorem for piecewise smooth manifolds can be treated on the basis of Lemma 1.1. Very recently,

N. Teleman has proved this theorem independently; (see [27]).

LEMMA 1.1. Let Y be as above and α be piecewise smooth. Then $\alpha \in \operatorname{dom} \bar{d}$ and $\bar{d}\alpha = \hat{d}\alpha$. Moreover, if $f : Y \to Y'$ is piecewise smooth riemannian equivalence, then f induces $f^*: \operatorname{dom} \bar{d}_{Y'} \to \operatorname{dom} \bar{d}_Y$, and for all $\alpha' \in \operatorname{dom} \bar{d}_{Y'}$, $\bar{d} f^*(\alpha') = f^*(\bar{d}\alpha')$.

PROOF. Since f^* induces a norm bounded isomorphism between the respective spaces of piecewise smooth forms, and $\hat{d}f^* = f^*\hat{d}$, it clearly suffices to establish that α piecewise smooth, implies $\alpha \in \operatorname{dom} \bar{d}$ and $\bar{d}\alpha = \hat{d}\alpha$. Let $\beta \in \Lambda_0^{n-i+1}(Y)$ be a smooth form having compact support, and $L \subset S$ a finite subcomplex containing the support of β. Let $\epsilon_2 < <\epsilon_3 \ldots <\epsilon_n$ be small positive numbers and let T_{j,ϵ_j} denote the set of points at distance $\leq \epsilon_i$ from Σ^{n-j}, with respect to some piecewise flat metric which is equivalent to g on L. Since α is piecewise smooth, if we apply Stokes' Theorem on the region $L - \bigcup_{j=2}^{n} (\Sigma^{n-j})$, the contributions from (n-1) dimensional faces cancel, (or $\beta \equiv 0$) and we get

$$(1.14) \quad \int_{L - \bigcup_{j=2}^{n} T_{j,\epsilon_j}} d\alpha \wedge *\beta = \int_{\bigcup_{j=2}^{n} \partial_{j,\epsilon_j}} \alpha \wedge *\beta + \int_{L - \bigcup_{j=2}^{n} T_{j,\epsilon_j}} (-1)^{i+1} \alpha \wedge d*\beta.$$

Here ∂_{j,ϵ_j} denotes the piece of $\partial(L - \bigcup_{j=2}^{n} T_{j,\epsilon_j})$ consisting of points on $\partial T_{j,\epsilon_j}$. (Note if Y is not orientable, $*$, and the orientation of the domain of integration, are chosen locally). In order to see that the boundary integrals in (1.14) vanish, we will use the following simple estimate whose proof will be recalled in Lemma 1.2 below. Let $h(r) \in L^1(0,1)$. Then there exists a sequence $\epsilon_s \to 0$ such that

$$(1.15) \quad |h(\epsilon_s)| = o(\epsilon_s^{-1} \cdot |\log \epsilon_s|^{-1}).$$

Now suppose we have already shown that if $\epsilon_{j_0},\ldots\epsilon_n$ are held fixed and $\epsilon_2,\ldots\epsilon_{j_0-1}$ are made successively to approach zero, then the contributions to (1.14) due to $\partial_{2,\epsilon_2},\ldots\partial_{j_0-1,\epsilon_{j_0-1}}$ vanish. Here we allow $j_0 = 2$ so as to start the argument. Since $\alpha \in L^2(Y)$, in particular, $\alpha \in L^2(T_{j_0,\epsilon})$ for some fixed $\epsilon > 0$. Then by Fubini's Theorem,

$$(1.16)\quad \int_0^{\epsilon}\int_{\partial_{j_0,r}} \|\alpha\|^2\, dA_r \cdot dr < \infty,$$

where dA_r denotes the element of area on $\partial_{j_0,r}$ and $A(\partial_{j_0,r})$ the area of $\partial_{j_0,r}$. By (1.15), there exists a sequence $\epsilon_s \to 0$, such that

$$(1.17)\quad \int_{\partial_{j_0,\epsilon_s}} \|\alpha\|^2\, dA_{\epsilon_s} = o(\epsilon_s^{-1}\cdot|\log \epsilon_s|^{-1}).$$

Let ϵ_{j_0} tend to zero through the sequence ϵ_s, and let K be an upper bound for the pointwise norm of β. Then

$$(1.18)\quad \left|\int_{\partial_{j_0,\epsilon_s}} \alpha\wedge*\beta\right| \le \int_{\partial_{j_0,\epsilon_s}} \|\alpha\|\cdot\|\beta\|\cdot dA_{\epsilon_s},$$

$$\le K\int_{\partial_{j_0,\epsilon_s}} \|\alpha\|\cdot dA_{\epsilon_s},$$

$$\le K\left(\int_{\partial_{j_0,\epsilon_s}} dA_{\epsilon_s}\right)^{\frac12}\left(\int_{\partial_{j_0,\epsilon_s}} \|\alpha\|^2\cdot dA_{\epsilon_s}\right)^{\frac12},$$

$$\le K(A(\partial_{j_0,\epsilon_s}))^{\frac12}\left(\int_{\partial_{j_0,\epsilon_s}} \|\alpha\|^2\cdot dA_{\epsilon_s}\right)^{\frac12},$$

$$\le K'\epsilon_s^{(j_0-1)/2}\cdot o(\epsilon_s^{-\frac{1}{2}}\cdot|\log\epsilon_s|^{-\frac{1}{2}}),$$

$$= K'\circ o(1),$$

where the last step follows because $j_0 \ge 2$. Since $\alpha, \hat{d}\alpha, \beta, \delta\beta \in L^2(Y)$, it follows that the limit is actually independent of the particular sequence chosen, and from (1.14) we conclude by induction that

(1.19) $\langle\hat{d}\alpha, \beta\rangle = \langle\alpha, \delta\beta\rangle$.

Thus $\alpha \in \mathrm{dom}\,\bar{d}$ and $\bar{d}\alpha = \hat{d}\alpha$, which suffices to complete the proof.

We now give the proof of (1.15) in the following lemma which will be used repeatedly.

LEMMA 1.2. Let $h(r) \in L^1(0,1)$ and let $k(r) \ge 0$ be any function such that $k|[a,1] \in L^1[a,1]$ for any a, but $k(r) \notin L^1(A)$ for some $A \subset (0,1)$. Then A contains a sequence $\epsilon_s \to 0$, for which

(1.20) $|h(\epsilon_s)| = o(k(\epsilon_s))$.

In particular, taking $k(r) = \frac{1}{r\cdot|\log r|}$, we have (1.15).

PROOF. Since $k(r) \in L^1[a,1]$ and $k(r) \notin L^1(A)$, it follows that for all a, $k(r) \notin L^1(A\cap(0,a])$. If we had $|h(r)| \ge \frac{1}{s}k(r)$ on $A\cap(0,\frac{1}{s}]$, it would follow that $h(r)|(0,\frac{1}{s}] \notin L^1(0,\frac{1}{s}]$. So choose $\epsilon_s \in A\cap(0,\frac{1}{s}]$, with $|h(\epsilon_s)| \le \frac{1}{s}k(\epsilon_s)$.

In the sequel we will employ the frame work of the present section to study analysis on $X - \Sigma$ where X is a riemannian space with singularities, Σ. By this we mean that X is a metric space such that for some closed subset Σ, $X - \Sigma$ is a smooth dense (incomplete) riemannian manifold in the induced metric. More specifically, in this paper X will be a closed pseudomanifold, (i.e. n-circuit), satisfying a certain local topological condition to avoid the issue of "ideal boundary conditions". Recall that a closed pseudomanifold is a compact simplicial complex with the property that each $(n-1)$-simplex is a face of exactly two n-simplices. The metrics we consider will include as a special case, metrics which are piecewise smooth, smooth off the codimension 2, skelton, and equivalent in the sense of (1.13) to the standard flat metric on each n-simplex. For such a metric, every point on a j-simplex σ^j has a neighborhood which is equivalent to $D^j \times C(L(\sigma^j))$, where D^j is a j-disk, $C(L(\sigma^j))$, denotes the cone on the link $L(\sigma^j)$ of σ^j, and $L(\sigma^j)$ carries an admissible metric for dimension $n-j$. More generally we will be able to treat the case in which $C(L(\sigma^j))$ is an equivalent to a "horn" (in the sense of Section 2), on a previously admissible $L(\sigma^j)$.

2. COLLARS AND HORNS, STOKES' THEOREM

In this section we prove Stokes' Theorem for "collars" and for isolated singularities which we call "horns". By a collar we mean simply a manifold riemannian equivalent to $(a,b) \times N^m$, where N^m is some riemannian manifold with metric g, and $(0,1) \times N^m$ carries the product metric. If $f : R^+ \to R^+$ is a smooth function with $f(r) = o(r)$, by an f-horn, $C^f(N)$ on N^m, we mean the manifold $R^+ \times N^m$ with a metric equivalent to

(2.1) $dr^2 + f^2(r)g$.

In case we want to indicate the precise metric (2.1), we will call $C^f(N)$ the metric f-horn on N. For $f(r) = r$ we have the metric cone. From now on we make the assumption that $f(r) = r^c$ and $c \geq 1$. In this paper, we will write $C(N)$ for $C^{r^c}(N)$ and

(2.2) $C_{r_1,r_2}(N) = \{(r,x) \in C(N) | r_1 < r < r_2\}$.

We begin with the simpler case of a collar. Let

$$(2.3)\quad \alpha = \phi + dr\wedge\omega$$

be an i-form on $(0,1) \times N$; (ϕ, ω do not involve dr). Then

$$(2.4)\quad d\alpha = \tilde{d}\phi + dr\wedge\left(\frac{\partial\phi}{\partial r} - \tilde{d}\omega\right),$$

where $\tilde{d}$ denotes exterior differentiation with respect to the x-variables.

THEOREM 2.1 (Stokes' Theorem for Collars). Let $(0,1) \times N^m$ be a collar and suppose that Stokes' Theorem holds for N^m in the L^2 sense. Let $\alpha \in \operatorname{dom} \bar{d}$ be an i-form and $\beta \in \operatorname{dom} \bar{\delta}$ be an (i+1)-form. Then for almost all $b \in (0,1)$, we have for almost all a,

$$(2.5)\quad \int_{(a,b)\times N} \bar{d}\alpha\wedge *\beta = \int_{(a,b)\times N} \alpha\wedge *\bar{\delta}\beta + \int_{(b,N)} \alpha\wedge *\beta$$
$$- \int_{(a,N)} \alpha\wedge *\beta.$$

PROOF. Consider first the case $\alpha \in$ dom, $\beta \in \operatorname{dom} \delta$. Set

$$(2.6)\quad \begin{aligned} \alpha &= \phi + dr\wedge\omega, \\ *\beta &= \theta + dr\wedge\psi. \end{aligned}$$

Since, $d\alpha$, $d*\beta \in L^2((0,1) \times N^m)$, the following forms are in $L^2((r,N))$ for almost all r.

$$(2.7)\quad \begin{aligned} &\phi, \tilde{d}\phi, \left(\frac{\partial\phi}{\partial r} - \tilde{d}\omega\right), \\ &\theta, \tilde{d}\theta, \left(\frac{\partial\theta}{\partial r} - \tilde{d}\psi\right). \end{aligned}$$

Since we can not assume $\tilde{d}\omega$, $\tilde{d}\psi \in L^2$ we must use regularization. Let $\rho(r,x)$ be a form on $(0,1) \times N^m$ (not involving dr). For our purposes $\rho(r,x)$ can actually be assumed <u>smooth</u>. Define the δ-regularization, ρ_δ of ρ, to be the form on $(\delta/2, 1-\delta/2) \times N^m$ given by

$$(2.8)\qquad \rho_\delta(r,x) = \frac{1}{\delta}\int_{r-\delta/2}^{r+\delta/2} \rho(s,x)ds.$$

Set

$$(2.9)\qquad \alpha_\delta = \phi_\delta + dr\wedge\omega_\delta.$$

$$(*\beta)_\delta = \theta_\delta + dr\wedge\psi_\delta.$$

Clearly, differentiation under the integral gives,

$$(2.10)\qquad \tilde{d}(\alpha_\delta) = (\tilde{d}\alpha)_\delta$$

Also,

$$(2.11\qquad (\frac{\partial\rho}{\partial r})_\delta = \frac{1}{\delta}\int_{r-\delta/2}^{r+\delta/2} \frac{\partial\rho}{\partial s}(s,x)ds$$

$$= \frac{1}{\delta}(\rho(r+\delta/2,x) - \rho(r-\delta/2,x))$$

$$= \frac{\partial}{\partial r}\frac{1}{\delta}\int_{r-\delta/2}^{r+\delta/2} \rho(s,x)ds$$

$$= \frac{\partial}{\partial r}(\rho_\delta).$$

Hence,

$$(2.12)\qquad d(\alpha_\delta) = (d\alpha)_\delta.$$

Moreover, by standard arguments, (see [1]), the L^2 norm of ρ_δ is finite and satisfies

$$(2.13)\qquad \|\rho_\delta\| \le \|\rho\|.$$

Similarly,

$$(2.14)\qquad \rho_\delta \xrightarrow{L^2} \rho.$$

Now by (2.10), (2.11),

$$(2.15)\qquad (\frac{\partial\phi}{\partial r} - \tilde{d}\omega)_\delta = \frac{\theta(r+\delta/2) - \theta(r-\delta/2)}{\delta} - \tilde{d}(\omega_\rho)$$

$$(\frac{\partial\theta}{\partial r} - \tilde{d}\psi)_\delta = \frac{\theta(r+\delta/2) - \theta(r-\delta/2)}{\delta} - \tilde{d}(\psi_\delta)$$

and it follows from (2.7), (2.13) that these forms are in $L^2((\delta/2, 1-\delta/2) \times N^m)$. Since ϕ, $\theta \in L^2((\delta/2, 1-\delta/2) \times N^m)$, we have

$$(2.16) \quad \tilde{d}(\omega_\delta),\ \tilde{d}(\psi_\delta) \in L^2((\delta/2, 1-\delta/2) \times N^m)$$

and we can proceed with the proof.

$$(2.17) \quad \int_{(a,b)\times N} d\alpha\wedge *\beta = \lim_{\delta\to 0} \int_{(a,b)\times N} (d\alpha)_\delta \wedge (*\beta)_\delta$$

$$= \lim_{\delta\to 0} \int_{(a,b)\times N} d(\alpha_\delta)\wedge(*\beta)_\delta$$

$$= \lim_{\delta\to 0} \int_{(a,b)\times N} (\tilde{d}\phi_\delta + dr\wedge\frac{\partial}{\partial r}(\phi_\delta) - \tilde{d}(\omega_\delta))$$

$$\wedge(\theta_\delta + dr\wedge\psi_\delta).$$

By Fubini's Theorem, for fixed δ, the right hand side of the last equation equals

$$(2.18) \quad \int_a^b \int_{N^m} (-1)^{i+1}\tilde{d}(\phi_\delta)\wedge\psi_\delta + (\frac{\partial}{\partial r}(\phi_\delta) - \tilde{d}(\omega_\delta))\wedge\theta_\delta.$$

Since by (2.16) $\tilde{d}(\omega_\delta)$, $\tilde{d}(\psi_\delta) \in L^2(\delta/2, 1-\delta/2) \times N^m)$, for almost all r we have

$$(2.19) \quad \tilde{d}(\omega_\delta),\ \tilde{d}(\psi_\delta) \in L^2(r\times N),$$

and by hypothesis, we can apply Stokes' Theorem on $(r\times N)$ to obtain

$$(2.20) \quad \int_a^b \int_N \phi_\delta\wedge\tilde{d}(\psi_\delta) + \frac{\partial}{\partial r}(\phi_\delta)\wedge\theta_\delta + (-1)^i\omega_\delta\wedge\tilde{d}(\psi_\delta).$$

Again by Fubini's Theorem, we can reverse the order of integration in the second term and integrate by parts to obtain

$$(2.21) \quad \int_{(b,N)} \phi_\delta\wedge\theta_\delta - \int_{(a,N)} \phi_\delta\wedge\theta_\delta - \int_N \phi_\delta\wedge \frac{\partial}{\partial r}(\theta_\delta).$$

Using this, as in (2.17), we find that (2.20) can be rewritten as,

$$(2.22)\quad (-1)^{i+1}\int_{(a,b)\times N}\alpha_\delta\wedge(d*\beta)_\delta + \int_{(b,N)}\alpha_\delta\wedge(*\beta)_\delta - \int_{(a,N)}\alpha_\delta\wedge(*\beta)_\delta .$$

Now choose a sequence $\delta_j \to 0$. Then the first term on the right hand side of (2.22) converges to,

$$(2.23)\quad \int_{(a,b)\times N}\alpha\wedge*\delta\beta .$$

After passing to a subsequence if necessary, for almost all $a,\ b \in (0,1)$,

$$(2.24)\quad \alpha_{\delta_j}|(b,N^m) \xrightarrow{L^2} \alpha|(b,N^m),$$

$$(*\beta)_{\delta_j}|(b,N^m) \xrightarrow{L^2} *\beta|(b,N^m),$$

$$\alpha_{\delta_j}|(a,N^m) \xrightarrow{L^2} \alpha|(a,N^m),$$

$$(*\beta)_{\delta_j}|(a,N^m) \xrightarrow{L^2} *\beta|(a,N^m),$$

This together (2.22), (2.23), completes the proof for α, β smooth. The general case follows by a simple regularization argument.

We now introduce some notation. If $\theta(r,x)$ is an i-form on the metric horn $C(N)$, let

$$(2.25)\quad \|\theta(r,x)\|_s$$

$$\|\theta(r,x)\|_{r_1,r_2}$$

denote respectively the L^2 norm of $\theta(r,x)$ regarded as a form on N equipped with metric induced on the submanifold (s,N), and the L^2 norm of $\theta|C_{r_1,r_2}(N)$. Thus, $\|\theta\|_1$ denotes the norm of θ regarded as a form on N^m with its given metric, and $\|\theta\|_{0,1}$ denotes the L^2 norm of θ. Moreover,

$$(2.26)\quad \|\theta\|_s = r^{c(m/2-i)}\|\theta\|_1$$

$$(2.27)\quad \|\theta\|_{r_1,r_2} = \left(\int_{r_1}^{r_2} r^{c(m-2i)}\|\theta(r)\|_1^2\right)^{\frac{1}{2}}.$$

LEMMA 2.1. Let $\theta(x)$ be an i-form on N^m and $\theta(1,x)$ the natural extension to $C_{0,1}(N)$, i.e. $\theta(1,x)$ is independent of r. Then $\theta(1,x) \in L^2(C_{0,1}(N))$ if and only if $i < \frac{m+1}{2}$.

PROOF.

$$(2.28)\quad \|\theta\|^2_{r_1,r_2} = \int_{r_1}^{r_2} r^{c(m-2i)}\|\theta(1,x)\|_1^2 dr$$

$$= (r_2^{c(m-2i)+1} - r_1^{c(m-2i)+1})\|\theta(1,x)\|_1^2.$$

In order for the limit as $r_1 \to 0$ to exist, we must have

$$(2.29)\quad c(m-2i)+1 > 0,$$

or

$$(2.30)\quad i < \frac{m}{2} + \frac{1}{2c}\ .$$

Since i, m, are integers and $c \geq 1$, this just amounts to

$$(2.31)\quad i < \frac{m+1}{2}\ .$$

LEMMA 2.2

1)

$$(2.32)\quad \left\|\int_{r_1}^{r_2}\theta(r,x)\right\|_1 \leq K\cdot\|\theta\|_{r_1,r_2}\begin{cases} |r_2^{2c(i-m/2)+1} - r_1^{2c(i-m/2)+1}|^{\frac{1}{2}} & i \neq \frac{m}{2} - \frac{1}{2} \\ |\log r_2 - \log r_1|^{\frac{1}{2}} & i = \frac{m-1}{2}, (c=1) \end{cases}$$

2) If $i < \frac{m+1}{2}$ and $\theta \in L^2(C_{0,1}(N^m))$,

$$(2.33)\quad \left\|\int_r^1 \theta\right\|_r \leq K\cdot\|\theta\|_{r,1}\begin{cases} r^{\frac{1}{2}} & i < \frac{m-1}{2} \\ r^{\frac{1}{2}}\,|\log r|^{\frac{1}{2}} & i = \frac{m-1}{2}, (c=1) \\ 1 & i = \frac{m}{2} \end{cases}$$

In particular, for $i < \frac{m+1}{2}$,

(2.34) $\int_r^1 \theta \in L^2(C_{0,1}(N))$.

3) If $i > \frac{m-1}{2}$, and $\theta \in L^2(C_{0,1}(N^m))$,

(2.35) $\lim_{\epsilon \to 0} \int_\epsilon^r \theta = \int_0^r \theta \in L^2$

exists in the sense of convergence in $L^2(N^m)$. Moreover, for $i > \frac{m-1}{2}$,

(2.36) $\|\int_0^r \theta\|_r \le K \cdot \|\theta\|_{0,r} r^{\frac{1}{2}}$,

so that

(2.37) $\int_0^r \theta \in L^2(C_{0,1}(N^m))$.

PROOF.

1)

$$(2.38)\quad \|\int_{r_1}^{r_2} \theta\|_1 \le \int_{r_1}^{r_2} \|\theta\|_1$$

$$= \int_{r_1}^{r_2} r^{c(i-m/2)} r^{c(m/2-i)} \|\theta\|_1,$$

$$\le |\int_{r_1}^{r_2} r^{2c(i-m/2)}|^{\frac{1}{2}} \cdot \|\theta\|_{r_1, r_2},$$

where we have used (2.27) and the Schwartz inequality. Since $c \ge 1$ and i, m are integers, the first estimate in (2.32) holds unless $c = 1$, $i = \frac{m-1}{2}$, in which case the second estimate holds.

2) If $i \le \frac{m}{2} - \frac{1}{2c}$, the term in (2.32) involving r_1 is dominant. In view of (2.26), this gives the first two estimates in (2.33), with the second a possibility only if $c = 1$. If $i > \frac{m}{2} - \frac{1}{2c}$, (i.e. $i > \frac{m-1}{2}$), the term in (2.32) involving r_2 is dominant. Thus, by (2.26), for $i > \frac{m}{2} - 1$,

$$(2.39)\quad \|\int_r^1 \theta\|_r \le K \cdot \|\theta\|_{r,1} r^{c(m/2-i)}.$$

For $i = \frac{m}{2}$, this gives the third estimate in (2.33). It is immediate from (2.33) that $\int_r^1 \theta \in L^2(C_{0,1}(N))$ for $i < \frac{m+1}{2}$.

(2.40) $\|\int_{r_1}^{r_2}\theta\|_1 < K\cdot\|\theta\|_{0,r_2}\cdot r_2^{c(i-m/2)+\frac{1}{2}}$

This implies that the limit on the right hand side of (2.35) exists, which, in view of (2.26), gives (2.36). Then (2.37) follows immediately from (2.36)

Note that even if θ is smooth, for $i > \frac{m-1}{2}$, the precise degree of smoothness of $\int_0^r\theta$ is not altogether apparent. Let $\alpha \in L^2(C_{0,1}(N^m))$ have the expression $\phi + dr\wedge\omega$ in polar coordinates. For almost all r, $\phi(r,x) \in L^2(N)$. Let

(2.41) $\phi = \phi_{\widetilde{ce}} + \phi_{\widetilde{e}} + \phi_{\widetilde{H}}$

denote the decomposition of Theorem 1.1 for $\phi(r,x)$ regarded as a form on N^m, and set $\phi_{\widetilde{cc}} = \phi_{\widetilde{e}} + \phi_{\widetilde{H}}$. We now prove a Sobolev type inequality.

LEMMA 2.3. Suppose that Stokes' Theorem holds for N^m in the L^2 sense. Let $\alpha = \phi + dr\wedge\omega$ be an i-form on $C_{0,1}(N^m)$ such that $\alpha \in \operatorname{dom} \bar{d}$. Then for almost all r

(2.42)
$$\|\phi_{\widetilde{cc}}(r)\|_r \leq \begin{cases} K(\|\alpha\|_{r,1} + \|\bar{d}\alpha\|_{r,1})\begin{cases} r^{\frac{1}{2}} & i < \frac{m-1}{2} \\ r^{\frac{1}{2}}|\log r|^{\frac{1}{2}} & i = \frac{m-1}{2}, (c=1) \\ 1 & i = \frac{m}{2} \end{cases} \\ K\|\bar{d}\alpha\|_{0,r}\cdot r^{\frac{1}{2}} \qquad\qquad i > \frac{m}{2} \end{cases}$$

PROOF. By a simple regularization argument, it suffices to consider the case α smooth, α, $d\alpha \in L^2(C_{0,1}(N)$. As in (2.4), $d\alpha \in L^2(C_{0,1}(N))$ implies

(2.43) $(\frac{\partial\phi}{\partial r} - \tilde{d}\omega) \in L^2(C_{0,1}(N^m))$

Assume first that $i < \frac{m+1}{2}$. By (2.33), for say $\frac{1}{2} < a < 1$,

$$(2.44)\quad \left\|\int_a^r\left(\frac{\partial\phi}{\partial r} - \tilde{d}\omega\right)\right\|_r \le K\circ\left\|\frac{\partial\phi}{\partial r} - \tilde{d}\omega\right\|_{r,1}\begin{cases} r^{\frac{1}{2}} & i < \frac{m-1}{2} \\ r^{\frac{1}{2}}|\log r|^{\frac{1}{2}} & i = \frac{m-1}{2}, (c=1) \\ 1 & i = \frac{m}{2}\end{cases}$$

For almost all a, $\phi(a) \in L^2(N)$ and we can assume (using (2.26)) that

$$(2.45)\quad \|\phi(a)\|_r \le K\circ\|\phi\|_{\frac{1}{2},1}r^{c(m/2-i)}.$$

Since

$$(2.46)\quad \int_r^a \frac{\partial\phi}{\partial r} - \tilde{d}\omega = \phi(a) - \phi(r) - \tilde{d}\int_r^a\omega,$$

and for almost all r, $\phi(r) \in L^2(N)$, it follows that for such r,

$$(2.47)\quad \tilde{d}\int_r^a\omega \in L^2(N).$$

Also by (2.32),

$$(2.48)\quad \int_r^a\omega \in L^2(N).$$

Since by hypothesis, Stokes' Theorem holds for N^m in the L^2 sense,

$$(2.49)\quad \langle\phi_{\tilde{c}c}(r), \tilde{d}\int_r^a\omega\rangle_N = 0,$$

for almost all r. Thus, by (2.44), (2.45),

$$(2.50)\quad \|\phi_{\tilde{c}c}(r)\|_r \le K(\|\alpha\|_{r,1} + \|\bar{d}\alpha\|_{r,1})\begin{cases} r^{\frac{1}{2}} & i < \frac{m-1}{2} \\ r^{\frac{1}{2}}|\log r|^{\frac{1}{2}} & i = \frac{m-1}{2}, (c=1) \\ 1 & i = \frac{m}{2}\end{cases}$$

where the term in (2.44) dominates except in case $i = \frac{m}{2}$, in which case both terms are of the same order of magnitude. For $i > \frac{m}{2}$, using (2.36), as above we have

$$(2.51)\quad \lim_{\epsilon\to 0}\|\phi(r) - \phi(\epsilon) - \int_\epsilon^r\tilde{d}\omega\|_r \le K\cdot\|\bar{d}\alpha\|_{0,r}\cdot r^{\frac{1}{2}}.$$

In particular, we see that the limit on the left hand side exists. Using (2.49), as above, for almost all ϵ, r we have

(2.52) $\lim_{\epsilon\to 0} \|\phi_{\widetilde{cc}}(r) - \phi_{\widetilde{cc}}(\epsilon)\|_r \le K\cdot\|\bar{d}\alpha\|_{0,r} r^{\frac{1}{2}}$

By Lemma 1.2, we can choose a sequence $\epsilon_s \to 0$ such that

(2.53) $\lim_{\epsilon_s\to 0} \|\phi(\epsilon_s)\|_{\epsilon_s} = o(\epsilon_s^{-\frac{1}{2}}\cdot|\log \epsilon_s|^{-\frac{1}{2}})$.

By (2.26),

(2.54) $\|\phi(\epsilon_s)\|_r = (\frac{r}{\epsilon_s})^{c(m/2-i)} \|\phi(\epsilon_s)\|_{\epsilon_s}$.

Since $\frac{m}{2} - i \le -\frac{1}{2}$, $c \ge 1$,

(2.54) $\lim_{\epsilon_s\to 0} \|\phi_{\widetilde{cc}}(\epsilon_s)\|_r \le \lim_{\epsilon_s\to 0} \|\phi(\epsilon_s)\|_r = 0$

This, together with (2.52), suffices to complete the proof.

Note that without the assumption that Stokes' Theorem holds for N^m, the above argument implies that (2.42) holds with $\phi_{\widetilde{cc}}$ replaced by $\phi_{\widetilde{ce}}$. But actually, the main point is to estimate $\phi_{\widetilde{H}}$. In case $i = \frac{m}{2}$, (2.42) is not good enough for our purposes and we will need an additional estimate for $\phi_{\widetilde{ce}}$ which requires the hypothesis, range $\bar{\bar{d}}$ is closed. The estimate holds for all i, but to avoid the stronger hypothesis, we use it only for $i = m/2$. It is the lack of control over $\phi_{\widetilde{H}}$ for $i = \frac{m}{2}$, which leads to the phenomenon of "ideal boundary conditions".

LEMMA 2.4. Suppose $\bar{\bar{d}}$ is closed. Let α_1, α_2 be forms in $L^2(C_{0,1}(N^m))$ such that $\alpha_1, \alpha_2 \in \operatorname{dom} \bar{d}$. Then any subset $A \subset (0,1)$ for which $\int_A r^{-1}\cdot|\log r|^{-1} = \infty$, contains a sequence $\epsilon_s \to 0$ such that

(2.55) $\|(\phi_1)_{\widetilde{ce}}(\epsilon_s)\|_{\epsilon_s}, \|(\phi_2)_{\widetilde{ce}}(\epsilon_s)\|_{\epsilon_s} = o(\epsilon_s^{\frac{1}{2}}|\log \epsilon_s|^{-\frac{1}{2}})$

PROOF. As usual, we can assume $\alpha_j \in \operatorname{dom} d$, $j = 1,2$. As in (2.4), $d\alpha_j \in L^2(C_{0,1}(N^m))$ implies $\tilde{d}\phi_j \in L^2(C_{0,1}(N^m))$. Thus

(2.56) $\|\tilde{d}\phi_1\|_r^2, \|\tilde{d}\phi_2\|_r^2 \in L^1([0,1])$,

so that

(2.57) $\|\tilde{d}\phi_1\|_r^2 + \|\tilde{d}\phi_2\|_r^2 \in L^1([0,1])$.

By Lemma 1.2, given A as above, there exists a sequence $A \ni \epsilon_s \to 0$ such that

(2.58) $\|\tilde{d}\phi_1\|_{\epsilon_s}, \|\tilde{d}\phi_2\|_{\epsilon_s} = o(\epsilon_s^{-\frac{1}{2}}|\log \epsilon_s|^{-\frac{1}{2}})$

Range $\tilde{\tilde{d}}$ closed, implies $\tilde{d}^{-1}$ is bounded with respect to $\| \|_1$. But $\tilde{d}^{-1}\phi_1$, $\tilde{d}^{-1}\phi_2$, have degree i_1-1, i_2-1 respectively. Thus,

$$(2.59)\quad \begin{aligned} \|(\phi_j)_{\tilde{ce}}\|_{\epsilon_s} &= \epsilon_s^{c(m/2-(i_j-1))}\|(\phi_j)_{\tilde{ce}}\|_1 \\ &\le K\cdot\epsilon_s^{c(m/2-(i_j-1))}\|\tilde{d}\phi_j\|_1 \\ &= K\cdot\epsilon_s^{c(m/2-(i_j-1))}\epsilon_s^{-c(m/2-i_j)}\|\tilde{d}\phi_j\|_{\epsilon_s} \\ &= K\cdot\epsilon_s^{c}\|d\phi_j\|_{\epsilon_s}. \end{aligned}$$

(2.55) follows from (2.58), (2.59)

We are now in a position to prove Stokes' Theorem.

THEOREM 2.2 (Stokes' Theorem for horns). Let N^m be a riemannian manifold for which Stokes' Theorem holds in the L^2 sense. Let $\alpha \in \operatorname{dom}\bar{d}$ be an i-form and $\beta \in \operatorname{dom}\bar{\delta}$ be an (i+1)-form. If m is even, $i = \frac{m}{2}$, assume further that range $\tilde{\tilde{d}}_{\frac{m}{2}-1}$ is closed and $\dim \mathfrak{H}^{m/2} = 0$, where $\mathfrak{H}^{m/2}$ is the space of harmonic $\frac{m}{2}$-forms on N^m. Then for almost all r,

$$(2.60)\quad \int_{C_{0,r}(N^m)} \bar{d}\alpha\wedge*\beta = \int_{C_{0,r}(N^m)} \alpha\wedge*\bar{\delta}\beta + \int_{(r,N)} \alpha\wedge*\beta$$

PROOF. According to Stokes' Theorem for collars, Theorem 2.1, for almost all r, we have for almost all ϵ,

$$(2.61)\quad \int_{C_{\epsilon,r}(N^m)} \overline{d}\alpha\wedge*\beta = \int_{C_{\epsilon,r}(N^m)} \alpha\wedge*\overline{\delta}\beta + \int_{(r,N)} \alpha\wedge*\beta - \int_{(\epsilon,N)} \alpha\wedge*\beta.$$

Set $\alpha = \phi_1 + dr\wedge\omega_1$, $*\beta = \phi_2 + dr\wedge\omega_2$. Then,

$$(2.62)\quad \int_{(\epsilon,N)} \alpha\wedge*\beta = \int_{(\epsilon,N)} ((\phi_1)_{\widetilde{cc}} + (\phi_1)_{\widetilde{e}})\wedge((\phi_2)_{\widetilde{cc}} + (\phi_2)_{\widetilde{e}})$$

$$= \int_{(\epsilon,N)} (\phi_1)_{\widetilde{cc}}\wedge(\phi_2)_{\widetilde{cc}} + (\phi_1)_{\widetilde{e}}\wedge(\phi_2)_{\widetilde{cc}} + (\phi_1)_{\widetilde{cc}}\wedge(\phi_2)_{\widetilde{e}} + (\phi_1)_{\widetilde{e}}\wedge(\phi_2)_{\widetilde{e}}.$$

The fourth term on the right hand side of (2.62) is automatically zero by Stokes' Theorem in the usual sense. Since we assume $\widetilde{\mathcal{H}}^{m/2} = 0$, by Lemmas 1.2, 2.3, 2.4, we can choose a sequence $\epsilon = \epsilon_s \to 0$ such that

$$(2.63)\quad \|(\phi_1)_{\widetilde{e}}(\epsilon_s)\|_{\epsilon_s},\ \|(\phi_2)_{\widetilde{e}}(\epsilon_s)\|_{\epsilon_s} = o(\epsilon_s^{-\frac{1}{2}}|\log\epsilon_s|^{-\frac{1}{2}})$$

and

$$(2.64)\quad \|(\phi_1)_{\widetilde{cc}}(\epsilon_s)\|_{\epsilon_s},\ \|(\phi_2)_{\widetilde{cc}}(\epsilon_s)\|_{\epsilon_s} \le K\cdot\begin{cases} \epsilon_s^{\frac{1}{2}} & i = \frac{m-1}{2},\frac{m}{2} \\ \epsilon_s^{\frac{1}{2}}|\log\epsilon_s|^{\frac{1}{2}} & i = \frac{m-1}{2} \\ \epsilon_s^{\frac{1}{2}}|\log\epsilon_s|^{-\frac{1}{2}} & i = \frac{m}{2} \end{cases}$$

The claim follows immediately from (2.61) - (2.64).

Note that since

$$(2.65)\quad \lim_{\epsilon\to 0}\int_{C_{\epsilon,r}(N^m)} \overline{d}\alpha\wedge*\beta = \int_{C_{0,r}(N^m)} \overline{d}\alpha\wedge*\beta,$$

$$\lim_{\epsilon\to 0}\int_{C_{\epsilon,r}(N^m)} \alpha\wedge*\overline{\delta}\beta) = \int_{C_{0,r}(N^m)} \alpha\wedge*\overline{\delta}\beta,$$

it follows that in fact, the boundary integral in (2.61) tends to zero independent of how $\epsilon \to 0$.

Even though the singularity of $C_{0,1}(N)$ has codimension m+1, it is easy to see that Theorem 2.1 is false if $\tilde{\mathcal{H}}^{m/2} \neq 0$. Let $0 \neq \phi \in \tilde{\mathcal{H}}^{m/2}$ and let the natural extension of ϕ to the metric horn $C_{0,1}(N)$ also be denoted by ϕ. Clearly, we have $d\phi = \delta(dr\wedge\phi) = 0$. Thus

$$(2.66) \quad \int_{C_{0,1}(N^m)} d\phi\wedge*(dr\wedge\phi) = \int_{C_{0,1}(N^m)} \phi\wedge*\delta(dr\wedge\phi) = 0,$$

but

$$(2.67) \quad \int_{(r,N^m)} \phi\wedge*(dr\wedge\phi) \neq 0.$$

We first encountered this phenomenon about four years ago, in our initial approach to the Ray Singer Conjecture. We considered a nondegenerate critical point $p \in f^{-1}(k)$, of index k on a manifold of dimension 2k+1. It appeared that in the limit, the behavior of eigenfunctions and eigenvalues on the submanifolds $f^{-1}(k+\epsilon)$, $f^{-1}(k-\epsilon)$ depended on the side from which ϵ approached to zero. Notice that a neighborhood of p in $f^{-1}(k)$ is the cone on $S^k \times S^k$ and $H^k(S^k\times S^k,\mathbb{R}) \neq 0$. The attempt to understand this situation was one source of motivation for our subsequent work on singularities. Another was the realization that certain arguments of [4], which were carried out in polar coordinates for the annulus $C_{r_1,r_2}(S^m) \subset \mathbb{R}^{m+1}$, did not depend in an essential way on the fact that the base was the unit sphere.

Now let the situation be as in Theorem 2.2 but in case m = 2k, do not assume $\tilde{\mathcal{H}}^k = 0$. Fix a subspace $V \subset \mathcal{H}^k$ and let $V^\perp$ denote its orthogonal compliment with respect to the natural inner product on $\tilde{\mathcal{H}}_k$. Let

$$\alpha_k = \phi_1 + dr\wedge\omega_1$$

$$*\beta_k = \phi_2 + dr\wedge\omega_2$$

and assume the ideal <u>boundary conditions</u>

(2.68) $\|(\phi_1)_{\tilde{\mathcal{H}}^\perp}(r)\|_r = o(1)$

(2.69) $\|(\phi_2)_{\tilde{\mathcal{H}}}(r)\|_r = o(1)$.

Then if $\alpha \in \operatorname{dom} \bar{d}$, $\beta \in \operatorname{dom} \bar{\delta}$, as in Theorem 2.2,

(2.70) $\langle \bar{d}\alpha, \beta\rangle = \langle \alpha, \bar{\delta}\beta\rangle$.

3. COLLARS AND HORNS, POINCARÉ LEMMA AND BOUNDED d^{-1}

In order to eventually establish the appropriate de Rahm Theorem which gives the isomorphism of $H^*_{(2)}(X)$ with the Goresky-MacPherson theory, in particular to show that $H^i_{(2)}(X)$ is finite dimensional and thus that $\bar{d}^{-1}$ is bounded, we must have a suitable Poincaré Lemma.

We begin with some elementary facts. Let $(0,1) \times N^m$ be a collar and $\alpha = \phi + dr\wedge\omega$ be a form in $L^2((0,1) \times N^m)$. Note that there exists $K > 0$, independent of α, such that if $S \subset (0,1)$ is any subset of measure zero, there exists $a \in (0,1)-S$ (a depends on α) such that

(3.1) $\|\phi(a)\|_1 \leq K\cdot\|\phi\|_{0,1}$.

Here the notation is analogous to that of (2.25). Of course, (2.26), (2.27) are replaced by,

(3.2) $\|\theta\|_s = \|\theta\|_1$,

(3.3) $\|\theta\|_{r_1,r_2} = (\int_{r_1}^{r_2}\|\theta\|_1^2)^{\frac{1}{2}}$.

Define the homotopy operator $\mathcal{K}^\circ$ by

(3.4) $\mathcal{K}^\circ\alpha = \int_a^r \omega$.

Note that by the argument of (2.32), $\mathcal{K}^\circ\alpha \in L^2((0,1) \times N^m$ and $\mathcal{K}^\circ$ is a <u>bounded operator</u>. It therefore extends to all of L^2. Note also that $\mathcal{K}^{\cdot}$ is <u>not a compact operator</u>.

LEMMA 3.1 (Poincaré Lemma, bounded d^{-1} for collars).

1) Let $\alpha \in \Lambda^i((0,1) \times N^m)$. Then,

$$(d\mathcal{K}^{\circ} + \mathcal{K}^{\circ}d)\alpha = \alpha - \phi(a).$$

If $\alpha \in \operatorname{dom} \bar{d}$, for almost all a, $\mathcal{K}^{\circ}\alpha \in \operatorname{dom} \bar{d}$ and

$$(\bar{d}\mathcal{K}^{\circ} + \mathcal{K}^{\circ}\bar{d})\alpha = \alpha - \phi(a).$$

2) If $\bar{d}\alpha = 0$, the restriction map $[\alpha] \to [\phi(a)]$ is well defined and independent of a, for almost all a. It induces isomorphisms

$$\begin{array}{ccc} H^i_{(2),\#}((0,1) \times N^m) & \simeq & H^i_{(2),\#}(N^m) \\ \| & & \| \\ H^i_{(2)}((0,1) \times N^m) & \simeq & H_{(2)}(N^m). \end{array}$$

If $\|[\alpha]\|_{0,1}$, $\|[\phi(a)]\|_1$ denote the natural pseudo norms on L^2-cohomology, then for some $c > 0$ and all α,

$$(3.5) \quad \frac{1}{c}\|[\phi(a)]\| \le \|[\alpha]\|_{0,1} \le c\|[\phi(a)]\|_1.$$

In particular $\bar{d}$ has closed range if and only if $\bar{\tilde{d}}$ has closed range. Thus $\bar{d}^{-1}$ is bounded if and only if $\bar{\tilde{d}}^{-1}$ is bounded.

PROOF.

1) Suppose α is smooth.

$$(3.6) \quad d\mathcal{K}^{\cdot}\alpha = d \int_a^r \omega$$

$$= \tilde{d} \int_a^r \tilde{d}\omega + dr\wedge \frac{\partial}{\partial r} \int_a^r \omega$$

$$= \int_a^r \tilde{d}\omega + dr\wedge\omega.$$

$$(3.7) \quad \mathcal{K}^{\circ}d\alpha = \int_a^r \frac{\partial\phi}{\partial s} - \tilde{d}\omega$$

$$= \phi(r) - \phi(a) - \int_a^r \tilde{d}\omega.$$

By adding (3.6), (3.7) we obtain the first equation above. Now let $\alpha \in \operatorname{dom} \bar{d}$ and $\alpha_s \to \alpha$, $d\alpha_s \to d\alpha$. After passing to a subsequence, we have $\phi_s(a) \to \phi(a)$ for almost all a. Moreover, for almost all a, $\phi_s(a) \in L^2(N^m)$ for all s. Thus for almost all a,

$$(3.8)\quad \lim_{s\to\infty} dK^{\cdot}\alpha_s = \lim_{s\to\infty} - K^{\cdot}d\alpha_s + \alpha_s - \phi_s(a)$$

$$= - K^{\cdot}\bar{d}\alpha + \alpha - \phi(a)$$

Since $K^{\cdot}$ is bounded, $K^{\cdot}\alpha_s \to K^{\cdot}\alpha$ and the second equation follows.

2) Since by the argument of (2.32), $K^{\circ}\alpha|(r,N^m)$ is in $L^2(N^m)$ for all r, by 1) $[\phi(r)] = [\phi(a)]$ for almost all r, i.e. whenever $\phi(r) \in L^2(N^m)$. Similarly, since $K^{\cdot}$ is a bounded operator, α is L^2-cohomologous to the form $\phi(a)$, regarded as a form on $(0,1) \times N^m$. Clearly, $\phi(a)$ is the restriction of that extension of $\phi(a)$ to $(0,1) \times N^m$ which is independent of r. If $\phi(a) = d\psi(a)$, then

$$(3.9)\quad \alpha = dK^{\cdot} + d\psi(a).$$

If we choose a as in (3.1), the rest of the argument becomes straightforward.

In dealing with horns, we must be more careful about the choice of a in the definition of $K^{\cdot}$. If $\alpha = (\phi + dr\wedge\omega) \in L^2(C_{0,1}(N^m))$, set

$$(3.10)\quad K\alpha = \begin{cases} \int_a^r \omega & i \le \frac{m}{2} + 1 \\ \int_0^r \omega & i > \frac{m}{2} + 1 \end{cases}$$

According to (2.33)-(2.37) of Lemma 2.2, K is a <u>bounded</u> operator but is <u>not compact</u>; (note that if α is an i-form then ω, which plays the role of θ of Lemma 2.2, is an i-1 form). For $i \le \frac{m}{2} + 1$, we choose $a \in (\frac{1}{2},1)$ depending on α, so that (3.1) is satisfied. If $d\alpha \in L^2(C_{0,1}(N^m))$, then $Kd\alpha \in L^2(C_{0,1}(N^m))$.

If $i \le \frac{m}{2}$, then by Lemma 2.1, the extension to $C_{0,1}(N^m)$ of

$\phi(a)$ is in $L^2(C_{0,1}(N^m))$ and as above,

$$(3.11)\quad (d\mathcal{K} + \mathcal{K}d)\alpha = \alpha - \phi(a),$$
$$(\bar{d}\mathcal{K} + \mathcal{K}\bar{d})\alpha = \alpha - \phi(a).$$

For $i > \frac{m}{2} + 1$, $(i > \frac{m}{2})$ the precise degree of smoothness of $\mathcal{K}\alpha$, $(\mathcal{K}d\alpha)$ in the x-directions is not immediately apparent. So we proceed by verifying that the counterpart of (3.11) holds in the weak sense. In order to treat the case $i = \frac{m}{2} + 1$ along with the cases $i > \frac{m}{2} + 1$, it is convenient to modify the definition of $\mathcal{K}$; the case $i = \frac{m+1}{2}$ must still be treated separately. Set

$$(3.12)\quad \mathcal{K}^\circ\alpha = \begin{cases} \int_0^r \omega & i = \frac{m}{2} + 1 \\ \mathcal{K}\alpha & i > \frac{m}{2} + 1 \end{cases}$$

According to (2.33), (2.34) of Lemma 2.2, we still have that $\alpha \in L^2(C_{0,1}(N))$ implies $\mathcal{K}^\circ\alpha \in L^2(C_{0,1}(N^m))$.

LEMMA 3.2. Let $i > \frac{m+1}{2}$ and $\alpha \in \operatorname{dom} \bar{d}$ be an i-form on $C_{0,1}(N^m)$. Then

$$(3.13)\quad (\bar{d}\mathcal{K}^\circ + \mathcal{K}^\circ\bar{d})\alpha = \alpha.$$

PROOF. Since $\mathcal{K}^\circ$ is bounded, as usual, it suffices to consider α smooth, α, $d\alpha \in L^2(C_{0,1}(N)$. Let β be a smooth form having compact support in $C_{\delta,1}(N^m)$. It is enough to verify

$$(3.14)\quad \langle\mathcal{K}\alpha, \delta\beta\rangle + \langle\mathcal{K}d\alpha, \beta\rangle = \langle\alpha, \beta\rangle$$

For $\epsilon > 0$, set

$$(3.15)\quad \mathcal{K}^\circ_\epsilon\alpha = \int_\epsilon^r \omega.$$

Then clearly $\mathcal{K}^\circ_\epsilon\alpha$, $\mathcal{K}^\circ_\epsilon d\alpha \in L^2(C_{\delta,1}(N^m))$ and on $C_{\delta,1}(N^m)$,

$$(3.16)\quad \lim_{\epsilon\to 0} \mathcal{K}^\circ_\epsilon\alpha = \mathcal{K}^\circ\alpha$$
$$\lim_{\epsilon\to 0} \mathcal{K}^\circ_\epsilon d\alpha = \mathcal{K}^\circ d\alpha$$

in L^2. As in (3.11)

$$(3.17)\quad (d\mathcal{K}^{\circ}_{\epsilon} + \mathcal{K}^{\circ}_{\epsilon}d)\alpha = \alpha - \phi(\epsilon).$$

Since in particular, $i \geq \frac{m+1}{2}$, by (2.54), there exists a sequence $\epsilon_s \to 0$ such that

$$(3.18)\quad \lim_{\epsilon_s \to 0} \|\phi(\epsilon_s)\|_r = 0,$$

uniformly for $r > \delta$. Thus, regarding $\phi(\epsilon_s)$ as a form on $C_{\delta,1}(N^m)$ we have also,

$$(3.19)\quad \lim_{\epsilon_s \to 0} \|\phi(\epsilon_s)\|_{\delta,1} = 0.$$

Define $<,>_\delta$ by

$$(3.20)\quad <\theta,\gamma>_\delta = \int_{C_{\delta,1}(N^m)} \theta\wedge *\gamma.$$

Since β is supported on $C_{\delta,1}(N^m)$,

$$(3.21)\quad <\mathcal{K}^{\circ}\alpha, \delta\beta> + <\mathcal{K}^{\circ}d\alpha, \beta> = <\mathcal{K}^{\circ}\alpha, \delta\beta>_\delta + <\mathcal{K}^{\circ}d\alpha, \beta>_\delta.$$

Using (3.16), we have

$$(3.22)\quad <\mathcal{K}^{\circ}\alpha, \delta\beta>_\delta + <\mathcal{K}^{\circ}d\alpha, \beta>_\delta = \lim_{\epsilon\to 0} <\mathcal{K}^{\circ}_{\epsilon}\alpha, \delta\beta>_\delta + <\mathcal{K}^{\circ}_{\epsilon}d\alpha, \beta>_\delta,$$

and by Stokes' Theorem,

$$(3.23)\quad <\mathcal{K}^{\circ}\alpha, \delta\beta>_\delta + <\mathcal{K}^{\circ}_{\epsilon}d\alpha, \beta>_\delta = <(d\mathcal{K}^{\circ}_{\epsilon} + \mathcal{K}^{\circ}_{\epsilon}d)\alpha, \beta>_\delta.$$

By (3.17),

$$(3.24)\quad <(d\mathcal{K}^{\circ}_{\epsilon} + \mathcal{K}^{\circ}_{\epsilon}d)\alpha, \beta>_\delta = <\alpha,\beta>_\delta - <\phi(\epsilon),\beta>.$$

So by (3.21)-(3.24),

$$(3.25)\quad <\mathcal{K}^{\circ}\alpha, \delta\beta> + <\mathcal{K}^{\circ}d\alpha, \beta> = \lim_{\epsilon\to 0} <\alpha,\beta>_\delta - <\phi(\epsilon),\beta>_\delta,$$

where the limit exists independent of how $\epsilon \to 0$. If we set $\epsilon = \epsilon_s \to 0$, with ϵ_s as in (3.19), we get

$$(3.26)\quad \lim_{\epsilon_s\to 0} <\alpha,\beta>_\delta - <\phi(\epsilon_s),\beta>_\delta = <\alpha,\beta>_\delta, \\ = <\alpha,\beta>,$$

which together with (3.25), suffices to complete the proof.

We have yet to deal with the case $i = \frac{m+1}{2}$.

LEMMA 3.3. Let $i = \frac{m+1}{2}$ and $\alpha \in \operatorname{dom} \overline{d}$ be an i-form. Assume that $\overline{\tilde{d}}_{\frac{m-1}{2}}$ has closed range. Then for almost all a, there exists $\psi(a) \in \operatorname{dom} \overline{\tilde{d}}$ such that

1) $\overline{\tilde{d}}\psi(a) = \phi(a) - (\mathcal{K}\overline{d}\alpha)(a)$

2) $\overline{d}(\mathcal{K}\alpha + \psi(a)) + \mathcal{K}\overline{d}\alpha = \alpha.$

PROOF.

1) We can, as usual, let α be smooth, α, $d\alpha \in L^2(C_{0,1}(N^m))$.

$$(3.27)\quad \mathcal{K}_\epsilon d\alpha = \phi(r) - \phi(\epsilon) - \int_\epsilon^r \tilde{d}\omega.$$

If we let $\epsilon = \epsilon_s \to 0$, where ϵ_s is as in (3.19), we get

$$(3.28)\quad \mathcal{K}d\alpha = \phi(r) - \overline{\int_0^r}\tilde{d}\omega,$$

where

$$(3.29)\quad \overline{\int_0^r}\tilde{d}\omega = \lim_{\epsilon_s \to 0} \int_{\epsilon_s}^r \tilde{d}\omega,$$

and in particular, the limit exists. Clearly, for all r, $\overline{\int_0^r}\tilde{d}\omega$ is in the closure of the range of $\tilde{d}$, and since $\overline{\tilde{d}}_{\frac{m-1}{2}}$ has closed range, we can write

$$(3.30)\quad \overline{\int_0^r}\tilde{d}\omega = \overline{\tilde{d}}\psi(r)$$

If we take r = a in (3.28), (3.30), then 1) follows.

2) We have

$$(3.31)\quad d\mathcal{K}\alpha = dr \wedge \omega - \int_a^r \tilde{d}\omega.$$

Adding (3.27), (3.31) gives

$$(3.32)\quad d\mathcal{K}\alpha + \mathcal{K}_\epsilon d\alpha = \alpha - \phi(\epsilon) - \int_\epsilon^a \tilde{d}\omega,$$

and by arguing as in Lemma 2.2 we obtain

$$(3.33)\quad \langle \mathcal{K}\alpha, \delta\beta\rangle_\delta + \langle \mathcal{K}d\alpha, \beta\rangle_\delta + \langle \overline{\int_0^a}\tilde{d}\omega, \beta\rangle_\delta = \langle \alpha, \beta\rangle_\delta .$$

From (3.28) and 1), we have as in Lemma 2.2,

$$(3.34)\quad \langle K\alpha, \delta\beta\rangle + \langle Kd\alpha, \beta\rangle + \langle \psi(a), \delta\beta\rangle = \langle \alpha, \beta\rangle$$

Note that since $i = \frac{m+1}{2}$, we can not conclude in general that $\int_0^a \omega \in L^2(N^m)$. For $i > \frac{m+1}{2}$ we have

$$(3.35)\quad \bar{\tilde{d}} \int_0^a \omega = \bar{\int_0^a} \tilde{d}\omega.$$

We can now prove the Poincaré Lemma.

LEMMA 3.4 (Poincaré Lemma, bounded d^{-1} for horns). Let N^m be a riemannian manifold and if m is odd assume that $\bar{\tilde{d}}_{\frac{m-1}{2}}$ has closed range.

1) For $i \le \frac{m}{2}$ the map $[\alpha] \to [\phi(a)]$ induces an isomorphism

$$(3.36)\quad \begin{array}{ccc} H^i_{(2),\#}(C_{0,1}(N^m)) & \simeq & H^i_{(2),\#}(N^m) \\ \wr\wr & & \wr\wr \\ H^i_{(2)}(C_{0,1}(N^m)) & \simeq & H^i_{(2)}(N^m) \end{array}$$

Moreover, there exists $c > 0$, such that

$$(3.37)\quad \frac{1}{c}\|[\phi(a)]\|_1 \le \|[\alpha]\|_{0,1} \le c\|[\phi(a)]\|_1 .$$

In particular $\bar{d}_{i-1}$ has closed range and thus $\bar{d}_{i-1}^{-1}$ is bounded, if and only if $\bar{\tilde{d}}_{i-1}$ has closed range.

2) For $i \ge \frac{m+1}{2}$,

$$(3.38)\quad H^i_{(2),\#}(C_{0,1}(N^m)) = H^i_{(2)}(C_{0,1}(N^m)) = 0.$$

In particular $\bar{d}_{i-1}$ has closed range and $\bar{d}_{i-1}^{-1}$ is bounded.

PROOF.

1) By (3.11), if $\alpha \in \operatorname{dom} \bar{d}$ is closed,

(3.39) $\alpha = \bar{d}\mathcal{K}\alpha + \phi(a)$

It follows from Lemma 3.1 that $[\alpha] \to [\phi(a)]$ is well defined and by Lemma 2.1 the map is surjective. The injectivity also follows from Lemma 2.1. The rest of 1) is straight forward.

2) This follows immediately from Lemmas 3.2, 3.3 and the discussion of Section 1.

One can also define relative groups and give a relative version of the Poincaré Lemma.

We mention that if in case $m = 2k$, we impose the ideal boundary conditions (2.68), (2.69), corresponding to a subspace $V \subset \tilde{\mathcal{H}}^k$, and if the Strong Hodge Theorem holds for N^m, then it is immediate that (3.36) gets replaced by

(3.40) $H^k_{(2)}(C_{0,1}(N^m)) \simeq V \subset H^k_{(2)}(N^m)$.

4. SOME STANDARD LEMMAS

In this section we prepare for the inductive argument of Section 5. We begin with the observation that if Stokes' Theorem holds locally, then it can be globalized by means of a partition of unity. We then check the exactness of the Mayer Vietoris sequence. Since we consider $\bar{d}$ on various subsets, we will use a subscript to indicate the precise domain.

LEMMA 4.1. Let Y^n be a riemannian manifold, $A_0, A_1, \ldots A_N$ an open cover, and $h_0, \ldots h_N$ a piecewise smooth partition of unity subordinate to $\{A_j\}$, such that for some K, the pointwise norms satisfy $\|\bar{d}h_j\| \le K$.

Suppose that for all α_j, β_j, $j = 1, \ldots N$, with α_j an i-form in dom $\bar{d}_{A_j}$ and β_j an (i+1) form in dom $\bar{\delta}_{A_j}$, we have

(4.1) $\langle \bar{d}\alpha_j, \beta_j \rangle_{A_j} = \langle \alpha_j, \bar{\delta}\beta_j \rangle_{A_j}$.

whenever,

(4.2) $\sup \alpha_j \cap \sup \beta_j \subseteq (A_j \cap A_0) \cup \sup h_j$.

Then if α is an i-form in dom $\bar{d}_{X^n}$ and β is an (i+1)-form in dom $\bar{\delta}_{X^n}$ with

(4.3) $\sup \alpha \cap \sup \beta \subseteq X - A_0$,

we have

(4.4) $\langle \bar{d}\alpha, \beta \rangle = \langle \alpha, \bar{\delta}\beta \rangle$.

PROOF. Clearly $h_j\alpha | A_j$ is in dom $\bar{d}_{A_j}$ and $\beta | A_j$ is in dom $\bar{\delta}_{A_j}$. Thus,

$$(4.5) \quad \langle \bar{d}\alpha, \beta \rangle = \langle \sum_{j=0}^{N} \bar{d}(h_j\alpha), \beta \rangle$$

$$= \sum_{j=0}^{N} \langle \bar{d}(h_j\alpha), \beta \rangle_{A_j}$$

By (4.3) the first term in the sum is zero and by (4.1), (4.2), the right hand side equals

$$(4.6) \quad \sum_{j=1}^{N} \langle h_j\alpha, \bar{\delta}\beta \rangle_{A_j} = \sum_{j=1}^{N} \langle h_j\alpha, \bar{\delta}\beta \rangle$$

$$= \langle \alpha, \beta \rangle - \langle h_0\alpha, \bar{\delta}\beta \rangle$$

$$= \langle \alpha, \beta \rangle$$

where the last step follows from (4.3).

In order to check the exactness of the Mayer Vietoris sequence we will use the following extension lemma.

LEMMA 4.2. Let $Y' \subset Y$ be an open submanifold whose closure is a manifold with piecewise smooth boundary $\partial Y'$. Suppose there exists a neighborhood $\mathfrak{U} \supset \partial Y'$, a riemannian manifold N^m, and a piecewise smooth riemannian equivalence f of $\mathfrak{U}$ with the collar $(-1,1) \times N^m$, which carries $(0, N^m)$ onto $\partial Y'$. Then there exists a linear map e: dom $\bar{d}_{Y'} \to$ dom $\bar{d}_Y$, a constant $K > 0$, and an open set $\mathfrak{V} \supset \partial Y'$, such that for $\alpha \in$ dom $\bar{d}_{Y'}$,

1) $e(\alpha)|Y' = \alpha$

2) $\bar{d}e(\alpha)|\mathcal{V} = e(\bar{d}\alpha)|\mathcal{V}$

3) $\|e(\alpha)\| \le K\cdot\|\alpha\|$

4) $\|\bar{d}e(\alpha)\| \le K(\|\alpha\| + \|d\alpha\|)$

PROOF. Let $h : (0,1) \to (0,1)$ be $\equiv 1$ near 0 and $\equiv 0$ near 1. Let $\gamma \in \operatorname{dom} \bar{d}$ be a form on $(-1,0) \times N^m$ and define $\hat{\gamma}$ on $(-1,1) \times N^m$ by

$$(4.7)\quad \hat{\gamma}(r,x) = \begin{cases} \gamma(r,x) & -1 < r < 0 \\ h\gamma(-r,x) & 0 < r < 1 \end{cases}$$

Let $\alpha \in \operatorname{dom} \bar{d}$ on Y' and set

$$(4.8)\quad e(\alpha) = \begin{cases} \alpha & y \in Y' \\ f^*\left(\widehat{f^{*-1}(\alpha)}\right) & y \in \mathcal{U} - Y' \\ 0 & y \in (Y'\cup\mathcal{U}) \end{cases}$$

Clearly it will suffice to show that if $\alpha \in \operatorname{dom} d_{Y'}$ then $e(\alpha) \in \operatorname{dom} \bar{d}_Y$ and $\bar{d}e(\alpha)$ can be calculated pointwise. For this it is enough to establish the corresponding facts for $\hat{\gamma}$. Define $\hat{\gamma}_s$ by

$$(4.9)\quad \hat{\gamma}_s(r,x) = \begin{cases} \gamma(r,x) & -1 < r < -s \\ \gamma\left(\frac{(s-1)r}{s+1} - \frac{2s}{s+1}, x\right) & -s < r < 1 \end{cases}$$

Then $\hat{\gamma}_s(r,x)$ is piecewise smooth so $\hat{\gamma}_s(r,x) \in \operatorname{dom} \bar{d}$. Moreover, it is easy to see that

$$(4.10)\quad \lim_{s\to 0} d\hat{\gamma}_s = \bar{d}\hat{\gamma}$$

This completes the proof.

If the condition of Lemma 4.2 holds, we will say _$\partial Y'$ is collared_.

LEMMA 4.3. Let A_1, A_2 be an open cover of Y and suppose that the boundary of $A_1 \cap A_2 \subset A_1$ is collared. Then the Mayer Vietoris sequence

$$(4.11)\quad \xrightarrow{\delta} H^i_{(2),\#}(Y) \xrightarrow{i} H^i_{(2),\#}(A_1) \oplus H^i_{(2),\#}(A_2) \xrightarrow{j} H^i_{(2),\#}(A_1 \cap A_2) \xrightarrow{\delta}$$

is exact. If the boundary of $A_1 \cap A_2$ in both A_1 and in A_2 is collared, then there exists $c > 0$ such that for all $u \in H^{i-1}_{(2),\#}(A_1 \cap A_2)$,

$$(4.12)\quad \|\delta(u)\| \leq c\|u\|.$$

PROOF. To show exactness of (4.15), it suffices to show exactness of the sequence of cochain complexes,

$$(4.13)\quad 0 \to \{\mathrm{dom}\ \bar{d}_{A_1 \cup A_2}\} \xrightarrow{i} \{\mathrm{dom}\ \bar{d}_{A_1}\} \oplus \{\mathrm{dom}\ \bar{d}_{A_2}\} \xrightarrow{j} \{\mathrm{dom}\ \bar{d}_{A_1 \cap A_2}\} \to 0.$$

It is obvious that i is injective and that its image is contained in the kernel of j. Conversely, if $\alpha_1 \oplus \alpha_2$ is in the kernel of j, there is a well defined form α on $A_1 \cup A_2$ which restricts to α_1, α_2 and a form γ on $A_1 \cup A_2$ which restricts to $\bar{d}\alpha_1$, $\bar{d}\alpha_2$. To see that $\gamma = \bar{d}\alpha$, note that if $\beta \in \Lambda_0^{n-i-1}(A_1 \cup A_2)$ then by use of a partition of unity, we can write $\beta = \beta_1 + \beta_2$ where sup $\beta_j \subset A_j$. Then

$$
\begin{aligned}
(4.14)\quad \langle\gamma,\beta\rangle &= \langle\gamma,\beta_1+\beta_2\rangle \\
&= \langle\gamma,\beta_1\rangle_{A_1} + \langle\gamma,\beta_2\rangle_{A_2} \\
&= \langle\bar{d}\alpha_1,\beta_1\rangle_{A_1} + \langle\bar{d}\alpha_2,\beta_2\rangle_{A_2} \\
&= \langle\alpha,\delta\beta_1\rangle_{A_1} + \langle\alpha,\delta\beta_2\rangle_{A_2} \\
&= \langle\alpha,\delta\beta_1\rangle + \langle\alpha,\delta\beta_2\rangle \\
&= \langle\alpha,\delta\beta\rangle.
\end{aligned}
$$

Finally, since the boundary of $A_1 \cap A_2 \subset A_1$ is collared, every form in dom $\bar{d}_{A_1 \cap A_2}$ is the restriction of a form in dom $\bar{d}_{A_1}$. In particular, the map j in (4.13) is surjective. By use of 4) of Lemma 4.2, (4.12) is straightforward.

Note that the boundedness of the maps i,j in (4.11) is obvious

5. HANDLEBODIES AND THE HODGE THEOREM

In this section we prove the Strong Hodge Theorem for spaces that can be built up out of pieces of the type considered in Sections 2 and 3.

Let X^n be a closed pseudomanifold as defined in Section 1, together with a piecewise flat metric. By this we mean that X^n is a metric space with distance function ρ, and each closed n-simplex σ^n is isometric to the convex hull of (n+1) points in general position in $\mathbb{R}^n$. Note that the restriction of ρ to $X^n - \Sigma^{n-2}$ induces the structure of a smooth flat riemannian manifold on $X^n - \Sigma^{n-2}$. Let $\hat{g}$ denote the corresponding riemannian metric.

As a consequence of Lemma 1.1, it follows easily that L^2-cohomology groups $H^i_{(2),\#}(X^n - \Sigma^{n-2})$ are independent of the particular choice of piecewise flat metric on X^n with the given triangulation T or on any subdivision S. For this we just apply Lemma 1.2 to any (infinite) smooth triangulation of the open manifold $X^n - \Sigma^{n-2}_S$ which includes $(X^n - \Sigma^{n-2}_S) \cap \Sigma^{n-1}_S$ as a subcomplex of the (n-1)-skeleton. Thus the groups $H^i_{(2),\#}(X^n - \Sigma^{n-2})$ are combinatorial invariants of X^n. It will follow from the Hodge Theorem that $H^i_{(2),\#}(X^n - \Sigma^{n-2}) = H^i_{(2)}(X^n - \Sigma^{n-2})$; (here we consider the case when ideal boundary conditions are not present).

We want to consider metrics which are more general than $\hat{g}$. For this purpose, and as a technical device, it is conenient to decompose $X^n - \Sigma^{n-2}$ as a union of handles. Let $C_{0,1}(N)$ denote the truncated cone on N. For $i > 1$, by an n-dimensional i-handle h^i, we mean a manifold which is piecewise smoothly equivalent to $I^{n-i} \times C_{0,1}(N^{i-1})$ with respect to some smooth triangulation. Here $I = (0,1)$. 0 and 1-handles are manifolds piecewise smoothly equivalent to I^n; for a 1 handle, we think of $I^{n-1} \times I$. Now choose a function r^c, $c \geq 1$ and let $C_{0,1}(N^{i-1})$ denote the corresponding metric horn for some choice of riemannian structure on N^{i-1}. A riemannian metric on h^i will be said to induce the structure of a riemannian i-handle, if the equivalence above is a piecewise smooth riemannian equivalence. Here $I^{n-i} \times C_{0,1}(N^{i-1})$ (or I^n) carries the product metric. Note that in the usual handlebody theory of piecewise linear manifolds, $C_{0,1}(N^{i-1})$ is replaced by I^i.

$X^n - \Sigma^{n-2}$ can be decomposed into handles by the following standard procedure. Fix a piecewise flat metric g as above. Let $T_{j,\epsilon} \subset X^n - \Sigma^{n-2}$ denote the subset of points at distance $< \epsilon$ from Σ^{n-j}, and $\overline{T}_{n,\epsilon}$ its closure in $X^n - \Sigma^{n-2}$. Choose ϵ_n so small that T_{n,ϵ_n} is a disjoint union of sets, each of which is equivalent to $C_{0,1}(\hat{L}(\sigma_j^0)) = h_j^0$, for some 0-simplex σ_j^0. Here $\hat{L}(\sigma_j^i)$ denotes the link of σ_j^i with the codimension 2 skeleton removed. Note that $L(\sigma_j^i)$ is a closed pseudomanifold with respect to the induced triangulation. Choose $\epsilon_n' < \epsilon_n$ and ϵ_{n-1} sufficiently small so that the intersection of $T_{n-1,\epsilon'_{n-1}}$ with the boundary of each h_j^0 is a disjoint union of subsets of the above type relative to the induced triangulation. Choose $\epsilon'_{n-1} < \epsilon_{n-1}$, and continue the process to obtain $\epsilon_1' < \epsilon_1 \ldots \epsilon'_n < \epsilon_n$. Set $T_{0,\epsilon_0} = X^n$, and $Z^{i-1} = X^n - \bigcup_{j \geq i} T_{j,\epsilon_i'}$. Then a simple standard argument (compare Section 6) shows that $T_{i,\epsilon_i} - \bigcup_{j>i} T_{j,\epsilon_j'}$ is a disjoint union of sets h_j^i such that for suitable piecewise smooth triangulations, there is a piecewise smooth equivalence $e_{i,j}$ of the pair

$(h^i_j, h^i_j \cap Z^{i-1})$ with the pair

(5.1) $(I^{n-i} \times C_{0,1}(\hat{L}(\sigma^{n-i})), I^{n-1} \times (\frac{1}{2},1) \times (\hat{L}(\sigma^{n-i})))$.

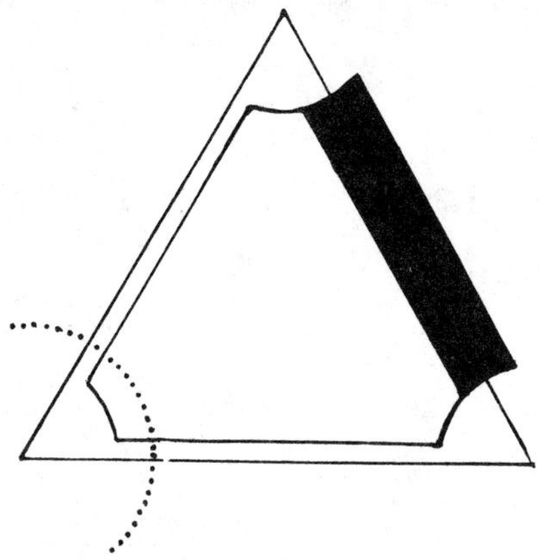

Fig. 5.1

We now define a class of admissible riemannian handles and pseudomanifolds (which we allow to be disconnected). 0 and 1 handles are admissible. A closed 1-dimensional pseudomanifold is a disjoint union of circles, and any such union is admissible. Suppose we have defined admissiblity for riemannian pseudomanifolds of dimension $< n$. A riemannian handle will be called admissible if it is equivalent to $I^{n-i} \times C_{0,1}(N^{i-1})$ where N^{i-1} is an admissible riemannian pseudomanifold with codimension 2 skelton deleted and (for the purposes of this paper) if in case $i - 1 = 2k$, $H^k_{(2),\#}(N^k) = 0$. X^n together with a riemannian metric g on $X^n - \Sigma^{n-2}$ will be called an admissible riemannian pseudomanifold, if with respect to the equivalences $e_{i,j}$, for some decomposition as above, g induces the structure of an admissible riemannian handle on each h^i_j. For example, the piecewise flat metric $\hat{g}$ is itself admissible as is any metric which is piecewise smoothly riemannian equivalent to it.

We point out that as a consequence of (5.1) the boundary of each $h^i_j \cap Z^{i-1}$ in h^i_j is collared. Also, if $k_1(y) : (0,1) \to (0,1)$ is smooth, has compact support, and is $\equiv 1$ except in some sufficiently small neighborhood, of the boundary and $k_2(r) : (0,1) \to (0,1)$ is smooth, $\equiv 1$ except in a sufficiently small neighborhood of 1, and vanishes near 1, then the supports of $(k_1^{n-i} \times k_2) \circ e_{i,j}$ will cover $X^n - \Sigma^{n-2}$. By the standard construction, we then have the existence of a piecewise smooth partition of unity subordinate to the cover $\{h_j^i\}$, and satisfying the hypothesis of Lemma 4.1.

THEOREM 5.1 (Strong Hodge Theorem). Let X^n be a closed n-dimensional admissible riemannian pseudomanifold. Then $H^i_{(2)}(X^n - \Sigma^{n-2})$ is finite dimensional and the Strong Hodge Theorem holds.

PROOF. We will verify properties A) and B) of Section 1. The proof is by induction over the handles. To begin with Z^0 is a disjoint union of sets which are piecewise smoothly riemannian equivalent to I^n. By repeated application of Lemma 3.1 it follows that $\bar{d}^{-1}_{Z^0}$ is bounded and that $H^i_{(2)}(Z^0)$ is finite dimensional, $(H^\ell_{(2)}(h_j^0) = 0$ for $\ell > 0$, $H^0_{(2)}(h_j^0) = R)$. The same properties clearly hold for any h_j^1 and $h_j^1 \cap Z^0$. Also, since a 1-dimensional closed riemannian pseudomanifold is a disjoint union of circles, the Strong Hodge Theorem holds. Assume by induction that the Strong Hodge Theorem has been established for closed admissible riemannian pseudomanifolds of dimension $< n$. It follows by repeated application of Lemma 3.1 and by the Poincaré Lemma 3.4, that for all n-dimensional i-handles h_j^i, $H^\ell_{(2)}(h_j^i) = H^\ell_{(2)}(C_{0,1}(\hat{L}(\sigma_j^{n-i}))$ is finite dimensional for all ℓ, and in particular that $\bar{d}^{-1}_{h_j^i}$ is bounded. In view of (5.1) these properties also hold for $h_j^i \cap Z^{i-1}$. Assume by induction that for some i, $H^\ell_{(2)}(Z^{i-1})$ is finite dimensional for all ℓ, and thus $\bar{d}^{-1}_{Z^{i-1}}$ is bounded. Since the boundary of $h_j^i \cap Z^{i-1}$ is collared in h^i_j, it follows from Lemma 4.3 that for all ℓ, $H_{(2)}(Z^i)$ is finite dimensional and thus that $\bar{d}^{-1}_{Z^i}$ is bounded. By induction, these properties hold for $X^n - \Sigma^{n-2}$, which establishes A).

Since by induction, the Strong Hodge Theorem holds for $\hat{L}(\sigma^i_j)$, by the definition of admissibility, the hypothesis of Theorem 2.2 is satisfied. If we also apply Theorem 2.1, it follows that the hypothesis of Lemma 4.1 is satisfied for the cover $\{h^i_j\}$. Thus (choosing A_0 empty), it follows that Stokes' Theorem holds for $X^n - \Sigma^{n-2}$ in the L^2 sense. This establishes B) and completes the proof.

From now on we write $H^i_{(2)}(X^n)$ for $H^i_{(2)}(X^n - \Sigma^{n-2})$. We mention that as an immediate corollary of the Hodge Theorem, it follows that if X^n is oriented, (i.e. $X^n - \Sigma^{n-2}$ is oriented) wedge product and integration induce a nonsingular pairing of $H^i_{(2)}(X^n)$ with $H^{n-i}_{(2)}(X^n)$. However, this could be established directly by using a generalization of the results of Section 3 and an inductive argument like the above; it also follows from the isomorphism with the Goresky-MacPherson groups to be discussed in the next section. Of course, without the admissibility condition on middle dimensional cohomology $H^i_{(2)}(X^n)$ and $H^{n-i}_{(2)}(X^n)$ do not have the same dimension in general. In suitable circumstances, Poincare duality can be restored by imposing $*$-invariant ideal boundary conditions. This means that for the subspace $V \subset \mathcal{H}$ satisfies $*V = V^{\perp}$, or equivalently that V is a maximal self-annihilating subspace under wedge product; compare also [5] and the work of Morgan [24].

For pseudomanifolds with collared boundary, the finite dimensionality of $H^i_2(X^n)$ and the boundedness of $\bar{d}^{-1}$ can be established by imbedding X^n in its double, and considering a metric invariant under the natural involution. A relative Poincaré duality can also be established in this way or by a direct inductive argument.

6. THE PAIRING WITH GORESKY-MACPHERSON

In this section we describe the dual pairing of $H^i_{(2)}(X^n)$ with the "middle intersection homology group" $IH^{\bar{m}}_i(X^n)$ introduced by Goresky and MacPherson in [18], [19]. Since we intend to discuss this more fully in [6], some details will be omitted. As mentioned in the introduction the connection of our work with that of Goresky-MacPherson was originally suggested by Dennis Sullivan (in conversations held during the summer of 1976). He observed that our calculation of L^2-cohomology for isolated conical singularities, agreed with the answers obtained by Goresky-MacPherson for their theory, and further that our *-invariant ideal boundary conditions corresponded precisely to the self-annihilating subspaces in the work of Morgan [24], (who had extended the definition of the middle intersection group).

We now recall the definition of $IH^{\bar{m}}_i(X^n)$; see [19] for more details. Let X^n be a pseudomanifold with triangulation T. A <u>stratification</u> of X^n is a filtration by closed subspaces

$$(6.1)\quad X^n = X_n \supset X_{n-1} = X_{n-2} \supset X_{n-3} \supset \ldots X_1 \supset X_0$$

such that for each point $p \in X_i - X_{i-1}$, there is a filtered space

$$(6.2)\quad V = V_n \supset V_{n-1} \supset \ldots V_i = \text{a point},$$

and a mapping $V \times B^i \to X$ which, for each integer j, takes $V_j \times B^i$ p.ℓ.-homeomorphically to a neighborhood of p in X_j. Here B^i is the p.ℓ. i-ball and p corresponds to $V_i \times$ interior point. In particular $X_i - X_{i-1}$ is an i-manifold (or is empty). Moreover, given T we have the associated stratification for which $X_j = \Sigma^j - \Sigma^{j-1}$, $j \le n - 2$.

The collection simplicial chains for all the various subdivisions of T can be divided into equivalence classes as follows. A chain on T' is equivalent to a chain on T" if their images coincide in some common refinement T". The resulting (direct limit) group $C_i(X)$ of equivalence classes is called the group of <u>p.l. geometric chains</u>; (here we will understand real coefficients). If $c_i \in C_i(X)$, the support $|c_i|$ of c_i is the union of all simplices for which c_i has a nonzero coefficient. Consider the space of i-chains c_i such that

$$(6.3) \quad i - \dim(|c_i| \cap X_{n-j}) \ge [\tfrac{j+3}{2}],$$

where $[\frac{j+3}{2}]$ denotes the greatest integer contained in $\frac{j+3}{2}$. Define $IC_i^{\overline{m}}(X^n)$ to be the associated complex, i.e. the complex of chains such that c_i, ∂c_i satisfy the above conditions.

DEFINITION 6.1. $IH_i^{\overline{m}}(X^n)$ is the i-th homology group of the complex $IC_i^{\overline{m}}(X^n)$.

In fact Goresky and MacPherson define a family of intersection homology groups $IH_i^{\overline{p}}(X^n)$ parameterized by what are called "perversities" $\overline{p}$; $\overline{m}$ represents the perversity (0,0,1,1,2,2,...). They show that groups corresponding to "complimentary" perversities, are dually paired in complimentary dimensions, under an intersection product. $\overline{m}$ is shown to be <u>self-complimentary</u> if X^n has a stratification for which the strata all have even codimension. This condition is generalized in [24] and [25]. The corresponding condition for self duality in the present paper is that of "admissibility" as defined in Section 5, or more generally, $*$-invariant, boundary conditions, (which we have described here only for the case of isolated conical singularities).

A fundamental result of [19], is that the groups $IH_i^{\bar{p}}(X^n)$ are <u>independent</u> of the particular stratification of X^n. It is an easy consequence that

(6.4) $IH_i^{\bar{m}}((0,1) \times N^m) \simeq IH_i^{\bar{m}}(N^m)$

(6.5) $$IH_i^{\bar{m}}(C^*_{0,1}(N^m)) \simeq \begin{cases} IH_i^{\bar{m}}(N^m) & i \leq \frac{m}{2} \\ 0 & i > \frac{m}{2} \end{cases}$$

Here $C^*_{0,1}(N^m)$ denotes the open pseudomanifold $C_{0,1}(N^m) \cup p$. These can be verified by using the natural product and cone stratifications of $(0,1) \times N^m$, $C^*_{0,1}(N^m)$; (see [19] for the generalization of (6.5) to arbitrary $\bar{p}$). Finally, it is immediate that the Mayer Vietoris sequence holds for p.l. open covers $A_1 \cup A_2$ of X^n.

By using the properties above, (or by proceeding directly) the groups $IH_i^{\bar{m}}(X^n)$ can be calculated for the basic example of a manifold with isolated conical singularities. Let $X^n = M^{m+1} \cup C^*_{0,1}(N^m)$, where M^{m+1} is a manifold with boundary N^m and the union is along the boundary.

(6.6) $m = 2k - 1$

$$IH_i^{\bar{m}}(X^{2k}) = \begin{cases} H_i(M^{2k}) & i < k \\ i(H_k(M^{2k})) \subset H_k(M^{2k}, N^{2k-1}) & i = k \\ H_k(M^{2k}, N^{2k-1}) & i > k \end{cases}$$

(6.7) $m = 2k$

$$IH_i^{\bar{m}}(X^{2k+1}) = \begin{cases} H_i(M^{2k+1}) & i < k+1 \\ i(H_{k+1}(M^{2k+1})) \subset H_{k+1}(M^{2k+1}, N^{2k}) & i = k+1 \\ H_i(M^{2k+1}, N^{2k}) & i > k+1 \end{cases}$$

The corresponding calculation for $H^i_{(2)}(X^{m+1})$ is stated in [5] and can be carried out on the basis of the Poincaré Lemma of Section 3 and the Mayer Vietoris sequence of Section 4.

More generally, in view of the formal properties (6.4), (6.5), and the corresponding Poincaré Lemma of Section 3, it will follow in general that

$$(6.8) \quad H^i_{(2)}(X^n) \simeq (IH^{\overline{m}}_i(X^n))^*$$

provided one can show that there exists a <u>natural map</u> $H^i_{(2)}(X^n)$ to the dual space $(IH^{\overline{m}}_i(X^n))^*$ of $IH^{\overline{m}}_i(X^n)$. In fact, by use of the handle decomposition of Section 5, (6.8) follows directly from the Mayer Vietoris sequence and induction over the handles.

The required map is given by integration. Note that if $\beta \in \operatorname{dom} d$ then for <u>arbitrary</u> chains c, $\int_c \beta$ is <u>not</u> defined in general. However, one can show that for <u>almost all</u> chains in $IC^{\overline{m}}_i(X^n)$, $\int_c \beta$ is defined and Stokes' Theorem holds.

There is another approach to the pairing which is also convenient for the purpose of defining products of classes in $H^*_{(2)}(X^n)$; namely, to pass to a subcomplex S^* of $\{\operatorname{dom} d_i\}$ which has the same cohomology, but whose members are nice enough so that integration over elements of $IC^{\overline{m}}_i(X^n)$ is <u>always</u> defined. To define S^*, one proceeds as follows.

Let Ω^* be a complex of smooth forms on the smooth manifold N^m. Assume that

$$(6.9) \quad \int_a^r \omega(x,s)ds \in \Omega^i,$$

whenever $\omega(x,s)$ is a smooth form on $(0,1) \times N^m$ with $\omega(s,x) \in \Omega^i$ for each fixed x. Consider the complex of forms $I \times \Omega^*$ on $(0,1) \times N^m$ associated to all smooth forms

$$(6.10) \quad \phi(x,r) + dr\wedge\omega(x,r)$$

such that $\phi(x,r) \in \Omega^i$, $\omega(x,r) \in \Omega^{i-1}$ for each fixed r. Then if we use (6.9) to define a homotopy operator $\mathcal{K}^{\bullet}$ as in Section 3, we see that restriction to (a,N^m) defines an isomorphism

$$(6.11) \quad H^i(\Omega^*) \simeq H^i(I \times \Omega^*).$$

Now define another complex $C(\Omega^*)$ on $(0,1) \times N^m$ (which we regard as $C_{0,1}(N^m)$, but with no metric specified), as follows. Consider smooth forms as in (6.10) which in addition satisfy the following conditions on $(0,1/4) \times N^m$

$$(6.12) \quad \beta_i \equiv \begin{cases} \phi(1/4) & i \le \frac{m}{2} \\ 0 & i > \frac{m}{2} \end{cases}$$

Let $C(\Omega^*)$ denote the associated complex, i.e. the space of forms α such that α, $d\alpha$ satisfy the above conditions. Then by use of the homotopy operator $\mathcal{K}$ of Section 3, it follows immediately that

$$(6.13) \quad H^i(C(\Omega^*)) = \begin{cases} H^i(\Omega^*) & i \le \frac{m}{2} \\ 0 & i > \frac{m}{2} \end{cases}$$

which is the analogue of the Poincaré Lemma 3.4. The corresponding result for the handle $I^i \times C_{0,1}(N^m)$ also follows.

Now suppose for example, $X^{m+1} = C^*_{0,1}(N^m) \cup M^{m+1}$ has a metric which is horn-like. Then if $\mathcal{S}^*$ denotes the complex of forms on $X^{n+1} - p$, whose restriction to $C_{0,1}(N)$ are in $C(\Lambda^*(N^m))$, it is easy to see that $\mathcal{S}^* \subset \{\text{dom } d\}$. Moreover, if $c \in IC^{\overline{m}}_{i+1}(X^n)$, $\alpha \in \mathcal{S}^i$ then

$$(6.14) \quad \int_c d\alpha = \int_{\partial c} \alpha.$$

In fact, X_0 must contain p. If $c_i \cap \neq \phi$ then $i \ge [\frac{m}{2} + 2]$, and it follows from (6.12) that α, $d\alpha$ are supported away from p. Thus (6.14) is clear.

For an arbitrary pseudomanifold X^n with triangulation T, we use a handle decomposition to define the complex $\mathcal{S}^*$. For this, we should be explicit about the compatibility of the various product structures on the intersections $h^i_j \cap Z^{i-1}$. We will use a variation of the construction of Section 5. The basic idea is that if σ^n is an n-simplex of X^n then $(\sigma^n)' - \cup T_{\epsilon_j}(\Sigma^{n-j})$ is a union of n-cubes. Here $(\sigma^n)'$ denotes the first barycentric subdivision of σ^n.

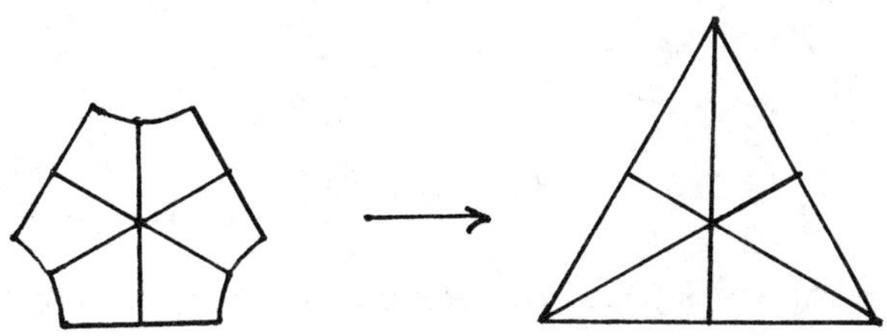

Fig. 6.1

Choose a piecewise flat metric on X^n. Let $\sigma^n \supset \sigma^{n-1} \ldots \sigma^0$ determine an n-simplex τ^n, in the first barycentric subdivision of σ^n. Let ℓ_j denote the distance from the barycenter of the face σ^{n-j+1}, to the face σ^{n-j}, $j = 1,\ldots n$. Then there is a natural map f_τ, which carries the image of the unit cube I^n under the collapsing map

(6.15) $g_\tau : (x_1, x_2, \ldots x_n) \to (\ell_1 x_1, \ell_2 x_1 x_2, \ldots, \ell_n x_1 x_2 \ldots x_n)$,

isometrically onto τ^n. Consider

(6.16) $\bigcup_\tau f_\tau \circ g_\tau((\frac{1}{3},1] \times \ldots (\frac{1}{3},1])$, $j = 0$

(6.17) $\bigcup_\tau f_\tau \circ g_\tau((\frac{1}{3},1] \times \ldots [0,\frac{2}{3}) \times [0,1])$, $j = 1$

(6.18) $\bigcup_\tau f_\tau \circ g_\tau((\frac{1}{3},1] \times \ldots [0,\frac{2}{3}) \times \overbrace{[0,1] \times \ldots [0,1]}^{j})$. $2 \le j \le n$

Here τ varies over <u>all</u> n-simplies in the first barycentric subdivision T', of T. The components of (6.16) and (6.17) are

0-handles and 1-handles respectively. The components of (6.17), with their intersections with Σ^{n-2} removed, are j-handles, $j \geq 2$.

A smooth i-form on $[0,1] \times [0,1] \times [0,1] \times \ldots [0,1]$ can be written as a sum of terms

(6.19) $h_{(\ell)}(x_1,\ldots x_n)dx_{\ell_1} \wedge \ldots dx_{\ell_i}$,

where $\ell = (\ell_1,\ldots \ell_i)$. We impose the following conditions which generalize (6.12).

For $j > 1$, $x_j < 1/4$ implies,

A) $h_{(\ell)}(x_1,\ldots x_n)$ is independent of x_j,

B) $j \notin (\ell_1,\ldots \ell_i)$,

C) at most $\frac{j}{2}$ of the $(\ell_1,\ldots \ell_i)$ are $< j$.

Define $\mathcal{S}^*$ to be the complex of piecewise smooth forms α on $X^n - \Sigma^{n-2}$ such that for all τ, $(f_\tau \circ g_\tau)^*(\alpha)$, $(f_\tau \circ g_\tau)^*(d\alpha)$ satisfy the above conditions. Then it is straightforward to show that (6.14) holds. By induction over the handles, the inclusion $\mathcal{S}^* \subset \{\text{dom } d\}$ induces an isomorphism

(6.20) $H^i(\mathcal{S}^*) \simeq H^i_{(2)}(X^n)$.

Thus, for metrics of the type considered in Section 5, but without assuming the admissibility condition, we have

THEOREM 6.1. $H^i_{(2)}(X^n) \simeq (IH^{\overline{m}}_i(X^n))^*$

In view of Theorem 6.1, the admissibility condition in the Hodge Theorem 5.1 can be replaced by the corresponding condition on $IH^{\overline{m}}_i(X^n)$.

Finally, we mention that if $\alpha_{i_1}, \alpha_{i_2}$ are i_1, i_2-forms in $\mathcal{S}^*$, then $\alpha_{i_1} \wedge \alpha_{i_2} \in \mathcal{S}^*$ provided either i_1 or $i_2 \geq \frac{n}{2}$. However, if various strata are empty, or if their links have vanishing middle intersection groups, the definition of S^* can be revised and products will be defined more frequently. This will be discussed further elsewhere, as will be the identification of wedge product as the dual of intersection product.

7. APPLICATIONS AND FURTHER RESULTS

As we mentioned in the introduction, the most interesting applications of our main results, Theorem 5.1 and 6.1, concern the topology of singular algebraic varieties which admit a Kahler metric of the type considered in Section 5. *For such varieties, the various standard consequences of Hodge Theory such as the Hodge structure and primitive decomposition of cohomology, the Hard Lefschetz Theorem, and the Hodge Index Theorem, are valid for the cohomology theory dual to the Goresky-MacPherson middle intersection groups*; (since X admits a stratification by strata of even codimension, X is admissible in the sense of Section 5). In fact, given Theorems 5.1 and 6.1, these properties follow by precisely the same arguments (applied to the open Kahler manifold $X^n - \Sigma^{n-2}$) as are standard in the compact case. The arguments depend only on the existence of a Hodge theory and the *local* properties of a Kahler metric.

The above properties were conjectured jointly with MacPherson. They will be discussed further in [8]. More generally, MacPherson and I have the following conjecture.

CONJECTURE 7.1. The standard consequences of Hodge Theory are valid for *all* projective algebraic varieties and the middle intersection group.

Of course, a natural approach to this conjecture is to attempt to establish directly, that the Hodge Theory we have described extends to the general case. It is planned to discuss this further in [8]. On the other hand, one can try to use the complex structure more vigorously, via resolution of singularities. For example, in [8], the existence of a Hodge structure on the middle intersection group for surfaces, will be derived from this viewpoint.

Finally, we mention that the Schubert varieties form a significant class of spaces admitting metrics for which the techniques of the present paper apply; see [8]. This seems to be of particular interest because, as I understand from D. Kazhdan and G. Lusztig, the consequences of the Hodge Theory in this case may be relevant for their recent work on representations of Coxeter groups and Hecke algebras; see [22].

By way of contrast, we now briefly discuss some developments arising out of the approach to the Laplacian on singular spaces which was described in [5]. This method is more closely tied to separation of variables and eigenfunction expansions than that of the present paper. Thus, it can be carried out most explictly in those situations where the metric has a standard form. Of course, the basic case is that of a metric cone and in [5], we gave a functional calculus for the Laplace operator on the cone. From a technical standpoint the main idea was to use the functional calculus for the Laplace operator on the base of the cone in order to "sum" the eigenfunction expansions which were obtained from the Hankel inversion formula; (this technique was also applied in [4]). Fundamental solutions for the Laplace, heat and wave operators on the cone were obtained in [6], and in [9], a detailed discussion of the diffraction of waves by conical singularities will be given.

We point out that at the level of an explicit functional calculus, (as opposed to that of the present paper), the more difficult case of general metric horns must be distinguished from the special case of metric cones. In fact, it is only for the case of cones that separation of variables leads to an ordinary differential equation with a regular singular point.

We now restrict attention to metrics on pseudomanifolds, for which the simplices have constant curvature and totally geodesic faces. Note that even in order to treat the piecewise flat case one must simultaneously consider the case of constant positive curvature, since the links of simplicies in a piecewise flat space inherit metrics of this type. There is functional calculus for "suspensions" which allows us to handle the positive case; see [7]. In the compact case, we get in particular, a complete orthonormal basis of eigenfunctions, and can calculate the asymptotic expansion of the trace of the heat kernel; (compare [6]). Note that for the case of admissible metrics compactness of the Green's operator in the piecewise constant curvature case, immediately implies the same result for riemannian equivalent metrics. In general the relation between these cases is like the relation between constant and variable coefficients.

Here are two applications of these ideas. Both were referred to in [5], [6], but were not stated explicitly. The first is a Bochner Vanishing Theorem in our context. Let X^n be an admissible riemannian pseudomanifold with a piecewise smooth metric of constant curvature. Assume that for each simplex, the link is connected. There is a natural notion of positive (nonegative) curvature at the singularities Σ^{n-2} of X^n. Namely, for each simplex of Σ^{n-2}, the link (which is a circle) has length $< 2\pi$ ($\leq 2\pi$). Although this condition is sometimes taken to be the analogue of positive <u>sectional curvature</u>, on the basis of the following result, Gromov has suggested that <u>positive cur-</u>

vature operator might be a more appropriate interpretation.

THEOREM 7.1. Let X^n be an admissible riemannian pseudomanifold such that the link of each simplex is connected.

1) If X^n has a piecewise flat metric such that the curvature at the singularities is nonnegative then X^n is a rational homology manifold whose i-the Betti number is $\leq \binom{n}{i}$. In fact, every harmonic form on $X^n - \Sigma^{n-2}$ is parallel.

2) If X^n as above has strictly positive curvature at the singularities then it is a rational homology sphere.

3) In fact, if X^n has a metric of piecewise positive constant curvature and the curvature at the singularities is nonnegative then X^n is a rational homology sphere.

It does not seem to be completely out of the question that a (perhaps strengthtened) version of Theorem 7.1 could be proved by more elementary means. In any event, we believe that our method of proof is useful in other contexts.

Now let X^n be an oriented pseudomanifold with piecewise flat metric. If X^n is not admissible we assume that it is possible to make an explicit choice of $*$-invariant boundary conditions (in an inductive sense that we won't make precise here); however, compare [24]. Then by use of Poincaré duality for the middle intersection group, L-classes can be defined for X^n; see [19], [24]. They are ordinary homology classes. These L-classes are given by an explicit local formula whose derivation follows from the heat equation approach to the index theorem. For each oriented simplex σ_j^{n-4k} of codimension 4k, we assert that the link $L(\sigma_j^{n-4k})$ with the induced orientation, has an η-invariant in the sense of Atiyah-Patodi-Singer, [2].

THEOREM 7.2 (Compare [12], [13]) $L_{4k}(X^n) = \sum_j \eta(L(\sigma_j^{n-4k}))\cdot\sigma_j^{n-4k}$

The related signature operator can also be generalized to the case of coefficients in an arbitrary vector bundle. This generalization comes most easily if one restricts attention to a suitable class of partially flat connections.

We refer to [7] and a planned sequel for a more detailed discussion of these results.

8. APPENDIX

We will briefly recall the construction of the regularization operators R_ϵ and the homotopy operators A_ϵ which were referred to in Section 1; compare [10], [15]. Consider the cube $(-1,1)^n \subset \mathbb{R}^n$ and let J_ϵ be the Friedrichs molifier on i-forms. Then for $\theta \in L^2((-1,1)^n)$, on any subcube $(a,b)^n$ with $[a,b] \subset (-1,1)$ we have, $\|J_\epsilon - I\| \to 0$, $J_\epsilon\theta \in \Lambda^i((a,b)) \cap L^2$, and $dJ_\epsilon\theta = J_\epsilon d\theta$ if $d\theta$ exists in the weak sense and ϵ is sufficiently small. In particular it follows that $\theta \in \operatorname{dom} \bar{d}_{(a,b)^n}$.

Write

$$(8.1) \quad \theta = \phi_i + dx_i \wedge \omega_i$$

where ϕ_i, ω_i do not contain dx_i. Then as in (3.4), we have bounded operators

$$(8.2) \quad K_{i,a}\theta = \int_a^{x_i} \omega(x_1,\ldots,u,x_{i+1}\ldots x_n)\,du$$

which satisfy,

$$(8.3) \quad (dK_{i,a} + K_{i,a}d)\theta = \theta - \phi_i(x_1,\ldots,a,x_{i+1},\ldots,x_n),$$

for $\theta \in \Lambda^i$. We can obtain bounded operators on the subcube $(-1/2,1/2)^n$, which are independent of a, by setting

$$(8.4)\quad \mathcal{K}_i\theta = \int_{-1/2}^{1/2} \mathcal{K}_{i,a}\theta da,$$

$$(8.5)\quad \mathcal{B}_i\theta = \int_{-1/2}^{1/2} \phi_i(x_1,\ldots,a,x_{i+1},\ldots x_n)da.$$

Then since d commutes with restriction, it follows that

$$(8.6)\quad d\mathcal{B}_i\theta = \mathcal{B}_i d\theta.$$

Using the fact that $\mathcal{K}_i$, $\mathcal{B}_i$, are bounded, from (8.3), (8.6) we get

$$(8.7)\quad (\bar{d}\mathcal{K}_i + \mathcal{K}_i\bar{d})\theta = \theta - \mathcal{B}_i\theta,$$

$$(8.8)\quad \bar{d}\mathcal{B}_i\theta = \mathcal{B}_i\bar{d}\theta,$$

for $\theta \in \operatorname{dom} \bar{d}_{(-1/2,1/2)^n}$. In particular $\mathcal{B}_i : \operatorname{dom} \bar{d}_{(-1/2,1/2)^n} \to \operatorname{dom} \bar{d}_{(-1/2,1/2)^n}$. It follows from (8.7), (8.8) that

$$(8.9)\quad (\bar{d}\mathcal{K}_j\mathcal{B}_{j-1}\cdots\mathcal{B}_1 + \mathcal{K}_j\mathcal{B}_{j-1}\cdots\mathcal{B}_1\bar{d})\theta = \mathcal{B}_{j-1}\cdots\mathcal{B}_1\theta - \mathcal{B}_j\cdots\mathcal{B}_1\theta.$$

Thus,

$$(8.10)\quad (\bar{d}\mathcal{K} + \mathcal{K}\bar{d})\theta = \theta - \mathcal{B}\theta,$$

where

$$(8.11)\quad \mathcal{K} = \mathcal{K}_1 + \sum_{j=2}^{n} \mathcal{K}_j\mathcal{B}_{j-1}\cdots\mathcal{B}_1,$$

$$(8.12)\quad \mathcal{B} = \mathcal{B}_n\cdots\mathcal{B}_1.$$

As is obvious by inspection, for $i > 0$,

$$(8.13)\quad \mathcal{B} \equiv 0,$$

while for $i = 0$,

$$(8.14)\quad \mathcal{B}\theta = \int_{(-1/2,1/2)^n}\theta.$$

As for example in Section 3, (or otherwise), a <u>bounded</u> homotopy operator could also be obtained by working directly with $\bar{d}$ on the ball, $B_1(0) = C^*_{0,1}(S^{n-1})$, but not as easily. This illustrates the essential difference between $\mathbb{R}^n = C^*(S_1^{n-1})$ and an arbitrary cone. Let f be a smooth positive function which is identically 1 on $(-1/4,1/4)^n$ and whose support is contained in the interior of $(-1/2,1/2)^n$. Let $\theta \in L^2(-1,1)^n$, and let $\mathcal{K}\theta$ denote $\mathcal{K}$ applied to the restriction of θ to $(-1/2,1/2)^n$. If we define the values of $\mathcal{R}_\epsilon$, $\mathcal{G}_\epsilon$ (ϵ sufficiently small) on the support of f to be given by

$$(8.15)\quad \mathcal{R}_\epsilon = fJ_\epsilon + (1 - f)I - df\wedge(I - J_\epsilon)\mathcal{K}\theta,$$

$$(8.16)\quad \mathcal{G}_\epsilon = f(I - J_\epsilon)\mathcal{K},$$

we have $\mathcal{R}_\epsilon = J_\epsilon$ on $(-1/4,1/4)^n$, and $\mathcal{R}_\epsilon = I$, $\mathcal{G}_\epsilon = 0$ on the boundary of supp f. If $d\theta$ exists in the weak sense on $(-1,1)^n$, then $\theta|(-1/2,1/2)^n \in \operatorname{dom} \bar{d}_{(-1/2,1/2)^n}$ and on the support of f,

$$\begin{aligned}(8.17)\quad (\bar{d}\mathcal{G}_\epsilon + \mathcal{G}_\epsilon\bar{d})\theta &= \bar{d}[f(I - J_\epsilon)\mathcal{K}\theta] + [f(I - J_\epsilon)\mathcal{K}]\bar{d}\theta,\\ &= df\wedge(I - J_\epsilon)\mathcal{K}\theta + f(I - J_\epsilon)\bar{d}\mathcal{K}\theta\\ &+ f(I - J_\epsilon)\mathcal{K}\bar{d}\theta,\\ &= df\wedge(I - J_\epsilon)\mathcal{K}\theta + f(I - J_\epsilon)(\theta - \mathcal{B}\theta),\\ &= (I - \mathcal{R}_\epsilon)\theta + f(I - J_\epsilon)\mathcal{B}\theta,\\ &= (I - \mathcal{R}_\epsilon)\theta,\end{aligned}$$

where the last step follows by using (8.13), (8.14). If we **define** the values of $\mathcal{R}_\epsilon$, $\mathcal{G}_\epsilon$ outside supp f to be given by $\mathcal{R}_\epsilon = I$, $\mathcal{G}_\epsilon = 0$, then,

$$(8.18)\quad \bar{d}\mathcal{G}_\epsilon + \mathcal{G}_\epsilon\bar{d} = I - \mathcal{R}_\epsilon,$$

holds on $(-1,1)^n$ and immediately implies

$$(8.19)\quad \bar{d}\mathcal{R}_\epsilon = \mathcal{R}_\epsilon\bar{d}.$$

Moreover, it is clear from (8.10), (8.15), (8.16), that if $d\theta$ exists in the weak sense, then so does $d\mathcal{R}_\epsilon\theta$ and that $\mathcal{A}_\epsilon : \Lambda^i \cap L^2 \to \Lambda^i \cap L^2$.

Now let Y^n be an arbitrary riemannian manifold and let $\{U_j\}$ be a locally finite open cover by images of $(-1,1)^n$ such that the images of $(-1/4,1/4)^n$ still cover. Then $\mathcal{R}_\epsilon$, $\mathcal{A}_\epsilon$ can be transplanted to U_i and extended to all Y^n by defining their values to be I, 0, respectively, outside of U_i. The operators $R_{\epsilon_i,i}$, $A_{\epsilon_i,i}$, so obtained, satisfy

$$(8.20)\quad \bar{d}A_{\epsilon_i,i} + A_{\epsilon_i,i}\bar{d} = I - R_{\epsilon_i,i},$$

$$(8.21)\quad \bar{d}R_{\epsilon_i,i} = R_{\epsilon_i,i}\bar{d}.$$

Since the cover is locally finite, the infinite composition

$$(8.22)\quad R_\epsilon = \cdots \circ R_{\epsilon_2,2} \circ R_{\epsilon_1,1},$$

and the sum

$$(8.23)\quad A_\epsilon = \sum_{i=1}^{\infty} A^i,$$

where

$$(8.24)\quad A^i = A_{\epsilon_i,i} \circ R_{\epsilon_{i-1},i-1} \circ \cdots \circ R_{\epsilon_1,1},$$

are well defined. If $\epsilon_1 > \epsilon_2 > \cdots \to 0$ are chosen sufficiently small, R_ϵ, $A_\epsilon : L^2 \to L^2$ and $\|R_\epsilon - I\|$, $\|A_\epsilon\|$ can be made arbitrarily small. A simple formal argument, (see [10], shows that

$$(8.25)\quad \bar{d}A_\epsilon + A_\epsilon\bar{d} = I - R_\epsilon,$$

(8.26) $\bar{d}R_\epsilon = R_\epsilon\bar{d}$,

and since the supports of the images of (-1/4,1/4) cover Y, it is clear that R_ϵ is infinitely smoothing. Clearly, it follows that if $d\theta$ exists in the weak sense, then $\theta \in \operatorname{dom} \bar{d}_Y$ and moreover that $H^i_{(2)}(Y) = H^i_{(2),\#}(Y)$.

BIBLIOGRAPHY

1. Agmon, S., Lectures on Elliptic Boundary Value Problems, D. Van Nostrand Co., 1965.

2. Atiyah, M., Patodi, V., Singer, I., Spectral Asymmetry and Riemannian Geometry I, Math. Proc. Comb. Phil. Soc. 77 (1975), 43-69.

3. Borel, A., Stable Real Cohomology of Arithmetic Groups, Ann. Scient. Ec. Norm. Sup 4e Série, t.7, 1974, 235-272.

4. Cheeger, J., Analytic Torsion and the Heat Equation, Ann. Math., 109 (1979), 259-322.

5. Cheeger, J., On the Spectral Geometry of Spaces with Cone-like Singularities, P.N.A.S., V. 76, No. 5, (1979), 2103-2106.

6. Cheeger, J., Spectral Geometry of Spaces with Cone-like Singularities, preprint, 1978.

7. Cheeger, J., Spectral Geometry of Singular Riemannian spaces, (to appear).

8. Cheeger, J., Goresky, M., MacPherson, R., (to appear).

9. Cheeger, J., Taylor, M., On the Diffraction of Waves by Conical Singularities, (to appear).

10. De Rahm, G., Varietés Différentiables, Hermann, 1960.

11. Dodziuk, J., De Rahm-Hodge Theory for L^2-Cohomology of Infinite Coverings, Topology, V. 16, 1977, 157-165.

12. Gabrielov, A. M., Gel'fand, I. M., Losik, A Combinatorial Calculation of Characteristic Classes, Funct. Annal. and App., V. 9, N. 2, (1975), 103-115.

13. Gabrielov, A. M., Combinatorial Computation of Characteristic Classes, Funct. Anal. and App. V. 12, N. 2, (1978), 75-80.

14. Gaffney, M., The Harmonic Operator for Exterior Differential Forms, Thesis, U. of Chicago, 1951.

15. Gaffney, M., The Harmonic Operator for Differential Forms, P.N.A.S., V. 37, (1951), 48-50.

16. Gaffney, M., A Special Stokes Theorem for Riemannian manifolds, Ann. Math., V. 60, No. 1, (1954), 140-145.

17. Gaffney, M., Hilbert Space Methods in the Theory of Harmonic Integrals, Trans. A.M.S., V. 78, (1955), 426-444.

18. Goresky, M., MacPherson, R., La Dualité de Poincaré pour les Espaces Singuliers, C. R. Acad. Sci. Paris, t. 284, (1977), 1549-1551.

19. Goresky, M., MacPherson, R., Intersection Homology, preprint, 1978.

20. Harder, G., O, On the Cohomology of Arthmetically Defined Groups, Bombay Colloquium on Discrete Subgroups Lie Groups and Applications to Moduli, Oxford, 1975.

21. Harvey, R. Polking, J., Removable Singularities of Solutions, Acta. Math., V. 125, (1970), 39-55.

22. Kazhdan, D., Lusztig, G., Representations of Coxeter Groups and Hecke Algebras, preprint, 1979.

23. Kodaira, K., Harmonic Fields in Riemannian Manifolds (Generalized Potential Theory), Ann. Math., 50, (1949), 587-665.

24. Morgan, J., A Geometric Proof of the Characteristic Variety Theorem, (to appear).

25. Siegel, P., Witt spaces: A Geometric Cycle Theory for KO Homology at Odd Primes, Thesis, M.I.T., 1979.

26. Singer, I., Future Extensions of Index Theory and Elliptic Operators, Ann. Math. Stud., 70, (1971), Princeton U. Press, 171-185.

27. Teleman, N., Combinatorial Hodge Theory and Signature Theorem, preprint, 1979.

28. Zucker, S., Hodge Theory with Degenerating Coefficients, I, preprint, 1978.

DEPARTMENT OF MATHEMATICS
SUNY, Stony Brook
Stony Brook, NY 11790

Proceedings of Symposia in Pure Mathematics
Volume 36, 1980

LIOUVILLE THEOREM FOR HARMONIC MAPS

Shiu-Yuen Cheng*

§0. INTRODUCTION.

The classical Liouville theorem states that

LIOUVILLE THEOREM: Any bounded harmonic function defined on the entire plane is a constant function.

In 1975 Yau [2] generalized the theorem to harmonic functions on Riemannian manifolds. He proved that

THEOREM: Suppose that M is a complete Riemannian manifold of nonnegative Ricci curvature. Then any bounded harmonic function defined on M is a constant function.

The main purpose of this paper is to prove a Liouville theorem for harmonic maps. We prove the following

THEOREM: Suppose that M is a complete Riemannian manifold of nonnegative Ricci curvature and N is a simply-connected complete Riemannian manifold of nonpositive sectional curvature. If the image of any harmonic map from M into N lies in a compact subset, then it is a constant map.

The above theorem is a consequence of the interior gradient estimate derived in §2. Actually we prove a stronger theorem. Roughly speaking we prove that if a harmonic map grows slower than linear growth, then it is necessarily a constant map. Since harmonic maps are solutions of a nonlinear elliptic system, the interior gradient estimate in §2 is of interest in its own right.

During the Hawaii symposium Professor James Eells informed us that he also proved a Liouville theorem for harmonic maps by using the Sampson maximum principle.

§1. NOTATIONS AND FORMULAS.

Suppose that M, N are Riemannian manifolds of dimension m, n respec-

---

1980 Mathematics Subject Classification 53C99.

*Research partially supported by Sloan Fellowship and a NSF grant.

tively. Let $\{\theta_i\}_{i=1}^m$, $\{\omega_\alpha\}_{\alpha=1}^n$, be local orthonormal frame fields of M, N respectively. There exist unique anti-symmetric matrices (θ_{ij}) , $(\omega_{\alpha\beta})$ of 1-forms satisfying

(1.1) $$d\theta_i = \sum_j \theta_j \wedge \theta_{ji} ,$$

(1.2) $$d\omega_\alpha = \sum_\beta \omega_\beta \wedge \omega_{\beta\alpha}$$

The curvature tensors of M, N are obtained as

(1.3) $$d\theta_{ij} - \sum_k \theta_{ik} \wedge \theta_{kj} = \Theta_{ij} = -\frac{1}{2} \sum_{k,\ell} R_{ijk\ell}\theta_k \wedge \theta_\ell ,$$

(1.4) $$d\omega_{\alpha\beta} - \sum_\gamma \omega_{\alpha\gamma} \wedge \omega_{\gamma\beta} = \Omega_{\alpha\beta} = -\frac{1}{2} \sum_{\gamma,\delta} K_{\alpha\beta\gamma\delta}\omega_\gamma \wedge \omega_\delta .$$

The curvature tensors $R_{ijk\ell}$, $K_{\alpha\beta\gamma\delta}$ satisfy the well-known symmetry relations. The Ricci tensor (R_{ij}) of M is defined by

$$R_{ij} = \sum_k R_{ikjk} ,$$

and one concludes easily from the symmetry relations of $R_{ijk\ell}$ that (R_{ij}) is a symmetric matrix. We then say that the Ricci curvature is nonnegative if (R_{ij}) is nonnegative semi-definite. We say that the sectional curvature of N is nonpositive if $\sum K_{\alpha\beta\gamma\delta}\xi_\alpha\xi_\gamma\eta_\beta\eta_\delta \le 0$ for any (ξ_α) , (η_α) .

Suppose there is a smooth mapping $f: M \longrightarrow N$. Then

(1.5) $$f^*(\omega_\alpha) = \sum_i a_{\alpha i}\theta_i .$$

$|\nabla f| = (\sum(a_{\alpha i})^2)^{1/2}$ is then the length of the gradient of f . Exterior differentiate (1.5), we have

(1.6) $$\sum_i (da_{\alpha i} - \sum_j a_{\alpha j}\theta_{ij} + \sum_\beta a_{\beta i}f^*(\omega_{\beta\alpha})) \wedge \theta_i = 0 .$$

Define

(1.7) $$\sum_j a_{\alpha ij}\theta_j = da_{\alpha i} - \sum_j a_{\beta j}\theta_{ij} + \sum_\beta a_{\beta i}f^*(\omega_{\beta\alpha}) .$$

(1.6) then says that $\sum_{i,j} a_{\alpha ij}\theta_j \wedge \theta_i = 0$. This is equivalent to

(1.8) $$a_{\alpha ij} = a_{\alpha ji} .$$

f is said to be harmonic iff $\sum_i a_{\alpha ii} = 0$.

Exterior differentiate (1.7) we obtain

(1.9) $$\sum_{k,j} a_{\alpha ijk}\theta_k \wedge \theta_j = \frac{1}{2} \sum a_{\alpha j}R_{ijk\ell}\theta_k \wedge \theta_\ell - \frac{1}{2} \sum a_{\beta i}K_{\beta\alpha\gamma\delta}a_{\gamma k}a_{\delta\ell}\theta_k \wedge \theta_\ell ,$$

where

$$\sum a_{\alpha ijk}\theta_k = da_{\alpha ij} - \sum_k a_{\alpha kj}\theta_{ik} - \sum_k a_{\alpha ik}\theta_{jk} + \sum_\beta a_{\beta ij} f^*(\omega_{\beta\alpha}) . \tag{1.10}$$

(1.9) implies the following commutation formula

$$a_{\alpha ik\ell} - a_{\alpha i\ell k} = -\sum a_{\alpha j} R_{ijk\ell} + \sum a_{\beta i} a_{\gamma k} a_{\delta\ell} K_{\beta\alpha\gamma\delta} . \tag{1.11}$$

Using (1.8) and (1.11) we derive the following Bochner type formula for harmonic maps

$$\frac{1}{2}\Delta\left(\sum (a_{\alpha i})^2\right) = \sum_{i,j} (a_{\alpha ij})^2 + \sum R_{ij} a_{\alpha i}\, a_{\alpha j} - \sum K_{\alpha\beta\gamma\delta} a_{\alpha i} a_{\beta j} a_{\gamma i} a_{\delta j} . \tag{1.12}$$

In applications we usually use the Schwarz inequality to estimate the

$$|d|\nabla f|^2|^2 \le 4|\nabla f|^2 \left(\sum (a_{\alpha ij})^2\right) . \tag{1.13}$$

Suppose $\phi : N \longrightarrow \mathbb{R}$ is a function on N. Then $\phi \circ f$ is a function on M. Then it is easy to see that

$$d(\phi\circ f)(x) = \sum_i a_{\alpha i}(x)\phi_\alpha(f(x))\theta_i , \tag{1.14}$$

$$\Delta(\phi\circ f) = \sum_{\alpha,i} a_{\alpha ii}\phi_\alpha + \sum a_{\alpha i} a_{\beta i}\phi_{\alpha\beta} . \tag{1.15}$$

So if f is harmonic and $(\phi_{\alpha\beta}) \ge \lambda(\delta\alpha\beta)$, then

$$\Delta(\phi\circ f) \ge \lambda \sum (a_{\alpha i})^2 . \tag{1.16}$$

§2. AN INTERIOR GRADIENT ESTIMATE.

Let $B_a(x_o)$ be the closed geodesic ball of radius a and center x_o in M and let r be the distance function from x_o. Let $f: M \longrightarrow N$ be a harmonic map into N which is simply-connected and is having nonpositive sectional curvature. Let y_o lie outside $f(B_a(x_o))$ and ρ denote the distance function from y_o. Consider the function on $B_a(x_o)$

$$\Phi = \frac{(a^2-r^2)^2|\nabla f|^2}{(b^2-\rho^2\circ f)^2} ,$$

where $b > \sup\{\rho(f(x)) : x \in B_a(x_o)\}$. Then Φ assumes its maximum at an interior point $\bar{x}$ of $B_a(x_o)$. At this point r may not be twice differentiable. We will remedy this by the following consideration. Let $\gamma:[0,r(\bar{x})] \longrightarrow M$ be the minimizing geodesic joining $\gamma(0) = x_o$ to $\gamma(r(\bar{x})) = \bar{x}$. Let ε be a small positive number. Along γ, $\gamma(\varepsilon)$ is not conjugate to $\bar{x}$. So there exists a geodesic cone C with vertex at $\gamma(\varepsilon)$

and contains a neighborhood of $\bar{x}$ such that r_ε, the distance function from $\gamma(\varepsilon)$ measured in C, is smooth inside C. Let $\tilde{r} = r_\varepsilon + \varepsilon$. Then triangle inequality states that

$$(2.1)\qquad \begin{aligned} r &\le \tilde{r} \\ r(\bar{x}) &= \tilde{r}(\bar{x}) . \end{aligned}$$

Then $\dfrac{(a^2-\tilde{r}^2)^2|\nabla f|^2}{(b^2-\rho^2 \circ f)^2}$ is smooth near $\bar{x}$ and also attains the same maximum value at $\bar{x}$. Letting $\varepsilon \longrightarrow 0$, we may assume without loss of generality that r is already smooth near $\bar{x}$. Then, at $\bar{x}$ we have

$$(2.2)\qquad 0 = \frac{-2dr^2}{(a^2-r^2)} + \frac{d|\nabla f|^2}{|\nabla f|^2} + 2\,\frac{d(\rho^2 \circ f)}{b^2-\rho^2 \circ f} ,$$

$$(2.3)\qquad 0 \ge \frac{-2\Delta r^2}{(a^2-r^2)} - \frac{2|dr^2|^2}{(a^2-r^2)^2} + \frac{\Delta|\nabla f|^2}{|\nabla f|^2} - \frac{|d|\nabla f|^2|^2}{|\nabla f|^4} + {} + 2\,\frac{\Delta(\rho^2 \circ f)}{(b^2-\rho^2 \circ f)} + \frac{2|d(\rho^2 \circ f)|^2}{(b^2-\rho^2 \circ f)^2} .$$

Suppose that the Ricci curvature of $B_a(x_o)$ is bounded from below by $-K$, $K \ge 0$. Then using (1.12), (1.13) and the curvature conditions we see that

$$(2.4)\qquad \frac{\Delta|\nabla f|^2}{|\nabla f|^2} \ge \frac{1}{2}\,\frac{|d|\nabla f|^2|^2}{|\nabla f|^4} - K .$$

Applying the Hessian comparison theorem in [1] to (1.16) when $\phi = \rho^2$ we have

$$(2.5)\qquad \Delta(\rho^2 \circ f) \ge |\nabla f|^2 .$$

Also (2.2) and (1.4) imply

$$(2.6)\qquad \frac{|d|\nabla f|^2|^2}{|\nabla f|^4} \le \frac{4|d(\rho^2 \circ f)|^2}{(b^2-\rho^2 \circ f)^2} + \frac{8|dr^2|\times|d\rho^2|\times|\nabla f|}{(a^2-r^2)(b^2-\rho^2{}_\circ f)} + \frac{4\ |dr^2|^2}{(a^2-r^2)^2} .$$

Putting (2.4), (2.5), (2.6) into (2.3), the result is

$$(2.7)\qquad 0 \ge \frac{-2\Delta r^2}{(a^2-r^2)} - \frac{6|dr^2|^2}{(a^2-r^2)^2} - \frac{8|dr^2|\times|d\rho^2|\times|\nabla f|}{(a^2-r^2)(b^2-\rho^2 \circ f)} + \frac{2|\nabla f|^2}{(b^2-\rho^2 \circ f)} - K .$$

The Gauss' lemma implies $|dr^2| = 2r$, $|d\rho^2| = 2\rho$. Also we know there is a constant c_m depending only on $m = \dim M$ such that $\Delta r^2 \le c_m(1+Kr)$. Using these two pieces of information in (2.7), it reduces to

$$(2.8) \qquad 0 \geq \frac{-2c_m(1+Ka)}{(a^2-r^2)} - \frac{24a^2}{(a^2-r^2)^2} - \frac{32ab|\nabla f|}{(a^2-r^2)(b^2-\rho^2 \circ f)} + \frac{2|\nabla f|^2}{(b^2-\rho^2 \circ f)} - K .$$

From (2.8), we conclude the following theorem.

THEOREM: Let M, N be complete Riemannian manifolds such that N is simply-connected and is having nonpositive sectional curvature. Let $f: M \longrightarrow N$ be a harmonic map. Let $x_o \in M$ and r be the distance function from x_o. Let $B_a(x_o) = \{x : r(x) \leq a\}$. Suppose that $y_o \notin f(B_a(x_o))$ and ρ is the distance function from y_o. Let $b > \sup\{\rho(f(x)) = x \in B_a(x_o)\}$. Then, on $B_a(x_o)$ we have

$$(2.9) \qquad \frac{(a^2-r^2)^2|\nabla f|^2}{(b^2-\rho^2 \circ f)^2} \leq c_m \max \left\{\frac{Ka^4}{\beta}, \frac{a^2(1+Ka)}{\beta}, \frac{a^2b^2}{\beta^2}\right\},$$

where c_m is a constant depending only on $m = \dim M$ and $\beta = \inf\{b^2-\rho^2 \circ f(x) : x \in B_a(x_o)\}$.

A simple application of (2.9) proves the following:

LIOUVILLE THEOREM FOR HARMONIC MAPS: Let M, N be complete Riemannian manifolds and $f: M \longrightarrow N$ be a harmonic map. Suppose that the Ricci curvature of M is nonnegative. Suppose that the sectional curvature of N is nonpositive and that N is simply-connected. Let $x_o \in M$, $y_o \in N$. Let r, ρ be the distance functions from x_o and y_o respectively. Suppose that

$$(2.10) \qquad \overline{\lim_{a\to 0}}\, a^{-1}\sup\{\rho(f(x)) : x \in B_a(x_o)\} = 0 .$$

Then, f is a constant map.

BIBLIOGRAPHY

1. Greene, R. and Wu, H., "Function theory on manifolds which possess a pole", Lecture Notes in Mathematics, Springer-Verlag, New York, 1979, #699.

2. Yau, S. T., "Harmonic functions on complete Riemannian manifolds", Communications on Pure and Applied Mathematics, Vol. XXVIII (1975).

DEPARTMENT OF MATHEMATICS
PRINCETON UNIVERSITY
PRINCETON, NJ 08544

Proceedings of Symposia in Pure Mathematics
Volume 36, 1980

Some inverse spectral results for negatively curved n-manifolds

Victor Guillemin and David Kazhdan

§1. This paper is a sequel to [1]. Let X be a compact, boundaryless, n-dimensional Riemannian manifold. We recall that X has simple length spectrum if there exist no two distinct closed geodesics such that the ratio of their lengths is a rational number. We will denote by Δ the Laplace-Beltrami operator on X. If q is a smooth real valued function on X we will denote by $\Delta + q$ the operator

$$u \to \Delta u + qu$$

on $C^\infty X$. It is well-known that this operator is self-adjoint and has discrete spectrum. In [1] we proved

Theorem 1. Suppose $n = 2$, X has simple length spectrum, and the curvature of X is everywhere negative. Let q_1 and q_2 be smooth functions on X. Then $\Delta + q_1$ and $\Delta + q_2$ have the same spectrum only if $q_1 = q_2$.

In this paper we will generalize this result to $n > 2$. Our technique of proof requires a pinching condition on X: Let x be a point of X. We will say that X is α-pinched at x if there exists a positive number A (depending on x) such that

(1.1) $$-\alpha < K/A + 1 < \alpha$$

1980 Mathematics Subject Classification 58G25, 58G30.

K being the sectional curvature associated with any two dimensional subspace of T_x. Notice that if $n = 2$ and $\alpha < 1$, all that (1.1) says is that X is negatively curved at x. We will prove the following generalization of Theorem 1.

Theorem 2. Suppose X has simple length spectrum and is $1/n$-pinched at all $x \in X$. Let q_1 and q_2 be smooth functions on X. Then $\Delta + q_1$ and $\Delta + q_2$ have the same spectrum only if $q_1 = q_2$.

Let S^*X be the cosphere bundle of X, and let Ξ be the vector field on S^*X which generates geodesic flow. In [2] Livcic proves the following:

Theorem 3. If g is a smooth function on S^*X with the property that its integral over every closed integral curve of Ξ is zero, there exists a C^1 function, f, on S^*X such that $\Xi f = g$.

The main ingredient in the proof of theorem 2 is a sharpening of this result which we will now describe. If S^{n-1} is the standard $n-1$ sphere we can write

$$L^2(S^{n-1}) = \sum H^i , \tag{1.2}$$

H^i being the restriction to S^{n-1} of the space of homogeneous harmonic polynomials on $\mathbb{R}^n$ of degree i. The sum, (1.2), is an orthogonal direct sum of Hilbert spaces, and $SO(n)$ acts irreducibly on each of the summands. There is also a decomposition

$$L^2(S^*X) = \sum H^i(X) \tag{1.3}$$

which, on each of the fibers over X, is just the decomposition (1.2). We will show in section 3 that

$$f \in H^i(X) \cap C^\infty(S^*X) \Rightarrow \Xi f \in H^{i-1}(X) + H^{i+1}(X) .$$

The main ingredient in the proof of Theorem 2 is the following:

Theorem 4. Let X be $1/n$-pinched. Let g be a smooth function on S^*X of the form

$$g = \sum_{i \le N} g_i , \qquad g_i \in H^i(X) .$$

Suppose the integral of g over every periodic integral curve of Ξ is zero. Then there exists a smooth function, f, on S^*X of the form

$$f = \sum_{i \le N-1} f_i , \qquad f_i \in H^i(X) ,$$

such that $\Xi f = g$.

The proof of "Thm 4 ⇒ Thm 2" is the same as the proof of the analogous assertion in [1]. We refer the reader to §4 of [1] for details. The first step in the proof of Theorem 4 is the proof of a Bochner-Yano type equality on the covariant derivative of a symmetric tensor which we will discuss in sections 2-3. In §4 we will show that if X is $1/n$-pinched, the curvature term in this equality is positive definite. In section 5 we will use this fact to get an a priori estimate on the operator "$f \to \Xi f$" similar to the estimate (3.6) of

[1]. From this, Theorem 4 follows easily. We have relegated to the appendix some elementary results about norms on the space of n-tensors. These results are probably in the literature somewhere, but we have not been able to find a reference for them.

Remark: The "spectral rigidity" result of [1] for manifolds which are 1/n-pinched is also an easy consequence of Theorem 4. (See [1], Thm. 1.)

§2. Let V be an n-dimensional vector space and $\langle\ ,\ \rangle$ an inner product on V. There is a corresponding inner product on $\overset{p}{\otimes} V$

$$(2.1)\quad \langle v_1 \otimes \dots \otimes v_p, w_1 \otimes \dots \otimes w_p\rangle = \langle v_1,w_1\rangle \dots \langle v_p,w_p\rangle ,$$

and this induces inner products on all subspaces of $\overset{p}{\otimes} V$, in particular on the subspace, S^pV, of symmetric p-th order tensors. Let G be the element of S^2V canonically associated with $\langle\ ,\ \rangle$ and G^* the dual element in S^2V^*. There is a canonical map, $\overset{p}{\otimes} V \to \overset{p+2}{\otimes} V$, sending F onto $F \otimes G$ and a canonical map in the other direction, $\overset{p+2}{\otimes} V \to \overset{p}{\otimes} V$ sending F onto the contraction of the tensor $F \otimes G^*$. Let

$$\sigma\colon \overset{p}{\otimes} V \to S^pV$$

be the orthogonal projection of $\overset{p}{\otimes} V$ onto S^pV, and let α and β be the composite maps

$$(2.2)\qquad S^pV \to \overset{p}{\otimes} V \to \overset{p+2}{\otimes} V \xrightarrow{\sigma} S^{p+2}V$$

and

$$(2.3)\qquad S^{p+2}V \to \overset{p+2}{\otimes} V \to \overset{p}{\otimes} V \xrightarrow{\sigma} S^pV .$$

Lemma 2.1. α and β are transposes of each other.

Proof: It is enough to verify that the maps: $\overset{p}{\otimes} V \to \overset{p+2}{\otimes} V$ and $\overset{p+2}{\otimes} V \to \overset{p}{\otimes} V$ are transposes, but this is obvious. Q.E.D.

Let $\Theta^p V$ be the kernel of $\beta \colon S^p V \to S^{p-2} V$. We will call $\Theta^p V$ the space of symmetric tensors of trace zero. Let $O(V)$ be the group of linear maps of V preserving $\langle\ ,\ \rangle$. See [4] for the following.

Lemma 2.2. $O(V)$ acts irreducibly on $\Theta^p V$.

Corollary. Under the action of $O(V)$, $S^p V$ breaks up into $[p/2] + 1$ distinct irreducible pieces:

$$(2.4) \qquad S^p V \cong \Theta^p + \Theta^{p-2} + \dots .$$

Proof: Since β is surjective $S^p V \cong \Theta^p V + S^{p-2} V$. Now apply induction. Q.E.D.

We will denote by $\tilde{\Theta}^i$ the component of $S^p V$ isomorphic to Θ^i.

Lemma 2.3. $\tilde{\Theta}^{p-2}$ is the image under $\alpha \colon S^{p-2} \to S^p$ of Θ^{p-2}, $\tilde{\Theta}^{p-4}$ is the image of $\alpha \circ \alpha$ etc. Also $\tilde{\Theta}^p + \tilde{\Theta}^{p-2}$ is the kernel of $\beta \circ \beta$, $\tilde{\Theta}^p + \tilde{\Theta}^{p-2} + \tilde{\Theta}^{p-4}$ the kernel of $\beta \circ \beta \circ \beta$, etc.

Proof: Corollary of lemma 2.1.

Lemma 2.4. $\Theta^p V \otimes V$ is the direct sum of three irreducible $O(V)$-invariant subspaces

(2.5) $$\Theta^p V \otimes V = W_+ + W_- + W_o \; .$$

Here W_+ is just the intersection

$$(\Theta^p \otimes V) \cap S^{p+1}V = \Theta^{p+1}V \; ,$$

W_o is the kernel of the symmetrization map,

$$\sigma\colon \Theta^p V \otimes V \to S^{p+1}V$$

and W_- is the preimage of $\tilde{\Theta}^{p-1}$ with respect to σ, i.e.

(2.5)' $$\sigma\colon W_- \to \tilde{\Theta}^{p-1}$$

is an $O(V)$-isomorphism.

<u>Proof</u>: The contraction map: $\Theta^p V \otimes V \to \Theta^{p-1}V$, is onto, and the diagram

$$\begin{array}{ccc} \Theta^p V \otimes V & \xrightarrow{\ \sigma\ } & S^{p+1}V \\ \big\downarrow & & \big\downarrow \beta \\ \Theta^{p-1}V & \longrightarrow & S^{p-1}V \end{array}$$

commutes up to a constant factor; so the existence of (2.5) is clear. The only non-obvious assertion is the irreducibility of W_o. Since we will not be using it, we will not prove it here. However, it follows easily from Clebsch-Gordan. (See [3].)

<u>Remark</u>: In dimension 2 $W_o = \{0\}$.

We will now state without proof some metric properties

of the mappings α , β and σ introduced above. (These statements will be proved in the appendix.)

Lemma 2.5. The map $\alpha\colon \Theta^p \to \widetilde{\Theta}^p$ satisfies

$$(2.6) \qquad \|\alpha(F)\|^2 = \alpha(p)\|F\|^2 \quad \text{for all} \quad F \in \Theta^p$$

where

$$(2.6)' \qquad \alpha(p) = \frac{4}{p+2}\left(1 + \frac{n-2}{2(p+1)}\right) .$$

Lemma 2.6. The map $\beta\colon \widetilde{\Theta}^p \to \Theta^p$ satisfies

$$(2.7) \qquad \|\beta(F)\|^2 = \alpha(p)\|F\|^2 \quad \text{for all} \quad F \in \widetilde{\Theta}^p .$$

Lemma 2.7. The map $\sigma\colon W_- \to \widetilde{\Theta}^{p-1}$ in (2.5)' satisfies

$$(2.8) \qquad \|\sigma(F)\|^2 = \frac{1}{p+1}\|F\|^2 \quad \text{for all} \quad F \in W_- .$$

Note that since α and β are transposes of each other, (2.7) is an automatic consequence of (2.6).

In section 3 we (regrettably) have had to use a certain amount of classical tensorial notation (i.e. indices). Therefore, it may be helpful at this point to indicate how some of the operations above look in indices.

Let $v_1,\ldots,v_n$ be basis of V and let

$$(2.9) \qquad F = \sum F^{i_1\ldots i_p} v_{i_1} \otimes \ldots \otimes v_{i_p}$$

be an element of S^pV . Then, with $g_{ij} = \langle v_i, v_j\rangle$,

(2.10) $$\beta(F) = \sum H^{i_1 \dots i_{p-2}} v_{i_1} \otimes \dots \otimes v_{i_{p-2}}$$

where

$$H^{i_1 \dots i_{p-2}} = \sum g_{ij} F^{i_1 \dots i_{p-2} ij} .$$

The g_{ij}'s and the dual g^{ij}'s can be used to convert subscripts to superscripts and visa-versa, e.g.

$$F^{i_1}{}_{i_2 \dots i_p} = \sum g^{i_1 j} F_{j i_2 \dots i_p}$$

and with this convention the inner product $\langle F,F \rangle$ can be writton

(2.11) $$\langle F,F \rangle = \sum F^{i_1 \dots i_p} F_{i_1 \dots i_p} .$$

Note that we have used a modified Einstein summation convention in expressions like (2.11); i.e. we have written the summation sign but not the indices which are being summed.

Given an element, F, of $S^p V$ of the form (2.9), we can associate with it a polynomial function, f, on V^*, viz.

$$f(\xi) = \sum F^{i_1 \dots i_p} \xi_{i_1} \dots \xi_{i_p}$$

where $\xi_i(w) = (v_i, w)$ for $w \in V^*$. This sets up a one-one correspondence between symmetric tensors of order p and homogeneous polynomials of order p on V^*. The following is trivial to verify:

<u>Lemma 2.8</u>. Let F be in $S^p V$ and let $F_1 = \alpha(F)$. Let f

and f_1 be the functions on V^* associated with f and f_1. Then $f_1 = gf$ where g is the function $\sum g^{ij}\xi_i\xi_j$.

Under this correspondence between tensors and functions, Θ^p gets associated with the space, H^p, of harmonic polynomials on V^* of degree p. We will also use the symbol, H^p, to denote the space of spherical harmonics on the unit sphere, S, in V^*. Let us denote by γ the map

$$\gamma \colon H^p \to \Theta^p \quad . \tag{2.12}$$

Since H^p is a subspace of $L^2(S)$ it has a canonical $O(V)$-invariant inner product, and Θ^p has, of course, the inner product (2.1). In the appendix we will prove

Lemma 2.9. For all $f \in H^p$

$$\|\gamma(f)\|^2 = \gamma(p)\|f\|^2 \tag{2.13}$$

where

$$\gamma(p) = 2^p \frac{\Gamma(p + n/2)}{\Gamma(p+1)} \quad . \tag{2.14}$$

§3. Let X be a Riemannian manifold and x a point of X. At x we have the decomposition (2.4)

$$S^pV = \Theta^pV + \widetilde{\Theta}^{p-2}V + \ldots$$

with $V = T_x$. We can identify Θ^pV with the spherical harmonics of degree p on the unit sphere in V^*. Let us denote by Θ^p the vector bundle over X whose fiber at x

is $\Theta^p T_x$. We will call sections of Θ^p <u>symmetric tensors of trace zero</u>. If $H^p(X)$ is the p-th summand in the Hilbert space decomposition (1.3), we get a canonical identification

$$(3.1) \qquad H^p(X) \cap C^\infty(S^*X) \cong C^\infty \text{ sections of } \Theta^p ,$$

which is given pointwise by (2.12). In addition, Θ^p is equipped with the inner product (2.1); and up to a universal constant (see Lemma 2.9)

$$(3.2) \qquad H^p(X) \cong L^2 \text{ sections of } \Theta^p ,$$

the isomorphism being an isomorphism of Hilbert spaces.

Let Ξ be the vector field on S^*X generating geodesic flow. If $f \in H^p(X) \cap C^\infty(S^*X)$, then by (3.1) there is associated with f a symmetric tensor, F , of trace zero on X . What about Ξf ? In terms of a local trivialization of the tangent bundle let

$$F = \{F^{i_1 \dots i_p}\} ,$$

and let

$$DF = \{F^{i_1 \dots i_p, i_{p+1}}\}$$

be its covariant derivative. We will prove

<u>Proposition 3.1</u>. The function Ξf is the restriction to S^*X of a function, h , on T^*X which is a homogeneous polynomial of degree $p+1$ on each of the fibers of T^*X over X .

Moreover the symmetric tensor, H, on X associated with h is just the symmetrization of DF :

$$(3.3) \quad H^{i_1 \dots i_{p+1}} =$$

$$\frac{1}{p+1}\left(F^{i_1 \dots i_p, i_{p+1}} + F^{i_{p+1} i_2 \dots i_p, i_1} + \dots + F^{i_1 \dots i_{p-1} i_{p+1}, i_p}\right) .$$

<u>Proof:</u> The vector field, Ξ, is the Hamiltonian vector field associated with the function

$$g(x,\xi) = \tfrac{1}{2} \sum g_{ij}(x)\xi_i\xi_j$$

on T^*X ; i.e.

$$\Xi = \tfrac{1}{2} \sum ((\partial g/\partial \xi_k)(\partial/\partial x_k)) - ((\partial g/\partial x_k)(\partial/\partial \xi_k)) .$$

The expression on the right is just

$$(3.4) \qquad \sum g_{ik}\xi_i(\partial/\partial x_k) - \Gamma_{ikj}\xi_i\xi_j\,(\partial/\partial \xi_k) ,$$

the Γ's being Christoffel symbols. Since $f(x,\xi)$ is

$$\sum F^{i_1 \dots i_p} \xi_{i_1} \dots \xi_{i_p} ,$$

when we differentiate f by the vector field (3.4) and write the resulting expression as a linear combination of the monomials $\xi_{i_1} \dots \xi_{i_{p+1}}$ with symmetric coefficients, we get (3.3).

Q.E.D.

Combining this proposition with Lemma 2.4 and Lemma 2.8, we get

Proposition 3.2. If $f \in H^p(X) \cap C^\infty(S^*X)$ then Ξf is in $H^{p-1}(X) + H^{p+1}(X)$.

Let D_+F be the Θ^{p+1}-component of the tensor (3.3) and D_-F the $\tilde{\Theta}^{p-1}$ component.

We can write DF as an orthogonal sum

$$DF = D_+F + D_-^{\#}F + D_oF \quad , \tag{3.5}$$

the summands corresponding to the terms in the sum (2.5), and we have

$$D_-F = \sigma(D_-^{\#}F) \quad , \tag{3.6}$$

σ being the symmetrization operation. The following theorem is the main result of this section.

Theorem 3.3. Let F be a symmetric tensor of order p and trace zero. Let $\Omega(F)$ be the quadratic form

(3.7)

$$-\int\{\sum R_{ij}F^{ii_2\ldots i_p}F^j_{i_2\ldots i_p} + (p-1)R_{ijk\ell}F^{iki_3\ldots i_p}F^{j\ell}_{i_3\ldots i_p}\}dx \quad ,$$

$R_{ijk\ell}$ being the curvature tensor of X and R_{ij} the Ricci tensor. Then

$$\|D_+F\|^2 = \kappa(p)\|D_-F\|^2 + \frac{1}{p}\|D_oF\|^2 + \Omega(F) \tag{3.8}$$

with

$$\kappa(p) = (p+1)(1 + \frac{n-2}{2p}) \quad . \tag{3.9}$$

Proof: Our starting point is the following identity of Bochner and Yano (see [5], page 331).

$$(3.10)\quad \Omega(F) = \int\Big\{\sum F^{ii_2\ldots i_p,j}F_{ji_2\ldots i_p,i} - F^{ii_2\ldots i_p}{}_{,i}F^{j}{}_{i_2\ldots i_p,j}\Big\}dx\ .$$

We will show that both terms on the right hand side of (3.10) can be expressed in terms of the norms of D_+F, D_-F and D_oF.

Lemma 3.4. Let $Q(F)$ be the first term on the right hand side of (3.10). Then

$$(3.11)\qquad (p+1)(\|D_+F\|^2 + \|D_-F\|^2) = pQ(F) + \|DF\|^2\ .$$

Proof: By (3.3)

$$(p+1)^2(\|D_+F\|^2 + \|D_-F\|^2) = \sum_I\Big(\sum_{s=1}^{p+1} F^{I_s,i_s}\Big)\Big(\sum_{t=1}^{p+1} F_{I_t,i_t}\Big)$$

where $I_s = \{i_1,\ldots,\hat{i}_s,\ldots,i_{p+1}\}$, and the sum is over all $p+1$ - tuples, I. Expanding the right hand side we get two terms:

$$(p+1)p\sum F^{ii_2\ldots i_p,j}F_{ji_2\ldots i_p,i}$$

and

$$(p+1)\sum F^{ii_2\ldots i_p,j}F_{ii_2\ldots i_p,j}\ .$$

The first term is $(p+1)pQ(F)$ and the second term is $(p+1)\|DF\|^2$. Q.E.D.

Lemma 3.5. For all sections, F , of Θ^p

$$\|D_+F\|^2 = Q(F) + \frac{1}{p}\|D_oF\|^2 \quad . \tag{3.12}$$

Proof: By (3.5)

$$\|DF\|^2 = \|D_+F\|^2 + \|D_-^\# F\|^2 + \|D_oF\|^2 \quad .$$

By (2.8) and (3.6)

$$\|D_-^\# F\|^2 = (p+1)\|D_-F\|^2 \quad .$$

Substituting this into (3.11) we get

$$(p+1)\|D_+F\|^2 + (p+1)\|D_-F\|^2 =$$

$$pQ(F) + \|D_+F\|^2 + (p+1)\|D_-F\|^2 + \|D_oF\|^2$$

or

$$p\|D_+F\|^2 = pQ(F) + \|D_oF\|^2 \; . \qquad \text{Q.E.D.}$$

Next we will examine the last term on the right hand side of (3.10) viz.

$$\int \left\{\sum F^{ii_2\ldots i_p}{}_{,i} F^j_{i_2\ldots i_p,j}\right\} dx \quad . \tag{3.13}$$

Lemma 3.6. The expression (3.13) is equal to

$$(3.14) \qquad (p+1)(1+\frac{n-2}{2p})\|D_-F\|^2 \ .$$

Proof: Let $\beta: \tilde{\Theta}^{p-1} \to \Theta^{p-1}$ be the contraction mapping (2.2). Since $\beta(D_+F) = 0$, $\beta(D_-F) = \beta(D_+F + D_-F)$; so by (3.3) the $(i_3 \ldots i_{p+1})$-component of $\beta(D_-F)$ is:

$$(3.15) \quad \frac{1}{p+1}\sum g_{ij}(F^{ii_3\ldots i_{p+1},j} + F^{ji_3\ldots i_{p+1},i} + \ldots)$$

the terms indicated by dots being of the form

$$\frac{1}{p+1}\sum g_{ij}F^{iji_3\ldots\hat{i}_s\ldots i_{p+1},i_s} \ .$$

Since the covariant derivative of the metric tensor is zero and F is of trace zero, these terms vanish and (3.15) reduces to

$$\frac{2}{p+1}\sum F^{ii_3\ldots i_{p+1}}{}_{,i} \ .$$

Thus (3.13) is equal to $(\frac{p+1}{2})^2\|\beta(D_-F)\|^2$. By Lemma (2.6), however,

$$\|\beta(D_-F)\|^2 = \frac{4}{p+1}(1+\frac{n-2}{2p})\|D_-F\|^2$$

so we finally get for (3.13) the formula (3.14). Q.E.D.

By Lemma 3.6 we can rewrite (3.10) in the form

$$\Omega(F) + (p+1)(1+\frac{n-2}{2p})\|D_-F\|^2 = Q(F) \ .$$

Substituting this formula for $Q(F)$ into (3.12), we get (3.8). This concludes the proof of Theorem 3.3.

As a corollary of Theorem 3.3 we will prove:

Proposition 3.7. The operator, D_+, is elliptic.

Proof: We have to show that for every $x_o \in X$ and $\xi_o \in T^*_{x_o} - 0$, the symbol map

$$\sigma(D_+)(\xi_o): \Theta^p_{x_o} \to \Theta^{p+1}_{x_o}$$

is injective. Suppose this were not the case for some x_o and ξ_o. Since the leading part of D_+ is locally an operator with constant coefficients, there exists a section, F, of Θ^p with support in a small neighborhood of x_o, and a real-valued function f such that $F_{x_o} \neq 0$, $df_{x_o} = \xi_o$ and such that, for all $x \in X$ and $\xi = df_x$,

$$\sigma(D_+)(\xi)F_x = 0 . \tag{3.16}$$

On the other hand by (3.8) there exists a constant C such that

$$C(\|D_+F\| + \|F\|) \geq \|DF\| \tag{3.17}$$

for all sections, F, of Θ^p. If we put $e^{i\lambda f}F$ into (2.24) where f and F are as in (2.23), the left hand side of (3.17) is bounded in λ for $\lambda \gg 0$. However, since D is elliptic the right hand side is of order, $O(\lambda)$. Contradiction

Remarks. 1. Bochner and Yano obtain from (3.10) an identity

analogous to (3.8) for skew-symmetric tensors and use this identity to prove vanishing theorems for the cohomology groups of X. It is curious that their identity seems to be most effective when X is positively curved while ours is most effective when X is negatively curved.

2. In two dimensions, the D_o term in (3.8) is zero. In higher dimensions it is possible that, by exploiting this term, one can improve the pinching assumptions in Theorem 4.

§4 In this section we will discuss the signature of the quadratic form $\Omega(F)$ in formula (3.8). Suppose first of all that X is a manifold of constant curvature. Then, if R is the scalar curvature,

$$R_{ij} = \frac{1}{n}R \quad \text{and} \quad R_{ijk\ell} = \frac{1}{n(n-1)}R(g_{kj}g_{i\ell} - g_{j\ell}g_{ik})$$

and

$$\Omega(F) = -\frac{(n+p-2)}{n(n-1)}R\|F\|^2 \; ; \tag{4.1}$$

so $\Omega(F)$ is positive definite only if $R < 0$. It is clear that if X is negatively curved, and its curvature is "close to being constant", $\Omega(F)$ is still positive definite. Unfortunately, at the moment, we do not have really satisfactory criteria which will guarantee the positivity of $\Omega(F)$. We will content ourselves here with proving

Theorem 4.1. $\Omega(F)$ is positive definite providing X is $1/n$-pinched.

Proof: Let x be a point of X, let $R_{ijk\ell}$ be the curvature tensor of X at x, R_{ij} the Ricci tensor at x and g_{ij}

the metric tensor. Let $V = T_x$ and consider the following quadratic form on $\circledS^2 V$:

$$Q_2(S) = -\sum R_{ijk\ell} S^{ik} S^{j\ell}$$

for $S \in \circledS^2 V$.

Lemma 4.2. If X is $1/n$ - pinched at x , Q_2 is positive definite.

Proof: With S fixed we can choose an orthonormal basis, $\{v_1, \ldots, v_n\}$ of V such that

$$S = \sum c_i v_i \otimes v_i \quad .$$

Note that since S is in $\circledS^2 V$

(4.2) $$\sum c_i = 0 \quad .$$

Note also that

(4.3) $$\|S\|^2 = \sum |c_i|^2 \quad .$$

In this basis

(4.4) $$Q_2(S) = -\sum R_{ijij} c_i c_j = \sum_{i \neq j} K_{ij} c_i c_j \quad ,$$

where K_{ij} is the sectional curvature of the two-plane spanned by v_i and v_j . If X is α-pinched at x , there exists a positive number, A , such that

$$-\alpha < (K_{ij}/A) + 1 < \alpha \qquad \text{for all } i,j .$$

From (4.4) we get

$$Q_2(S) = \sum_{i\neq j} K_{ij}c_i c_j = \sum_{i\neq j} (K_{ij} + A)c_i c_j - A \sum_{i\neq j} c_i c_j .$$

However, by (4.2)

$$- \sum_{i\neq j} c_i c_j = \sum c_i^2 ;$$

whereas

$$\left| \sum_{i\neq j} (K_{ij} + A)c_i c_j \right| \leq A\alpha \sum |c_i||c_j| \leq A\alpha\left(\sum |c_i| \right)^2 .$$

The term on the right is bounded by

$$A\alpha n \|S\|^2$$

by (4.3); so if $\delta = 1 - \alpha n$

$$Q(S) \leq A\delta \|S\|^2$$

proving the lemma.

Next consider the quadratic form on $V = T_x$:

$$Q_1(s) = - \sum R_{ij} s^i s^j .$$

<u>Lemma 4.3</u>. Q_1 is positive definite providing the sectional curvature at x is negative in all directions.

Proof: We can choose an orthonormal basis, $\{v_1, \ldots, v_n\}$ of T_x such that, in this basis, Q_1 is in diagonal form, i.e. $R_{ij} = 0$ for $i \neq j$. If K_{ij} is, as above, the sectional curvature of the plane spanned by v_i and v_j, then

$$R_{ii} = \sum_{j \neq i} K_{ij} \quad ;$$

so the assumption that $K_{ij} < 0$ for all i,j implies that Q_1 is positive definite. Q.E.D.

Consider now the integrand of (3.7). Applying Lemma 4.3 to the first part of this integrand and Lemma 4.2 to the second, we see that both these terms are pointwise positive definite. Therefore, since X is compact, the integral is positive definite. This proves the theorem.

§5. Let f be in $H^p(X) \cap C^\infty(S^*X)$ and let

$$\Xi f = \Xi_+ f + \Xi_- f$$

where $\Xi_+ f$ is the H^{p+1}-component of Ξf and $\Xi_- f$ is the H^{p-1}-component. Let F be the section of Θ^p associated with f. By Proposition 3.1, Lemma 2.8 and Lemma 2.9:

$$(5.1) \qquad D_+F = \gamma(\Xi_+ f) \qquad \text{and} \qquad D_-F = \alpha \circ \gamma(\Xi_- f) \quad .$$

If X is $1/n$-pinched, the last term on the right in (3.8) is positive, and

$$\|D_+F\|^2 \geq (p+1)\left(1 + \frac{n-2}{2p}\right)\|D_-F\|^2 \quad .$$

By (5.1):

$$\gamma(p+1)\|\Xi_+ f\|^2 \geq (p+1)(1+\frac{n-2}{2p})\alpha(p-1)\gamma(p-1)\|\Xi_- f\|^2$$

or

$$\|\Xi_+ f\|^2 \geq a(p)\|\Xi_- f\|^2 \tag{3.2}$$

where

$$a(p) = (p+1)(1+\frac{n-2}{2p})\alpha(p-1)\frac{\gamma(p-1)}{\gamma(p+1)} .$$

By (2.6)' and (2.14)

$$a(p) = (1+\frac{n-2}{2p})^2\frac{p(p+1)}{(p+n/2)(p+n/2-1)} = 1 + O(1/p^2) .$$

In particular we have

<u>Lemma 5.1</u>. For p sufficiently large and for all $s > 0$ the product

$$a(p)\, a(p+1) \ldots a(p+s)$$

is bounded away from zero uniformly in s .

We will now prove Theorem 4. Let

$$h = \sum_{i=0}^{N} h_i$$

where h_i is in $H^i(X) \cap C^\infty(S^*X)$, and the integral of h

over every periodic integral curve of Ξ is zero. By Theorem 3 there exists a C^1-function, f, on S^*X such that $\Xi f = h$. Let

$$f = \sum_{i=0}^{\infty} f_i$$

with $f_i \in H^i$. The equation $\Xi f = h$ is equivalent to the system of equations

$$\Xi_- f_{i+1} + \Xi_+ f_{i-1} = h_i \quad , \quad i = 0,1,2,\ldots .$$

Since $h_i = 0$ for $i > N$ we have

$$\Xi_+ f_{i-1} + \Xi_- f_{i+1} = 0$$

for $i > N$, and we get from (5.2)

$$(5.3) \qquad \|\Xi_+ f_{i+1}\|^2 \geq a(i+1)\|\Xi_+ f_{i-1}\|^2$$

for all $i > N$. Since f is in C^1, the functions, $\Xi_+ f_i$, tend to zero in the L_2-topology as $i \to \infty$. Therefore by Lemma 5.1, $\Xi_+ f_i = 0$ for all $i \geq N$. By (3.8), $f_i = 0$ for all $i \geq N$. For $i \leq N$

$$\Xi_+ f_{i-1} = h_i - \Xi_- f_{i+1} .$$

For $i = N$ the right hand side of this equation is smooth; so by Proposition 3.7 f_{N-1} is smooth. By induction f_i is smooth for all i.

Appendix

Let V be an n-dimensional vector space and $\langle\ ,\ \rangle$ an inner product on V. Consider the maps

(A.1) $$\alpha\colon \circledcirc^p \to \tilde{\circledcirc}^p ,$$

(A.2) $$\sigma\colon W_- \to \tilde{\circledcirc}^{p-1} , \text{ and}$$

(A.3) $$\gamma\colon H^p \to \circledcirc^p$$

described in §2. (See Lemma 2.5, Lemma 2.7 and Lemma 2.9.) Since these are maps between irreducible $O(V)$-spaces, it is clear that they are unitary, up to constants of conformality. Our goal is to compute these constants of conformality. We will begin with the map (A.1). First of all assume that V is two-dimensional. The inner product on V extends to a Hermetian inner product on $V \otimes \mathbb{C}$. Let $\{v_1, v_2\}$ be an orthnormal basis of V, and let $\omega = v_1 + \sqrt{-1}\, v_2$ and $\bar{\omega} = v_1 - \sqrt{-1}\, v_2$. Clearly $\|\omega\|^2 = \|\bar{\omega}\|^2 = 2$ and $\langle \omega, \bar{\omega} \rangle = 0$. The inner product on V corresponds to a unique $O(V)$ invariant element of S^2V which we can express as

$$v_1 \otimes v_1 + v_2 \otimes v_2 = \tfrac{1}{2}(\omega \otimes \bar{\omega} + \bar{\omega} \otimes \omega) .$$

Consider the element, ω^p, of $\circledcirc^p V$. $\alpha(\omega^p)$ is the symmetrization of

$$\omega^p \otimes \tfrac{1}{2}(\omega \otimes \bar{\omega} + \bar{\omega} \otimes \omega) ,$$

which is

$$(A.4) \qquad \frac{1}{p+2} \sum_{r+s=p+1} \omega^r \otimes \bar{\omega} \otimes \omega^s$$

Since $\|\omega\|^2 = 2$, $\|\omega^p\|^2 = 2^p$. By (A.4)

$$\|\alpha(\omega^p)\|^2 = \frac{2^{p+2}}{p+2}$$

since the individual terms in (A.4) are mutually perpendicular. Thus we have proved

Lemma A.1. Let $\alpha(p)$ be the constant of conformality (2.6). Then in dimension two

$$\alpha(p) = \frac{4}{p+2} .$$

Now let V be of dimension $n > 2$. Let $\{v_1, v_2, \ldots, v_n\}$ be an orthonormal basis of V , and as above let $\omega = v_1 + \sqrt{-1}\, v_2$. Then ω^p is in Θ^p , and $\alpha(\omega^p)$ is the symmetrization of

$$(A.5) \qquad \omega^p \otimes (v_1 \otimes v_1 + v_2 \otimes v_2 + \ldots + v_n \otimes v_n) .$$

Symmetrizing (A.5) we get $n-1$ terms: the term (A.4) plus the terms

$$(A.6)_k \qquad \rho_k = \frac{2}{(p+2)(p+1)} \sum_{r+s+t=p} \omega^r \otimes v_k \otimes \omega^s \otimes v_k \otimes \omega^t ,$$

$k = 3, \ldots, n$. There are $\binom{p+2}{2}$ mutually orthogonal summands in $(A.6)_k$ each of $(\text{norm})^2$ 2^p ; so

$$\|\sigma_k\|^2 = \frac{2^{p+1}}{(p+2)(p+1)} .$$

Since $\|\omega^p\|^2 = 2^p$, we have proved

<u>Proposition A.2</u>. Let $\alpha(p)$ be the constant of conformality in (2.6). Then in dimension n

$$\alpha(p) = \frac{4}{p+2}(1 + \frac{n-2}{2(p+1)}) .$$

Consider next the map (A.2). As above let $\{v_1,\ldots,v_n\}$ be an orthonormal basis of V , and let $\omega = v_1 + \sqrt{-1}\,v_2$. It is clear that $\omega^p \otimes \bar{\omega}$ is in W_- . Its image, $\sigma(\omega^p \otimes \bar{\omega})$, in $\widetilde{\Theta}^p$ is

$$\text{(A.7)} \qquad \frac{1}{p+1} \sum_{r+s=p} \omega^r \otimes \bar{\omega} \otimes \omega^s .$$

The individual terms in (A.7) are mutually orthogonal and the $(\text{norm})^2$ of each is 2^{p+1} ; so

$$\|\sigma(\omega^p \otimes \bar{\omega})\|^2 = \frac{2^{p+1}}{p+1}$$

Since $\|\omega^p \otimes \bar{\omega}\|^2 = 2^{p+1}$ we have proved

<u>Proposition A.3</u>. Let $\sigma(p)$ be the constant of conformality in (2.8). Then (in all dimensions)

$$\sigma(p) = \frac{1}{p+1} .$$

Finally we must compute the constant of conformality, $\gamma(p)$, in (2.13). As above let $\{v_1,\ldots,v_n\}$ be an orthonormal basis of V and let x_i be the coordinate function

$$w \to (v_i, w)$$

on V^*. The function, $f = (x_1 + \sqrt{-1}\, x_2)^p$, is in H^p and $\gamma(f) = (v_1 + \sqrt{-1}\, v_2)^p = \omega^p$. We have already shown that $\|\gamma(f)\|^2 = 2^p$. We will next show that

$$\text{(A.8)} \qquad \int_{S^{n-1}} |f|^2 ds = \frac{\Gamma(p+1)}{\Gamma(p + n/2)} .$$

From this it will follow that

$$\gamma(p) = 2^p \frac{\Gamma(p + n/2)}{\Gamma(p+1)} .$$

Remark: The left hand side of (A.8) is somewhat ambiguously defined since we have not specified what our choice is of an $O(n)$-invariant measure, ds, on S^{n-1}. Strictly speaking, what we will show is that we can normalize ds so that (A.8) is correct.

Proof of (A.8): Let Q_1 be the function, $x_1^2 + x_2^2$, let Q be the function, $x_1^2 + \ldots + x_n^2$ and let Q_2 be the function, $x_3^2 + \ldots + x_n^2$. The expression, (A.8), is given by the integral

$$\text{(A.9)} \qquad \int_0^1 s^p h(s) ds$$

where $h(s)$ is the total volume of the surface $Q_1 = s$, $Q = 1$ with respect to the Leray form, μ, defined by

$$\text{(A.10)} \quad dx_1 \wedge \ldots \wedge dx_n = dQ_1 \wedge dQ \wedge \mu = dQ_1 \wedge dQ_2 \wedge \mu .$$

The surface, $Q_1 = s$, $Q = 1$, is the product of the circle, $x_1^2 + x_2^2 = s$, and the $(n-3)$-sphere, $x_3^2 + \dots + x_n^2 = 1 - s$. The standard volume forms on these spaces are

$$\mu_1 = (1/\sqrt{s})(x_1 dx_2 - x_2 dx_1)$$

and

$$\mu_2 = (1/\sqrt{1-s})\left(\sum_{i=3}^{n} (-1)^{i-1} x_i \, dx_3 \wedge \dots \wedge \widehat{dx_i} \wedge \dots \wedge dx_n\right) .$$

Setting

(A.11) $$\mu = M\mu_1 \wedge \mu_2 ,$$

we get

$$dx_1 \wedge \dots \wedge dx_n = M(dQ_1 \wedge dQ_2) \wedge \mu_1 \wedge \mu_2$$

$$= (M/\sqrt{s}\sqrt{1-s})(x_1^2 + x_2^2)(x_3^2 + \dots + x_n^2) dx_1 \wedge \dots \wedge dx_n$$

$$= M \sqrt{s} \sqrt{1-s} .$$

Thus

(A.12) $$M = (1 / \sqrt{s} \sqrt{1-s}) .$$

The integral of μ_1 over $q_1 = s$ is $2\pi\sqrt{s}$ and the integral of μ_2 over $Q_2 = 1-s$ is (up to a universal constant depending on n) $(1-s)^{n-2/2}$. Ignoring this universal constant, we get from (A.11) and (A.12):

$$h(s) = (1-s)^{n/2 - 2} .$$

Thus the integral (A.9) is

$$\int s^p (1-s)^{n/2 - 2} ds = \frac{\Gamma(p+1)\Gamma(n/2 - 1)}{\Gamma(n/2 + p)} .$$

Bibliography

1. V. Guillemin and D. Kazhdan, "Some inverse spectral results for negatively curved 2-manifolds". Topology (1979).
2. A. N. Livčic, "Some homological properties of U-systems". Mat. Zametki, 10 (1971), 555-564.
3. R. Steinberg, "A general Clebsch-Gordan theorem", Bull. AMS, 67 (1961), 406-407.
4. H. Weyl, The classical groups, their invariants and representations. Princeton University Press, Princeton, NJ (1946).
5. K. Yano, "Some remarks on tensor fields and curvature". Annals of Math. 55 (1952), 328-347.

Department of Mathematics
Massachusetts Institute of Technology
Cambridge, Massachusetts 02139

Department of Mathematics
Harvard University
Cambridge, Massachusetts 02138

Proceedings of Symposia in Pure Mathematics
Volume 36, 1980

ON THE SPECTRUM OF THE LAPLACE OPERATOR ON COMPACT RIEMANN SURFACES

Heinz Huber

1. Let F be a compact Riemann surface of genus $g_F > 1$. F carries exactly one metric of curvature -1 compatible with the conformal structure of F, i.e. the Poincaré metric. Let $0 = \lambda_0 < \lambda_1 \leqq \lambda_2 \leqq \ldots$ be the eigenvalues of the Laplace operator associated with this metric and $\varphi_0, \varphi_1, \varphi_2, \ldots$ a sequence of corresponding orthonormal eigenfunctions. φ_0 is constant. Hence, by the Gauss-Bonnet theorem we have $\varphi_0 = (4\pi(g_F-1))^{-1/2}$.

Define $$A_F(x) = \sum_{\lambda_n \leqq x} 1, \quad x \geqq 0,$$

$$Q_F(x) = A_F(x)/(g_F-1)x, \quad x > 0.$$

By Weyl's asymptotic law $Q_F(x)$ converges to 1 as x tends to infinity. The velocity of convergence will depend on the surface F. We are interested in the behavior of the function $q(x) = \inf_F Q_F(x)$ where the greatest lower bound is taken with respect to <u>all</u> surfaces of arbitrary genus.

THEOREM. (i) $q(x) = 0$ for $0 < x \leqq \frac{1}{4}$;
(ii) $q(x) > 0$ for $x > \frac{1}{4}$.

Of course (i) could happen to be true in a rather trivial way. For instance, if $A_F(\frac{1}{4})$ would be bounded, then $Q_F(\frac{1}{4})$ would tend to zero as g_F tends to infinity. But B. Randol [4] showed in 1974 that $A_F(\frac{1}{4})$ may be arbitrarily large for certain surfaces F. Theorem (i) is a consequence of the following inequality [2]:

(A) $$Q_F(\tfrac{1}{4}) \leqq (3\pi^2/2)(\log\cosh \ell_F/4)^{-3},$$

where ℓ_F denotes the neck of F, i.e. the length of the shortest closed geodesic on F. Now (i) follows from (A) if there exist surfaces of arbitrarily large neck. Because of the negative curvature it is not at all obvious that such surfaces do exist. However, one can construct nice examples by taking the quotient of the upper half-plane by certain

---

1980 Mathematics Subject Classification Primary 58G25, 35P15, 35P20; Secondary 30F10, 30F35.

arithmetically defined groups of noneuclidean motions [2]. Recently P. Buser [1] has shown in a more geometric way that there is always a surface of prescribed neck $\ell > 0$ and given genus $g > \exp(\ell^2/4)$.

If we combine inequality (A) with Theorem (ii) we meet a rather interesting fact. Take x slightly greater than $\frac{1}{4}$ and compare $A_F(x)$ with $A_F(\frac{1}{4})$:

$$\frac{A_F(x)}{A_F(\frac{1}{4})} = 4x\frac{Q_F(x)}{Q_F(\frac{1}{4})} \geqq \frac{8}{3\pi^2} xq(x) (\log\cosh\frac{\ell_F}{4})^3$$

Hence, if the neck of F is very large, then the eigenvalues between 0 and x concentrate in the small interval $[\frac{1}{4}, x]$. One should keep in mind that there are surfaces of arbitrarily large neck having as many eigenvalues in $(0, \frac{1}{4})$ as one wishes. Indeed, take one surface F_0 of large neck. Then there is a Galois covering F of F_0 having the required properties [4].

Theorem (ii) is a consequence of an inequality contained essentially in [3]: For all $\mu > \frac{1}{4}$, $n \geqq 0$, $p \in F$ we have

$$\text{(B)} \qquad 1 \leqq \frac{\mu}{\lambda_{n+1}} + 2\pi s(\mu) \sum_{k=0}^{n} \left(1 - \frac{\lambda_k}{\lambda_{n+1}}\right) |\varphi_k(p)|^2$$

To define the function $s(\mu)$ consider Legendre's function $F_\mu(t) = P_\nu(t)$, $\nu(\nu+1) = -\mu$, which has infinitely many zeros in $(1, +\infty)$ if $\mu > \frac{1}{4}$. Now

$$s(\mu) = \left(\int_1^{a(\mu)} F_\mu \, dt\right)^2 / \int_1^{a(\mu)} F_\mu^2 \, dt,$$

where $a(\mu) > 1$ is the first zero of F_μ. One can show that

(1) $\quad s(\mu) \to \infty, \ \mu \downarrow \frac{1}{4},$

(2) $\quad s(\mu) \sim 2/\mu, \ \mu \to +\infty$

Integrating inequality (B) over the surface F we get

$$\text{(B)}' \qquad 1 \leqq \frac{\mu}{\lambda_{n+1}} + \frac{s(\mu)}{2(g-1)} \sum_{k=0}^{n} \left(1 - \frac{\lambda_k}{\lambda_{n+1}}\right) \leqq$$

$$\leqq \frac{\mu}{\lambda_{n+1}} + \frac{s(\mu)}{2(g-1)} (n+1), \ n \geqq 0, \ \mu > \tfrac{1}{4}.$$

Now take $x > \frac{1}{4}$ and choose $n \geqq 0$ such that $\lambda_n \leqq x$, $\lambda_{n+1} > x$. We then have $1 < \mu/x + (s(\mu))/2 \times Q_F(x)$. Choosing now $\mu = \frac{1}{4} + \delta(x - \frac{1}{4})$, $0 < \delta < 1$, we get

$$Q_F(x) > \frac{2(x-\frac{1}{4})}{x^2} \frac{1-\delta}{s(\frac{1}{4} + \delta(x - \frac{1}{4}))}$$

Because of (2) the right-hand side tends to $\delta(1-\delta)$ as $x \to +\infty$. Hence we choose $\delta = \frac{1}{2}$ and arrive finally at

(C) $$q(x) \geqq \frac{x - \frac{1}{4}}{x^2 s(1/8 + x/2)}, \quad x > \tfrac{1}{4},$$

the right-hand side tending to $\frac{1}{4}$ as $x \to +\infty$.

2. Let me now point out some further consequences of inequalities (B) and (B'). One can, not quite easily, show that

(3) $$s(\mu) < 2/(\mu - 1), \; \mu > 1.$$

Combining this with (B') we get by optimal choice of $\mu > 1$:

$$\left[\sum_{k=0}^{m-1}\left(1 - \frac{\lambda_k}{\lambda_m}\right)\right]^{1/2} \geqq \tfrac{1}{2}(\lambda_m^{1/2} - \lambda_m^{-1/2}), \quad m \geqq 1.$$

From this inequality one derives easily

(D) $$\sum_{k=0}^{m-1} \lambda_k < \frac{m^2}{g-1} + m, \quad m \geqq 1,$$

which is rather remarkable in view of

$$\sum_{k=0}^{m-1} \lambda_k \sim \frac{m^2}{2(g-1)}, \quad m \to \infty.$$

Furthermore one gets an estimate for λ_m:

(E) $$\lambda_m < \frac{2m}{g-1} + 1 + 2\sqrt{\frac{m}{g-1}\left(\frac{m}{g-1} + 1\right)}$$

$$< \frac{4m}{g-1} + 2 - \frac{g-1}{4m + 2(g-1)} < \frac{4m}{g-1} + 2, \quad m \geqq 1.$$

In particular, since $g \geqq 2$, we always have

$$\lambda_1 < 3 + 2\sqrt{2} < 5.83.$$

One can get a better estimate, even for λ_2, by the following observation. Since φ_1 is orthogonal to the constant eigenfunction φ_0 there is a point $p \in F$ where φ_1 vanishes. If we now apply inequality (B) at this point, taking $n = 1$, we get

(4) $$1 \leqq \frac{\mu}{\lambda_2} + \frac{s(\mu)}{2(g-1)} \leqq \frac{\mu}{\lambda_2} + \frac{s(\mu)}{2}.$$

Using an estimate better than (3) in the vicinity of $\mu = 3$ one can by suitable choice of μ infer from (4) that

(5) $$\lambda_2 < 5 \cdot 2.$$

Assume now that the surface F admits an isometry I with a fixpoint p. I generates a cyclic group of finite order N and we can choose the sequences of eigenfunctions φ_n such that

$$\varphi_n \circ I = \exp\left(2\pi i \frac{k_n}{N}\right) \cdot \varphi_n, \quad 0 \leqq k_n < N.$$

φ_n is an <u>invariant</u> eigenfunction if $k_n = 0$; φ_0 is trivially invariant. Let now φ_{n+1}, $n \geqq 0$, be the first nontrivial invariant eigenfunction. Applying inequality (B) at the fixpoint p we get again

$$1 \leq \frac{\mu}{\lambda_{n+1}} + \frac{s(\mu)}{2(g-1)} \leq \frac{\mu}{\lambda_{n+1}} + \frac{s(\mu)}{2},$$

because the noninvariant $\varphi_1, \ldots, \varphi_n$ vanish all at the fixpoint p. Hence there is always a nontrivial invariant eigenfunction with an eigenvalue less then 5.2.

BIBLIOGRAPHY

1. P. Buser, "Riemannsche Flächen mit grosser Kragenweite", Comment. Math. Helv. 53 (1978).

2. H. Huber, "Ueber die Eigenwerte des Laplace-Operators auf kompakten Riemannschen Flächen", Comment. Math. Helv. 51 (1976), 215-231.

3. ---, "Ueber die Eigenwerte des Laplace-Operators auf kompakten Riemannschen Flächen II", Comment. Math. Helv. 53 (1978), 458-469.

4. B. Randol, "Small eigenvalues of the Laplace operator on compact Riemann surfaces", Bulletin Amer. Math. Soc. 80 (1974), 998-1001.

MATHEMATISCHES INSTITUT DER UNIVERSITÄT BASEL
RHEINSPRUNG 21, CH-4000 BASEL

Proceedings of Symposia in Pure Mathematics
Volume 36, 1980

Schubert varieties and Poincaré duality.

David Kazhdan
Harvard University

and

George Lusztig
Massachusetts Institute of Technology

This paper is a continuation of [12]. In the appendix to [12] we have shown that there is a very close connection between the failure of local Poincaré duality on a Schubert variety and the non-triviality of certain polynomials $P_{y,w}$ defined purely in terms of the corresponding Weyl group. The purpose of this paper is to give a much more precise form to this relationship. We shall give a cohomological interpretation for the coefficients of $P_{y,w}$ in terms of a new cohomology theory introduced by Goreski and MacPherson [8,9] (or, rather,its algebraic version due to Deligne [7]). This shows, in particular, that the coefficients of $P_{y,w}$ are non-negative, in accordance with a conjecture in [12]. We also consider a generalization of Schubert varieties; these are varieties associated to elements in an affine Weyl group. The question of positivity of the coefficients of $P_{y,w}$ for an arbitrary Coxeter group is still open.

We wish to thank MacPherson and Cheeger for some useful discussions. Our paper contains only formal manipulations with deep unpublished results of Deligne, which he kindly explained to us.

1980 Mathematics Subject Classification 14M15, 57P10.

Both authors were partially supported by the National Science Foundation.

§1. Schubert varieties.

1.1 Throughout this paper, k will denote an algebraic closure of the field $\mathbb{F}_p$ with p elements. If G is any algebraic k-group and A is a k-algebra, G(A) will denote the group of A-points of G. We shall identify G with G(k).

1.2 Let G be a semisimple, simply connected algebraic k-group. We fix a pair B,B* of opposed Borel subgroups in G. Their intersection $T = B \cap B^*$ is a maximal torus of G. Let N be the normalizer of T and let $W = N/T$ be the Weyl group. We have the Bruhat decompositions (see for example [3])

$$G = \coprod_{w \in W} BwB = \coprod_{w \in W} B^* wB .$$

Let $\mathcal{B}$ be the variety of Borel subgroups of G. For each $w \in W$, let

$$\mathcal{B}_w = \{B' \in \mathcal{B} \mid B' = gBg^{-1} \text{ for some } g \in BwB\}$$

$$\mathcal{B}^w = \{B' \in \mathcal{B} \mid B' = gBg^{-1} \text{ for some } g \in B^*wB\} .$$

We have the cell decompositions

$$\mathcal{B} = \coprod_{w \in W} \mathcal{B}_w = \coprod_{w \in W} \mathcal{B}^w .$$

Given two elements y,w in W, we say that $y \leq w$ if one of the following equivalent conditions is satisfies: (a) $\overline{\mathcal{B}}_y \subset \overline{\mathcal{B}}_w$, (b) $\overline{\mathcal{B}}^w \subset \overline{\mathcal{B}}^y$, (c) $\mathcal{B}_w \cap \mathcal{B}^y \neq \emptyset$. (The bar denotes Zariski closure.) The varieties $\overline{\mathcal{B}}_w, \overline{\mathcal{B}}^w$ are called Schubert varieties. They are projective, irreducible and $\dim \overline{\mathcal{B}}_w = \operatorname{codim}_{\mathcal{B}} \overline{\mathcal{B}}^w = \ell(w)$, where ℓ is the length function on W . Note that $\overline{\mathcal{B}}^w$ is isomorphic to $\overline{\mathcal{B}}_{w_o w}$, where w_o is the longest element of W .

1.3 We shall now make some remarks on the local structure of $\overline{\mathcal{B}}_w$. First, note that $\overline{\mathcal{B}}_w$ is the union of strata $\mathcal{B}_y$ $(y \leq w)$ and each $\mathcal{B}_y$ is non-singular as it is an orbit of B , which acts on $\overline{\mathcal{B}}_w$ by conjugation. Actually, $\mathcal{B}_y$ is isomorphic to the affine space $U \cap yU^*y^{-1}$ where U, U^* denote the unipotent radicals of B, B^* . The isomorphism is given by $U \cap yU^*y^{-1} \ni u \mapsto {}^{uy}B \in \mathcal{B}_y$. (we use the notation ${}^{g}B' = gB'g^{-1}$, $g \in G$, $B' \in \mathcal{B}$.)

We now describe an affine open neighborhood of $\mathcal{B}_y$ in $\overline{\mathcal{B}}_w$. Let $\mathcal{A}^y = \{B' \in \mathcal{B} \mid B' \text{ opposed to } {}^{y}B^*\}$. Then $\overline{\mathcal{B}}_w \cap \mathcal{A}^y$ is an affine open neighborhood of $\mathcal{B}_y$ in $\overline{\mathcal{B}}_w$. At the same time, it is a neighborhood of $\overline{\mathcal{B}}_w \cap \mathcal{B}^y$. We note the following fact:

1.4 <u>If</u> $y \leq w$, <u>the map</u> $({}^{uy}B, B') \mapsto {}^{u}B'$ $(u \in U \cap yU^*y^{-1})$ <u>defines a</u> T-<u>equivariant isomorphism</u> $\mathcal{B}_y \times (\overline{\mathcal{B}}_w \cap \mathcal{B}^y) \xrightarrow{\approx} \overline{\mathcal{B}}_w \cap \mathcal{A}^y$ <u>carrying</u> $\mathcal{B}_y \times \{{}^{y}B\}$ <u>onto</u> $\mathcal{B}_y$ <u>and</u> $\{{}^{y}B\} \times (\overline{\mathcal{B}}_w \cap \mathcal{B}^y)$ <u>onto</u> $\overline{\mathcal{B}}_w \cap \mathcal{B}^y$.

(This is the restriction of an isomorphism $\mathcal{B}_y \times \mathcal{B}^y \xrightarrow{\approx} \mathcal{A}^y$ defined by the same formula.)

Thus, $\overline{\mathcal{B}}_w \cap \mathcal{B}^y$ can be regarded as a transversal cross section to the stratum $\mathcal{B}_y$ in $\overline{\mathcal{B}}_w$. In particular, it is irreducible, of dimension $\ell(w) - \ell(y)$.

1.5 The variety $\mathcal{A}^y$ is isomorphic to k^N for some N ; the isomorphism can be chosen so that the induced action of T on k^N is linear and there exists an imbedding $\mathbb{G}_m \to T$ such that the corresponding action of $\mathbb{G}_m$ on k^N is given by

$$\lambda\cdot(z_1,\dots,z_n) = (\lambda^{a_1}z_1,\dots,\lambda^{a_n}z_n)$$

where $a_1 > 0,\dots,a_n > 0$. (See [12] Lemma A6.)
Since $\bar{\mathcal{B}}_w \cap \mathcal{B}^y$, $\bar{\mathcal{B}}_w \cap \mathcal{U}^y$ are closed subvarieties of $\mathcal{U}^y$, we see that they can be regarded as $\mathbb{G}_m$-stable closed subvarieties of k^N .

1.6 We now give some examples of Schubert varieities. Let $G = SL(V)$, where V is a 4-dimensional k-vector space. We identify the variety $\mathcal{B}$ with the set of flags $V_1 \subset V_2 \subset V_3$ ($\dim V_i = i$) in V . We fix a flag $V_1^0 \subset V_2^0 \subset V_3^0$. The 4-dimensional variety

$$\{(V_1 \subset V_2 \subset V_3) \in \mathcal{B} \mid V_1 \subset V_3^0,\ V_3 \supset V_1^0\}$$

and the 5-dimensional variety

$$\{(V_1 \subset V_2 \subset V_3) \in \mathcal{B} \mid \dim(V_2 \cap V_2^0) \geq 1\}$$

are examples of singular Schubert varieties. In fact, they fail to satisfy Poincaré duality. If G has rank ≤ 2 , the Schubert varieties corresponding to G may be singular, but are rationally smooth in the sense of [12], A1 and hence satisfy Poincaré duality. (This follows easily from [12], A2).

§2. The polynomials $P_{y,w}$.

2.1 In this section we review some results of [12]. Let (W,S) be a Coxeter group with length function ℓ . Let $\leq$ be the standard partial order on W . The Hecke algebra $\mathcal{H}$

is the algebra over the ring $\mathbb{Z}[q^{1/2},q^{-1/2}]$ with basis elements T_w (one for each element $w \in W$) and multiplication defined by the rules

$$(2.1.1)\qquad T_wT_{w'} = T_{ww'} \quad \text{if} \quad \ell(ww') = \ell(w) + \ell(w')$$

$$(T_s+1)(T_s-q) = 0 \quad \text{if} \quad s \in S .$$

The elements T_w are invertible in $\mathcal{H}$. More precisely, to any pair $y \leq w$ in W, one can associate a well defined polynomial $R_{y,w} \in \mathbb{Z}[q]$ of degree $\ell(w) - \ell(y)$ such that

$$(2.1.2)\qquad T^{-1}_{w^{-1}} = \sum_{y\leq w} \epsilon_y\epsilon_w R_{y,w}(q)q^{-\ell(w)}T_y .$$

Here, $\epsilon_w = (-1)^{\ell(w)}$.

We note the following identity valid for any $y \leq w$:

$$(2.1.3)\qquad q^{\ell(w)-\ell(y)}R_{y,w}(q^{-1}) = \epsilon_y\epsilon_w R_{y,w}(q) .$$

Next, to any pair $y \leq w$ in W, one can associate in a unique way a polynomial $P_{y,w} \in \mathbb{Z}[q]$ of degree $\leq \frac{1}{2}(\ell(w) - \ell(y) - 1)$ (if $y < w$) and equal to 1 (if $y = w$) in such a way that the identity:

$$(2.1.4)\qquad \sum_{y\leq w} \epsilon_y\epsilon_w q^{(\ell(w)/2)-\ell(y)}P_{y,w}(q^{-1})T_y$$

$$= \sum_{y\leq w} \epsilon_y\epsilon_w q^{-(\ell(w)/2)+\ell(y)}P_{y,w}(q)T^{-1}_{y^{-1}}$$

is satisfied for all $w \in W$. Using (2.1.2), one can rewrite (2.1.4) in the equivalent form

(2.1.5) $$q^{\ell(w)-\ell(y)}P_{y,w}(q^{-1}) = \sum_{y\leq z\leq w} R_{y,z}(q)P_{z,w}(q)\ , \quad (\forall y \leq w)\ .$$

Finally, we define, for each $y \leq w$ in W, a polynomial $Q_{y,w} \in \mathbb{Z}[q]$, by the requirement that

(2.1.6) $$\sum_{y\leq z\leq w} \epsilon_y\epsilon_z P_{y,z}Q_{z,w} = \delta_{y,w}\ , \quad (\forall y \leq w)\ .$$

It is clear that $Q_{w,w} = 1$ and that, for $y < w$, $Q_{y,w}$ has degree $\leq \frac{1}{2}(\ell(w) - \ell(y) - 1)$.

From (2.1.3), (2.1.5), and (2.1.6) it follows easily that

(2.1.7) $$q^{\ell(w)-\ell(y)}Q_{y,w}(q^{-1}) = \sum_{y\leq z\leq w} Q_{y,z}(q)R_{z,w}(q)\ , \quad (\forall y < w)\ .$$

§3. Definition of $\mathbb{H}^*(X)$, after Deligne.

3.1 In their papers [8,9] Goreski and Mac Pherson defined a new cohomology theory which associates to a complex algebraic variety, a graded $\mathbb{Q}$-vector space, called middle intersection cohomology. It is the usual cohomology when the variety is non-singular, and it satisfies Poincaré duality for an arbitrary compact variety. Their definition is topological and applies to a class of spaces more general than algebraic varieties.

J. Cheeger [2] has considered L^2-cohomology spaces of certain riemannian spaces with singularities. His results show that there is a strong resemblance between these L^2-cohomology spaces and the middle intersection cohomology of Goreski and MacPherson.

Deligne [7] found an algebraic version of the middle

intersection cohomology, which is applicable to algebraic varieties over an algebraically closed field of arbitrary characteristic.

Let X be a variety of pure dimension n over k and let ℓ be a prime $\neq p$. Deligne associates to X a complex ${}^{\pi}\mathbb{Q}_{\ell}$ of ℓ-adic sheaves, uniquely defined as an object in the corresponding derived category ([4], Rapport, 4.6), by the following properties.

(a) The cohomology sheaf $\mathcal{H}^i(X)$ of ${}^{\pi}\mathbb{Q}_{\ell}$ is zero for $i < 0$, and its support has dimension $\leq n-i-1$, for $i > 0$.

(b) ${}^{\pi}\mathbb{Q}_{\ell}$ is self dual in the derived category.

(c) The restriction of ${}^{\pi}\mathbb{Q}_{\ell}$ to the smooth part of X is equivalent to the complex reduced to the constant sheaf $\mathbb{Q}_{\ell}$ in degree 0.

3.2 Deligne constructs ${}^{\pi}\mathbb{Q}_{\ell}$ in terms of a stratification $X = X^n \supset X^{n-1} \supset X^{n-2} \supset \dots \supset X^0$ where X^i is closed, $X^i - X^{i-1}$ is smooth of pure dimension i and X is equisingular (in a suitable sense) along it, together with its stratification.

Let $U_i = X - X^{n-i}$ and let $j_i\colon U_i \hookrightarrow U_{i+1}$ be the inclusion. Then,

$$ {}^{\pi}\mathbb{Q}_{\ell} = \tau_{\leq n-1} R_{j_n *} \tau_{\leq n-2} \cdots \tau_{\leq 0} R_{j_1 *}\mathbb{Q}_{\ell} $$

where $R_{j_1 *},\dots,R_{j_n *}$ denote direct image (in derived category) and $\tau_{\leq i}$ denotes truncation (see [5] (1.4.6)); $\tau_{\leq i}$ has the effect of keeping the cohomology of a complex unchanged in degrees $\leq i$ and replacing it by zero in degrees $> i$). Note that $\mathcal{H}^i(X)$ is zero on U_{i+1} (except for $i = 0$).

Following Deligne, we define $\mathbb{H}^i(X)$ (resp. $\mathbb{H}_c^i(X)$) to be the hypercohomology space with arbitrary support (resp. with compact support) of X with coefficients in $\pi_{\mathbb{Q}_\ell}$.

It follows from (b) that

(3.2.1) <u>There is a natural non-singular Poincaré duality pairing</u> $\mathbb{H}^i(X) \times \mathbb{H}_c^{2n-i}(X) \to \mathbb{Q}_\ell(-n)$.

We also note the following consequence of the definitions:

(3.2.2) $$\mathcal{H}^i(X) = 0 \quad \text{if} \quad i \geq n > 0 .$$

3.3 Here are two examples. If X is non-singular, then $\mathbb{H}^i(X)$ coincides with the usual ℓ-adic cohomology $H^i(X)$ of X; if X is the (affine) cone with vertex 0 over a non singular projective variety Y, we have

$$\mathbb{H}^i(X) = \begin{cases} H^i(X-0), & \text{if } i < \dim(X) \\ 0, & \text{if } i \geq \dim(X) \end{cases} .$$

3.4 We shall use the following notation. If x is a point of X, $\mathcal{H}_x^i(X)$ will denote the stalk of $\mathcal{H}^i(X)$ at x.

§4. <u>The main results.</u>

4.1 In this section we shall assume that G is provided with an $\mathbb{F}_p$-rational structure, such that B, B^* are defined over $\mathbb{F}_p$ and T is a $\mathbb{F}_p$-split torus. Let $F: G \to G$ be the corresponding Frobenius map. It acts naturally on $\mathcal{B}$, $\overline{\mathcal{B}}_w$, $\overline{\mathcal{B}}^w$ $(w \in W)$, on the sheaves $\mathcal{H}^i(\overline{\mathcal{B}}_w)$, etc.

Theorem 4.2. *Let* $w \in W$. *The sheaf* $\mathcal{H}^i(\overline{\mathcal{B}}_w)$ *is zero for odd* i . *If* i *is even and* $B' \in \overline{\mathcal{B}}_w$ *is defined over* $\mathbb{F}_{p^r}$, *then all eigenvalues of* $(F^r)^*$ *on* $\mathcal{H}^i_{B'}(\overline{\mathcal{B}}_w)$ *are equal to* $p^{ir/2}$.

Theorem 4.3. *Let* $y \leq w$ *in* W . *Then*

$$P_{y,w}(q) = \sum_i \dim \mathcal{H}^{2i}_{y_B}(\overline{\mathcal{B}}_w)q^i \quad .$$

For the proof, we shall need two Lemmas; the first one is a deep results of Deligne, while the second one is a simple application of the first. We shall say that an irreducible algebraic variety X defined over $\mathbb{F}_p$ is pure (resp. very pure) if for any $x \in X(\mathbb{F}_{p^r})$ and for any i , all the eigenvalues of $(F^*)^r$ on $\mathcal{H}^i(X)$ have complex absolute value $\leq p^{ri/2}$ (resp. $= p^{ri/2}$) .

Lemma 4.4 (Deligne [6], [7]). *If* X *is projective and pure, then all the eigenvalues of* F^* *on* $\mathbb{H}^i(X)$ *have complex absolute value* $p^{i/2}$, *and the hard Lefschetz theorem (see the proof of next Lemma) holds for* X .

Lemma 4.5. *Let* Y *be an irreducible closed* $\mathbb{F}_p$*-subvariety of* k^N *invariant under the* $\mathbb{G}_m$*-action defined by* $\lambda(z_1,\ldots,z_n) = (\lambda^{a_1}z_1,\ldots,\lambda^{a_n}z_n)$, *where* $a_1 > 0,\ldots,a_n > 0$. *Then*

(a) $\mathbb{H}^i(Y) = \mathcal{H}^i_0(Y)$, *for all* i , *where* 0 *is the origin of* k^N .

(b) *If* $Y - 0$ *is very pure, then* Y *is very pure.*

The proof of (a) will be left to the reader. (Its analogue over complex numbers can be easily proved by topological considerations.) We now turn to the proof of (b).

The action of G_m on $Y-0$ and k^N-0 is locally standard, in the sense of [12], Def. A8. Therefore, we can consider the geometric quotient $Y-0 \to X = \mathbb{G}_m\backslash(Y-0)$ (set of G_m-orbits) as a projective subvariety of $P = \mathbb{G}_m\backslash(k^N-0)$. Moreover, as $Y-0$ is very pure, we see that X is very pure. Consider the "Wang exact sequence"

$$\cdots \longrightarrow \mathbb{H}^{i-2}(X)(-1) \xrightarrow{\wedge\omega} \mathbb{H}^i(X) \longrightarrow \mathbb{H}^i(Y-0) \longrightarrow$$

$$\mathbb{H}^{i-1}(X)(-1) \xrightarrow{\wedge\omega} \mathbb{H}^{i+1}(X) \longrightarrow \cdots$$

where $\omega \in H^2(X)(1)$ is the restriction of the corresponding class in $H^2(P)(1)$. By Lemma 4.4, the Lefschetz hard theorem holds on X: the map $\mathbb{H}^{i-1}(X)(-1) \xrightarrow{\wedge\omega} \mathbb{H}^{i+1}(X)$ is injective for $i \leq \dim(X)$. It follows that $\mathbb{H}^i(Y-0) = \operatorname{coker}(\mathbb{H}^{i-2}(X)(-1) \to \mathbb{H}^i(X))$, for $i \leq \dim(X)$. Using Lemma 4.4, we see that all the eigenvalues of F^* on $\mathbb{H}^i(Y-0)$ have complex absolute value equal to $p^{i/2}$, if $i \leq \dim(X)$. We must prove that all the eigenvalues of F^* on $\mathcal{H}^i_0(Y)$ have complex absolute value equal to $p^{i/2}$, for all i. Taking into account (a) and the fact that $\mathcal{H}^i_0(Y) = 0$ if $i \geq \dim(Y) = \dim(X)+1$, (cf. (3.2.2)) we see that it is enough to show that $\mathbb{H}^i(Y) \xrightarrow{\approx} \mathbb{H}^i(Y-0)$, for $i < \dim(Y)$. By Poincaré duality (3.2.1), this is equivalent to showing that $\mathbb{H}^i_c(Y-0) \xrightarrow{\approx} \mathbb{H}^i_c(Y)$ for $i > \dim(Y)$. But this follows from the long exact sequence associated with the partition $Y = (Y-0) \cup \{0\}$:

$$\cdots \to \mathcal{H}^{i-1}_0(Y) \to \mathbb{H}^i_c(Y-0) \to \mathbb{H}^i_c(Y) \to \mathcal{H}^i_0(Y) \to \cdots$$

4.6 Proof of Theorem 4.1. We may assume $w \neq e$. For each $z < w$, we consider the property:

(P(z)): For any $B' \in \mathcal{B}_z$ defined over $\mathbb{F}_{p^r}$, we have $\mathcal{H}^i_{B'}(\overline{\mathcal{B}}_w) = 0$ for odd i and all eigenvalues of $(F^*)^r$ on $\mathcal{H}^i_{B'}(\overline{\mathcal{B}}_w)$ (i even) are equal to $p^{ri/2}$.

It is clear that the property (P(w)) holds. We now fix an element $y < w$. We shall assume that the property (P(z)) holds for all z, $y < z \leq w$. We shall deduce from this assumption that property (P(y)) must hold also.

Our assumption implies that $(\overline{\mathcal{B}}_w \cap \mathcal{U}^y) - \mathcal{B}_y$ is very pure. Using the isomorphism $(\overline{\mathcal{B}}_w \cap \mathcal{U}^y) - \mathcal{B}_y \approx \mathcal{B}_y \times ((\overline{\mathcal{B}}_w \cap \mathcal{B}^y) - \{{}^yB\})$ (1.4) and the fact that $\mathcal{B}_y$ is smooth, we deduce that $(\overline{\mathcal{B}}_w \cap \mathcal{B}^y) - \{{}^yB\}$ is very pure. Using 1.5 and 4.5(b), it follows that $\overline{\mathcal{B}}_w \cap \mathcal{B}^y$ is very pure. Using the isomorphism $\overline{\mathcal{B}}_w \cap \mathcal{U}^y \approx \mathcal{B}_y \times (\overline{\mathcal{B}}_w \cap \mathcal{B}^y)$ (see (1.4)) and the fact that $\mathcal{B}_y$ is smooth, we see that $\overline{\mathcal{B}}_w \cap \mathcal{U}^y$ is very pure. We can now write $\overline{\mathcal{B}}_w \cap \mathcal{U}^y$ as a disjoint union $\coprod_{y \leq z \leq w} (\mathcal{B}_z \cap \mathcal{U}^y)$. We consider an integer $r \geq 1$. By [12], Lemmas A3, A4, we see that for any z, $y \leq z \leq w$, we have $\#(\mathcal{B}_z \cap \mathcal{U}^y)^{F^r} = R_{y,z}(p^r)p^{r\ell(y)}$ (see (2.1.2)). By the Lefschetz fixed point formula for F, applied to $\overline{\mathcal{B}}_w \cap \mathcal{U}^y$ and $\pi_{\mathbb{Q}_\ell}$, we have

$$(4.6.1)\quad \mathrm{Tr}\left((F^*)^r, \mathbb{H}^*_c(\overline{\mathcal{B}}_w \cap \mathcal{U}^y)\right) = \sum_{y \leq z \leq w} \sum_{B' \in (\mathcal{B}_z \cap \mathcal{U}^y)^{F^r}} \mathrm{Tr}\left((F^*)^r, \mathcal{H}^*_{B'}(\overline{\mathcal{B}}_w)\right)$$

$$= \sum_{y \leq z \leq w} p^{r\ell(y)} R_{y,z}(p^r) \mathrm{Tr}\left((F^*)^r, \mathcal{H}^*_{{}^zB}(\overline{\mathcal{B}}_w)\right)$$

where $\mathbb{H}^*_c$ denotes the formal alternating sum $\sum (-1)^i \mathbb{H}^i_c$, and $\mathcal{H}^*_{B'}$ has a similar meaning.

By Poincaré duality (3.2.1) on $\overline{\mathcal{B}}_w \cap \mathcal{U}^y$, we have

$$(4.6.2)\quad \mathrm{Tr}\left((F^*)^r, \mathbb{H}^*_c(\overline{\mathcal{B}}_w \cap \mathcal{U}^y)\right) = p^{r\ell(w)} \mathrm{Tr}\left((F^*)^{-r}, \mathbb{H}^*(\overline{\mathcal{B}}_w \cap \mathcal{U}^y)\right).$$

We now apply Lemma 4.5(a) to the variety $\mathcal{B}_w \cap \mathcal{a}^y$ and its $\mathbb{G}_m$-action given by 1.5. It follows that

$$(4.6.3) \quad \mathrm{Tr}\left((F^*)^{-r}, \mathbb{H}^*(\mathcal{B}_w \cap \mathcal{a}^y)\right) = \mathrm{Tr}\left((F^*)^{-r}, \mathcal{H}^*{}_{y_B}(\mathcal{B}_w)\right) .$$

Comparing (4.6.1), (4.6.2), (4.6.3), we see that

$$(4.6.4) \quad p^{r\ell(w)}\mathrm{Tr}\left((F^*)^{-r}, \mathcal{H}^*_{y_B}(\mathcal{B}_w)\right)$$

$$= \sum_{y \le z \le w} p^{r\ell(y)} R_{y,z}(p^r)\mathrm{Tr}\left((F^*)^r, \mathcal{H}^*_{z_B}(\mathcal{B}_w)\right)$$

Using the assumption $(P(z))$ for $y < z \le w$, we see that there exists a polynomial $\Pi \in \mathbb{Z}[q]$, independent of r, such that

$$(4.6.5) \quad p^{r(\ell(w)-\ell(y))} \sum_i (-1)^i \mathrm{Tr}\left((F^*)^{-r}, \mathcal{H}^i_{y_B}(\mathcal{B}_w)\right)$$

$$- \sum_i (-1)^i \mathrm{Tr}\left((F^*)^r, \mathcal{H}^i_{y_B}(\mathcal{B}_w)\right) = \Pi(p^r)$$

for all $r \ge 1$.

If λ' (resp. λ'') is an eigenvalue of F^* on $\mathcal{H}^i_{y_B}(\mathcal{B}_w)$ (resp. on $\mathcal{H}^j_{y_B}(\mathcal{B}_w)$) we have $|\lambda'| = p^{i/2}$, $|\lambda''| = p^{j/2}$, since $\mathcal{B}_w \cap \mathcal{a}^y$ is very pure. In particular, λ' cannot be equal to λ'', unless $i = j$. Moreover, λ' cannot be equal to $p^{\ell(w)-\ell(y)}\lambda''^{-1}$, since both i and j are $\le \dim(\mathcal{B}_w \cap \mathcal{B}^y) - 1 = \ell(w) - \ell(y) - 1$ (see (3.2.2)) so that

$$|\lambda'| = p^{i/2} \ne p^{\ell(w)-\ell(y)-j/2} = |p^{\ell(w)-\ell(y)}\lambda''^{-1}| .$$

It follows that the identity (4.6.5) must split into several identities of the form

$$\mathrm{Tr}\Big((F^*)^r, \mathcal{H}^i_{y_B}(\overline{\mathcal{B}}_w)\Big) = \Pi_i(p^r) , \quad r \geq 1 , \; i \geq 0 ,$$

where $\Pi_i \in \mathbb{Z}[q]$ is independent of r. It follows that property $(P(y))$ holds. This completes the proof of Theorem 4.2

4.7 <u>Proof of Theorem 4.3</u>. Let us define

$$\widetilde{P}_{y,w}(q) = \sum_i \dim \mathcal{H}^{2i}_{y_B}(\overline{\mathcal{B}}_w) q^i \in \mathbb{Z}[q] .$$

Then Theorem 4.2 and (4.6.4) show that

$$q^{\ell(w)-\ell(y)} \widetilde{P}_{y,w}(q^{-1}) = \sum_{y \leq z \leq w} R_{y,z}(q) \widetilde{P}_{z,w}(q) .$$

It is clear that $\widetilde{P}_{w,w} = 1$. Moreover, if $y < w$, we have

$$\mathcal{H}^{2i}_{y_B}(\overline{\mathcal{B}}_w) \approx \mathcal{H}^{2i}_{y_B}(\overline{\mathcal{B}}_w \cap \mathcal{Q}^y) \approx \mathcal{H}^{2i}_{y_B}(\overline{\mathcal{B}}_w \cap \mathcal{B}^y) = 0$$

whenever $2i \geq \ell(w) - \ell(y) = \dim(\overline{\mathcal{B}}_w \cap \mathcal{B}^y)$ (cf. (3.2.2)). Therefore, $\deg(\widetilde{P}_{y,w}) \leq \frac{1}{2}(\ell(w) - \ell(y) - 1)$. Thus $\widetilde{P}_{y,w}$ satisfy the defining identities for $P_{y,w}$ (see §2). It follows that $\widetilde{P}_{y,w} = P_{y,w}$ for all $y \leq w$.

<u>Corollary 4.8</u>. <u>If</u> $y \leq w$ <u>in</u> W, <u>the polynomial</u> $P_{y,w}$ <u>has</u> ≥ 0 <u>coefficients</u>.

<u>Corollary 4.9</u>. <u>Let</u> $w \in W$. <u>We have</u> $\mathbb{H}^i(\overline{\mathcal{B}}_w) = 0$ <u>for</u> i <u>odd</u>. <u>If</u> i <u>is even, the eigenvalues of</u> F^* <u>on</u> $\mathbb{H}^i(\overline{\mathcal{B}}_w)$ <u>are equal to</u> $p^{i/2}$. <u>We have</u>

$$\sum_i \dim \mathbb{H}^{2i}(\overline{\mathcal{B}}_w) q^i = \sum_{y \leq w} q^{\ell(y)} P_{y,w}(q) .$$

We use the Lefschetz fixed point formula for F, applied to $\mathcal{B}_w$ and ${}^{\pi}Q_\ell$, taking also into account Theorems 4.2 and 4.3:

$$\sum_i (-1)^i \mathrm{Tr}\big((F^*)^r, H^i(\overline{\mathcal{B}}_w)\big) = \sum_{y \le w} p^{r\ell(y)} P_{y,w}(p^r), \quad (\forall r \geq 1),$$

since $\#(\mathcal{B}_y^{F^r}) = p^{r\ell(y)}$. It remains to apply to $\overline{\mathcal{B}}_w$, the first part of Lemma 4.4.

§5. A generalization.

5.1 We now introduce a more general kind of Schubert varieties; they will be associated to elements in an affine Weyl group. Let t be an indeterminate. Consider the k-algebras $A = k[t,t^{-1}]$, $A^+ = k[t]$, $A^- = k[t^{-1}]$. The natural inclusions $k \begin{smallmatrix} \subset A^+ \subset \\ \subset A^- \subset \end{smallmatrix} A$ induce inclusions $G \begin{smallmatrix} \subset G(A^+) \subset \\ \subset G(A^-) \subset \end{smallmatrix} G(A)$. The natural projections $A^+ \to k$ (taking t to 0) and $A^- \to k$ (taking t^{-1} to 0) define homomorphisms $\pi^+ : G(A^+) \to G$, $\pi^- : G(A^-) \to G$. Let $\widetilde{B} = (\pi^+)^{-1}(B)$, $\widetilde{B}^* = (\pi^-)^{-1}(B^*)$, $\widetilde{W} = N(A)/T$. $\widetilde{W}$ is in a natural way a Coxeter group on $\ell+1$ generators (ℓ = rank G); it is the affine Weyl group of G (see [11]). We have the decompositions (see [1], [10], [11], [13])

$$G(A) = \coprod_{w \in \widetilde{W}} \widetilde{B}\, w\, \widetilde{B} = \coprod_{w \in \widetilde{W}} \widetilde{B}^*\, w\, \widetilde{B}.$$

Let $\widetilde{\mathcal{B}}$ (resp. $\widetilde{\mathcal{B}}^*$) be the set of subgroups of $G(A)$ which are conjugate to $\widetilde{B}$ (resp. to $\widetilde{B}^*$). Note that the sets $\widetilde{\mathcal{B}}, \widetilde{\mathcal{B}}^*$ are disjoint.

For each $w \in W$, let

$$\tilde{\mathcal{B}}_w = \{\tilde{B}' \in \tilde{\mathcal{B}} \mid \tilde{B}' = g\tilde{B}g^{-1} \text{ for some } g \in \tilde{B}w\tilde{B}\}$$

$$\tilde{\mathcal{B}}^w = \{\tilde{B}' \in \tilde{\mathcal{B}} \mid \tilde{B}' = g\tilde{B}g^{-1} \text{ for some } g \in \tilde{B}^*w\tilde{B}\} .$$

We have the "cell" decompositions

$$\tilde{\mathcal{B}} = \coprod_{w\in\tilde{W}} \tilde{\mathcal{B}}_w = \coprod_{w\in\tilde{W}} \tilde{\mathcal{B}}^w .$$

5.2 The set $\tilde{\mathcal{B}}$ can be regarded in a natural way as an infinite dimensional algebraic variety over k. More precisely, there exist (finite dimensional) projective varieties X_i $(i = 0,1,2,\ldots)$ and algebraic imbeddings $X_0 \subset X_1 \subset X_2 \subset \ldots$ such that $\tilde{\mathcal{B}}$ is the union $\bigcup_{i=0}^{\infty} X_i$. Let us consider, for example, the case where $G = SL(V)$ where V is an n-dimensional k-vector space. Then $\tilde{\mathcal{B}}$ can be identified with the set of all sequences $L_0 \supset L_1 \supset \ldots \supset L_{n-1} \supset L_n = tL_0$ of A^+-lattices (A^+-submodules of rank n) of $V \otimes A$ such that $\dim L_j/L_{j+1} = 1$ $(0 \leq j \leq n-1)$ and L_0 is in the $G(A)$-orbit of $V \otimes A^+$. (The stabilizer in $G(A)$ of such a sequence is in $\tilde{\mathcal{B}}$.) For each integer $i \geq 0$, let X_i be the set consisting of all sequences $L_0 \supset L_1 \supset \ldots \supset L_{n-1} \supset L_n = tL_0$ in $\tilde{\mathcal{B}}$ such that

$$V \otimes A^+ \supset t^iL_0 \supset t^iL_1 \supset \ldots \supset t^iL_n \supset t^{n(i+1)}V \otimes A^+ .$$

It is clear that $X_0 \subset X_1 \subset X_2 \subset \ldots$ and that the union of the X_i's is $\tilde{\mathcal{B}}$. On the other hand, X_i can be identified with the set of all sequences $\tilde{L}_0 \supset \tilde{L}_1 \supset \ldots \supset \tilde{L}_n = t\tilde{L}_0$ of t-stable subspaces of the $n^2(i+1)$-dimensional k-vector space $V \otimes A^+/(t^{n(i+1)}V \otimes A^+)$ (on which t acts naturally as a nilpotent endomorphism), such that $\operatorname{codim} \tilde{L}_j = ni+j$

$(j = 0,1,\dots,n)$. Thus, X_i has a natural structure of projective k-variety. In the case where G is arbitrary, the corresponding varieties X_i can be defined using an imbedding $G \subset SL(V)$. It follows from what we have said that the Zariski closure $\overline{\tilde{\mathcal{B}}}_w$ of $\tilde{\mathcal{B}}_w$ in $\tilde{\mathcal{B}}$ is well defined. It is a projective irreducible k-variety of dimension $\ell(w)$, where ℓ is the length function on $\tilde{W}$. The usual partial order $\leq$ on the affine Weyl group $\tilde{W}$ can be described geometrically by the equivalent conditions:

$$y \leq w \Leftrightarrow \overline{\tilde{\mathcal{B}}}_y \subset \overline{\tilde{\mathcal{B}}}_w \Leftrightarrow \mathcal{B}_w \cap \mathcal{B}^y \neq \emptyset .$$

We define $\overline{\tilde{\mathcal{B}}}^y$ to be the infinite union $\bigcup_{z \geq y} \tilde{\mathcal{B}}^z$. It can be regarded as a closed subvariety of codimension $\ell(y)$ of the infinite dimensional variety $\tilde{\mathcal{B}}$; in other words, its intersection with X_i for $i >> 0$ has codimension $\ell(y)$ in X_i . Moreover, for any $w \in \tilde{W}$, $\overline{\tilde{\mathcal{B}}}_w \cap \overline{\tilde{\mathcal{B}}}^y$ is either empty or is a closed subvariety of codimension $\ell(y)$ of the projective variety $\overline{\tilde{\mathcal{B}}}_w$.

$\overline{\tilde{\mathcal{B}}}_w$ and $\overline{\tilde{\mathcal{B}}}^w$ are called the Schubert varieties corresponding to $w \in \tilde{W}$.

5.3 Let us consider the case where $G = SL(V)$ where V is a 2-dimensional k-vector space. Let e,f be a basis of V . We identify again $\tilde{\mathcal{B}}$ with the set of all sequences $L_0 \supset L_1 \supset L_2 = tL_0$ of A^+-lattices of $V \otimes A$ such that $\dim L_0/L_1 = 1$ and L_0 is in the $G(A)$-orbit of $V \otimes A^+$. Let $\mathcal{P}$ be the set of all A^+-lattices in $V \otimes A$ which are in the same orbit as $V \otimes A^+$. There is a natural map $\pi: \tilde{\mathcal{B}} \to \mathcal{P}$, $L_0 \supset L_1 \supset L_2 \to L_0$, all of whose fibres are isomorphic to the projective line over k . We wish to describe the images under π of the Schubert varieities in $\tilde{\mathcal{B}}$. (We take $\tilde{B}$ to be the stabilizer of the A^+-lattices

$L_{(0)} = A^+e + A^+f$, $L_{(1)} = A^+te + A^+f$; we take $\tilde{B}^*$ to be the stabilizer of the A^--lattices $L^{(0)} = A^-e + A^-f$, $L^{(1)} = A^-e + A^-t^{-1}f$.) For each integer $n \geq 0$, there is exactly one such image $\mathcal{O}_n$ of dimension n and exactly one such image $\mathcal{O}^n$ of codimension n . We have

$$\mathcal{O}_n = \{L \in \mathcal{O} | \ t^{\frac{1}{2}(n+\epsilon)}L \subset L_{(\epsilon)}\} ,$$

$$\mathcal{O}^n = \{L \in \mathcal{O} | \ t^{\frac{1}{2}(n+\epsilon)}L \cap L^{(1-\epsilon)} \neq 0\} ,$$

where $\epsilon = 0$ if n is even, $\epsilon = 1$ if n is odd. One checks easily that $\mathcal{O}_0 \subset \mathcal{O}_1 \subset \mathcal{O}_2 \ \ldots$, $\bigcup_{n=0}^{\infty} \mathcal{O}_n = \mathcal{O}$, $\mathcal{O} = \mathcal{O}^0 \supset \mathcal{O}^1 \supset \mathcal{O}^2 \supset \ldots$, $\bigcap_{n=0}^{\infty} \mathcal{O}^n = \emptyset$, $\mathcal{O}_n \cap \mathcal{O}^{n'} \neq \emptyset$ if and only if $n' \leq n$, $\mathcal{O}_n \cap \mathcal{O}^n$ = one point.

Although the varities $\mathcal{O}_n$ have, in general, singularities, they are rationally smooth in the sense of [12]A1 and hence satisfy Poincaré duality.

5.4 The Schubert varieties $\bar{\tilde{\mathcal{B}}}_w$ (for $w \in \tilde{W}$) are a generalization of the classical Schubert varieties $\bar{\mathcal{B}}_w$ $(w \in W)$. Indeed, W is in a natural way a subgroup of $\tilde{W}$ $(N/T \subset N(A)/T)$. If $w \in W$, the map $B' \to (\pi^+)^{-1}B'$ defines an isomorphism of $\bar{\mathcal{B}}_w$ onto $\bar{\tilde{\mathcal{B}}}_w$.

Theorems 4.2 and 4.3 describing the sheaves $\mathcal{H}^i$ on the Schubert varieties $\bar{\mathcal{B}}_w$ $(w \in W)$ extend in an obvious way to the case of the generalized Schubert varieities $\bar{\tilde{\mathcal{B}}}_w$ $(w \in \tilde{W})$. The proof in this more general case is essentially the same as that in 4.6, 4.7. We shall formulate here only the following version:

<u>Theorem 5.5. Let</u> $y \leq w$ <u>in</u> $\tilde{W}$. <u>Then</u> $\mathbb{H}^i(\bar{\tilde{\mathcal{B}}}_w \cap \tilde{\mathcal{B}}^y) = 0$ <u>for</u> i <u>odd, and</u> $P_{y,w} = \sum_{i \geq 0} \dim \mathbb{H}^{2i}(\bar{\tilde{\mathcal{B}}}_w \cap \tilde{\mathcal{B}}^y)q^i$, <u>where</u> $P_{y,w}$ <u>are the polynomials defined in §2.</u>

Corollary 5.6. *If* $y \leq w$ *in* $\widetilde{W}$, *the polynomial* $P_{y,w}$ *has* ≥ 0 *coefficients.*

It should not be difficult to prove the following

Proposition 5.7. *Let* $y \leq w$ *in* $\widetilde{W}$. *Then* $\mathbb{H}^i(\widetilde{\mathcal{B}}_w \cap \overline{\widetilde{\mathcal{B}}^y}) = 0$ *for* i *odd, and* $Q_{y,w}(q) = \sum_{i \geq 0} \dim \mathbb{H}^{2i}(\widetilde{\mathcal{B}}_w \cap \overline{\widetilde{\mathcal{B}}^y})q^i$, *where* $Q_{y,w}$ *are the polynomials defined in* §2. *In particular,* $Q_{y,w}$ *has* ≥ 0 *coefficients.*

References

1. G. D. Birkhoff, A theorem on matrices of analytic functions, Math. Ann. 74 (1913) 122-133.

2. J. Cheeger, On the Hodge theory of Riemannian pseudomanifolds, preprint.

3. C. Chevalley, Classification des groupes de Lie algébriques, Paris 1956-58.

4. P. Deligne, Cohomologie étale, SGA 4½, Lectures Notes in Math. 569, Springer Verlag.

5. P. Deligne, Théorie de Hodge II, Publ. Math. IHES, No. 40.

6. P. Deligne, La conjecture de Weil II, to appear.

7. P. Deligne, letter to the authors, 20 April 1979.

8. M. Goreski, R. MacPherson, La dualité de Poincaré pour les espaces singuliers. C. R. Acad. Sci. Paris, t. 284 serie A (1977), 1549-1551.

9. M. Goreski, R. MacPherson, Intersection homology theory, to appear.

10. A. Grothendieck, Sur la classifications des fibrés

holomorphes sur la sphère de Riemann, Am. J. Math. 79 (1957) 121-138.

11. N. Iwahori, H. Matsumoto, On some Bruhat decomposition and the structure of the Hecke rings of p-adic Chevalley groups, Publ. Math. IHES, No. 25.

12. D. Kazhdan, G. Lusztig, Representations of Coxeter groups and Hecke algebras, to appear in Invent. Math.

13. A. Pressley, Decompositions of the space of loops on a Lie group, to appear.

Proceedings of Symposia in Pure Mathematics
Volume 36, 1980

ESTIMATES OF EIGENVALUES OF A COMPACT RIEMANNIAN MANIFOLD

Peter Li and Shing-Tung Yau*

Let M be a compact manifold with (possibly empty) boundary. Then in this paper, we study the eigenvalues of M with respect to various boundary conditions.

Since the Poincaré inequality plays a very important role in analysis and since a lower bound of the first eigenvalue gives an upper bound of the constant in the Poincaré inequality, it is very desirable to find a good lower estimate of the first eigenvalue. For domains in euclidean space, there are classical works of Faber-Krahn, Polyá-Szegö, Payne, Weinberger, etc. The works of these authors are not only beautiful and important, but also give a deep impact to estimate eigenvalues on curved spaces. For many geometric problems, we often need to estimate the Poincaré inequality for domains on a curved space. Thus in this paper, we concentrate our attention to this case. The first major result in this direction was due to Lichnerowicz [10] and Obata [11]. In their beautiful work, they assumed the Ricci curvature of the compact manifold (without boundary) is greater than a positive constant and they estimated the first eigenvalue from below in terms of this constant. It is remarkable that this constant is sharp. This estimate of Lichnerowicz-Obata was generalized later by Reilly [14] to manifolds with boundary where he treated

1980 Mathematics Subject Classification 53-XX.
*This research was supported in part by the NSF Grant MCS78-04872

the Dirichlet boundary valued problem. Besides the assumption of Lichnerowicz-Obata, he also assumed the boundary of the manifold has non-negative mean curvature.

After this work of Lichnerowicz-Obata, J. Cheeger [3] studied the first eigenvalue by estimating it from below by a constant which is involved in certain type of isoperimetric inequality. In [1], Aubin gave a lower estimate of the first eigenvalue in terms of a lower bound of the volume, an upper bound of the diameter, a lower bound of the sectional curvature, an upper bound of the Ricci curvature and a lower bound of the injectivity radius. For the purpose of applications, it is important to relax the dependency of the lower bound on the geometric quantities. For this purpose, the second author [15] showed that one can estimate the first eigenvalue from below by lower bound of the volume, an upper bound of the diamter and a lower bound of the Ricci curvature. Basing on an upper estimate of Cheng [5], the second author conjectured that one should be able to drop the dependency of the volume in the above estimate. Combining with Cheng's result, this would give the best possible estimate of the first eigenvalue for a general compact manifold.

It turns out that the first author [9] was able to demonstrate the above conjecture in all the major cases. His method depended on a gradient estimate of the first eigenfunction. In this paper, we indicate how to make a slight modification of this method to give a complete demonstration of the above conjecture. For the special case of compact manifolds with non-negative Ricci curvature, we show that $\lambda_1 \geq \frac{\pi^2}{4d^2}$ where d is the diameter of the manifold. On the other hand, Cheng's estimate gives $\lambda_1 \leq \frac{n\pi^2}{d^2}$ where n is the dimension of the manifold. (Cheeger [2] had a weaker estimate prior to Cheng's result.) We also extend these estimates to compact manifolds with boundary. For Dirichlet boundary valued problem, the estimate also depends on the lower bound of the mean curvature of the boundary. For Neumann boundary valued problem, we have to assume the second fundamental form of the boundary is positive semidefinite. In the Dirichlet boundary valued problem, the diameter d can be replaced by the radius of the largest geodesic ball that can be inscribed into the manifold.

In the second part of the paper, we demonstrate how to use the variational principle to obtain both upper and lower estimates of higher eigenvalues λ_m. Let V be the volume of the compact manifold. Then the famous estimate of H. Weyl shows that when m tends to infinity, $\lambda_m \left(\frac{V}{m}\right)^{\frac{2}{n}}$ approaches to a constant $C(n)$ depending only on the dimension of the manifold. We demonstrate how to find an upper and lower estimate of $\lambda_m \left(\frac{V}{m}\right)^{\frac{2}{n}}$. For example, when the Ricci curvature of M is non-negative, we show that $\frac{\lambda_m}{C(n)}\left(\frac{V}{m+1}\right)^{\frac{2}{n}}$ is bounded from above by an absolute constant. When the Ricci curvature is bounded from below by $(n-1)K$, then we prove that $\lambda_m \leqslant C_1 \left(\frac{m+1}{V}\right)^{\frac{2}{n}} + C_2$ where C_1 depends on K, d and n; C_2 depends on K and n. Our method depends on Cheng's result [5].

The lower estimate of $\lambda_m \left(\frac{V}{m}\right)^{\frac{2}{n}}$ is more complicated. When the sectional curvature is non-negative, $\lambda_m \left(\frac{V}{m}\right)^{\frac{2}{n}}$ has a lower bound depending only on n when m is greater than a constant depending on the upper bound of sectional curvature and the radius of convexity. When the sectional curvature is allowed to be negative, then the lower bound of $\lambda_m \left(\frac{V}{m}\right)^{\frac{2}{n}}$ should be replaced by a positive constant depending on n, d and the lower bound of the sectional curvature.

Finally we should mention that Professor Gromov has pointed out that in a classical book of P. Levy [8] there was an indication of an estimate of the first eigenvalue. This estimate depends on a very non-trivial analysis of the regularity of certain hypersurface with constant mean curvature. Up to now, it is not known whether this analysis can be carried out when dimension $n \geqslant 7$. Furthermore, the constant involved in this analysis is not sharp whereas our estimate is sharp for manifolds whose Ricci curvature is non-negative and whose boundary has non-negative mean curvature. We should also mention that for convex domain in euclidean space, Payne-Weinberger [13] has already estimated λ_1 for Neumann problem. During the conference, Chavel-Feldman generalized the result of Payne-Weinberger to convex surfaces with non-negative curvature. Our theorem is more general while our method is very different. After we indicated our result to Professor Protter, he pointed out a paper of Payne and Stakgold [12] on the estimate of the first eigenvalue of a domain in euclidean space. It turns out that part of our estimates is similar to theirs

1. GRADIENT ESTIMATE. In this section we consider the solution of the equation

$$\Delta u = F(u) \tag{1.1}$$

defined on a compact manifold M of dimension n. In the case if M is a manifold with boundary ∂M, we imposed one of the following boundary conditions:

$$u \equiv 0 \quad \text{on} \quad \partial M \tag{1.2}$$

$$\frac{\partial u}{\partial \nu} \equiv 0 \quad \text{on} \quad \partial M \tag{1.3}$$

The first one is known to be the Dirichlet boundary condition and the latter is the Neumann boundary condition, where $\frac{\partial}{\partial \nu}$ denotes the outward normal to ∂M.

Suppose the Ricci curvature of M is bounded below by $(n - 1)K$. In the next three theorems we will derive an upper estimate for the gradient of the solution of (1.1)(with different boundary conditions) in terms of K, u, $F(u)$ and its derivative.

THEOREM 1. Let M be a compact manifold. If u is a solution of (1.1) and $\mu > 1$ is any constant, then

$$|\nabla u|^2 \leqslant \max\left\{4(n-1)\left[-(n-1)K - F_u + \left\|\frac{F(u)}{\mu \sup u - u}\right\|_\infty\right],\right.$$
$$\left.\sqrt{8}\left\|\frac{F(u)}{\mu \sup u - u}\right\|_\infty\right\}(\mu \sup u - u)^2$$

where F_u denotes the derivative of F with respect to u.

PROOF. For $\mu > 1$, we define the function

$$f = \frac{|\nabla u|^2}{(\mu \sup u - u)^2} \tag{1.4}$$

By the compactness of M, there is a point $x_o \in M$ such that f achieves its supremum. Hence at x_o

$$\nabla f = 0$$

(1.5) and

$$\Delta f \leqslant 0$$

This gives

$$\sum_j u_j u_{ji} + \frac{|\nabla u|^2 u_i}{\mu \sup u - u} = 0 \tag{1.6}$$

and

$$0 \geqslant \frac{\sum_{i,j} u_{ij}^2 + \sum_{i,j} u_j u_{jii}}{(\mu \sup u-u)^2} \tag{1.7}$$

$$+ \frac{4\sum_{i,j} u_i u_{ij} u_j}{(\mu \sup u-u)^3} + \frac{|\nabla u|^2 \sum_i u_{ii}}{(\mu \sup u-u)^3} + \frac{3|\nabla u|^4}{(\mu \sup u-u)^4}$$

By choosing suitable orthonormal frame at x_o, we may assume $u_\alpha = 0$ for $\alpha > 1$. Substituting (1.6) into (1.7) and using the Ricci formula, we have

$$\sum_{\alpha>1} u_{\alpha\alpha}^2 + \sum_i u_i (\Delta u)_i + \sum_{i,j} R_{ij} u_i u_j + \frac{|\nabla u|^2 \Delta u}{\mu \sup u-u} \leqslant 0 \tag{1.8}$$

Clearly

$$\begin{aligned} \sum_{\alpha>1} u_{\alpha\alpha}^2 &\geqslant \frac{1}{n-1}\left(\sum_{\alpha>1} u_{\alpha\alpha}\right)^2 \\ &= \frac{1}{n-1}(\Delta u - u_{11})^2 \\ &\geqslant \frac{u_{11}^2}{2(n-1)} - \frac{(\Delta u)^2}{n-1} \end{aligned} \tag{1.9}$$

By (1.6)

$$\sum_{\alpha>1} u_{\alpha\alpha}^2 \geqslant \frac{1}{2(n-1)} \frac{|\nabla u|^4}{(\mu \sup u-u)^2} - \frac{(\Delta u)^2}{n-1} \tag{1.10}$$

Therefore

$$\begin{aligned} &\frac{1}{2(n-1)} \frac{|\nabla u|^4}{(\mu \sup u-u)^2} - \frac{F^2(u)}{n-1} \\ &+ F_u|\nabla u|^2 + (n-1)K|\nabla u|^2 \\ &+ \frac{|\nabla u|^2 F(u)}{\mu \sup u-u} \leqslant 0 \end{aligned} \tag{1.11}$$

Hence

$$\frac{1}{2(n-1)} f^2(x_o) + \left(F_u + (n-1)K + \frac{F(u)}{\mu \sup u - u}\right) f(x_o) \leq \frac{1}{n-1}\left(\frac{F(u)}{\mu \sup u - u}\right)^2 \tag{1.12}$$

and

$$f(x_o) \leq \max\left\{4(n-1)\left[-(n-1)K - F_u + \left\|\frac{F(u)}{\mu \sup u - u}\right\|_\infty\right], \sqrt{8}\left\|\frac{F(u)}{\mu \sup u - u}\right\|_\infty\right\}$$

This proves theorem 1.

THEOREM 2. Let M be a compact manifold with boundary. Suppose H denotes the mean curvature of ∂M, and if u is a non-negative solution of (1.1) and (1.2). Then for $\mu > 1$,

either

$$|\nabla u| \leq \max\left\{-2(n-1)H, 2\left\|\frac{F(u)}{\mu \sup u - u}\right\|_{\partial M}^{1/2}\right\} \times (\mu \sup u - u)$$

or

$$|\nabla u|^2 \leq \max\left\{4(n-1)\left[-(n-1)K - F_u + \left\|\frac{F(u)}{\mu \sup u - u}\right\|_\infty\right], \sqrt{8}\left\|\frac{F(u)}{\mu \sup u - u}\right\|_\infty\right\} (\mu \sup u - u)^2$$

where $\| \ \|_{\partial M}$ denotes the supremum norm on ∂M.

PROOF. Again, we consider the function defined by (1.4). If $x_o \in \partial M$ is a point where f attains its supremum. By the strong maximal principle

$$\frac{\partial f}{\partial \nu}(x_o) > 0 \tag{1.14}$$

Hence

$$\sum_i u_i u_{i\nu} + \frac{|\nabla u|^2 u_\nu}{\mu \sup u - u} > 0 \tag{1.15}$$

By a suitable choice of orthonormal frame $e_1, \ldots, e_n$ such that $e_n = \frac{\partial}{\partial \nu}$ and e_α are tangential to ∂M for $\alpha < n$, since $u|_{\partial M} \equiv 0$

$$u_\alpha = 0 \quad \text{for } \alpha < n \tag{1.16}$$

Direct computation then gives

$$(n-1)Hu_\nu = \sum_{\alpha<n} u_{\alpha\alpha} = \Delta u - u_{\nu\nu} \tag{1.17}$$

Therefore (1.15) becomes

$$u_\nu \Delta u - (n-1)Hu_\nu^2 + \frac{u^3}{\mu \sup u - u} > 0 \tag{1.18}$$

Since $u_\nu < 0$,

$$f^{3/2}(x_o) < -(n-1)Hf(x_o) - \frac{\Delta u}{\mu \sup u - u} f^{1/2}_{(x_o)} \tag{1.19}$$

Hence

$$f^{1/2}(x_o) < \max\left\{-2(n-1)H,\ 2\left\|\frac{F(u)}{\mu \sup u - u}\right\|^{1/2}_{\partial M}\right\} \tag{1.20}$$

This gives the first part of the theorem. On the other hand, if x_o is in the interior of M, theorem 1 gives the second half of the estimate.

THEOREM 3. Let M be a compact manifold with boundary. Suppose the principal curvatures on ∂M are non-negative (i.e. ∂M is convex). If u is a solution of (1.1) and (1.3), then for $\mu > 1$.

$$|\nabla u|^2 \leqslant \max\left\{4(n-1)\left[-(n-1)K - F_u + \left\|\frac{F(u)}{\mu \sup u - u}\right\|_\infty\right],\ \sqrt{8}\left\|\frac{F(u)}{\mu \sup u - u}\right\|_\infty\right\}(\mu \sup u - u)^2$$

PROOF. In view of theorem 1, it suffices to show that the function

$$f = \frac{|\nabla u|^2}{(\mu \sup u - u)^2}$$

attains its supremum in the interior of M.

Assuming the contrary if $x_o \in \partial M$ is the maximum point of f. Then

$$\sum_i u_i u_{i\nu} + \frac{|\nabla u|^2 u_\nu}{\mu \operatorname{sip} u - u} > 0 \tag{1.21}$$

If $h_{\alpha\beta}$ are the second fundamental form elements of ∂M, then by direct computation one shows that

$$u_{\alpha\nu} = -h_{\alpha\beta}u_\beta \qquad 1 \leqslant \alpha,\beta < n \tag{1.22}$$

where we used the fact that $u_\nu \equiv 0$ on ∂M. Together with (1.21), we have

$$-\sum_{\alpha,\beta} h_{\alpha\beta}u_\alpha u_\beta > 0$$

which is a contradiction to the convexity of ∂M. The theorem follows.

2. ANOTHER GRADIENT ESTIMATE. In this section, another form of gradient estimate for the solution of (1.1) is obtained. In the case when the Ricci curvature of M is non-negative, the estimates in the following theorems are much sharper than those of the previous section. However for general manifolds theorem 1-3 are more effective as will be demonstrated in the next section.

THEOREM 4. Let M be a compact manifold. Suppose u is a solution of (1.1). If α is any constant and $\beta^2 \geq \sup(\alpha+u)^2$, then

$$|\nabla u|^2 \leq \sup_{x\in M}\left[\frac{-(F_u + (n-1)K)(\beta^2 - (\alpha+u)^2) - (\alpha+u)F(u)}{\beta^2}\right] \times (\beta^2 - (\alpha+u)^2)$$

PROOF. Consider the function

$$g = \frac{|\nabla u|^2}{\beta^2 - (\alpha+u)^2} \tag{2.1}$$

where α = constant and $\beta^2 \geq \sup(\alpha+u)^2$. If $x_o \in M$ is a point where g achieves its supremum, then at x_o, g satisfies

$$0 = \frac{g_i}{2} = \frac{u_j u_{ji}}{\beta^2 - (\alpha+u)^2} + \frac{|\nabla u|^2(\alpha+u)u_i}{(\beta^2-(\alpha+u)^2)^2} \tag{2.2}$$

and

$$0 \geq \frac{\Delta g}{2} = \frac{u_{ji}^2 + u_j u_{jii}}{\beta^2 - (\alpha+u)^2} + \frac{4u_j u_{ji} u_i(\alpha+u)}{(\beta^2-(\alpha+u)^2)^2} + \frac{|\nabla u|^4}{(\beta^2-(\alpha+u)^2)^2} + \frac{|\nabla u|^2(\alpha+u)F(u)}{(\beta^2-(\alpha+u)^2)^2} + \frac{4|\nabla u|^4(\alpha+u)^2}{(\beta^2-(\alpha+u)^2)^3} \tag{2.3}$$

Substituting (2.2) and using the Ricci identity, we have

$$0 \geq \frac{u_{ji}^2}{\beta^2-(\alpha+u)^2} + \frac{F_u|\nabla u|^2}{\beta^2-(\alpha+u)^2} + \frac{(n-1)K|\nabla u|^2}{\beta^2-(\alpha+u)^2} + \frac{|\nabla u|^4}{(\beta^2-(\alpha+u)^2)^2} + \frac{|\nabla u|^2(\alpha+u)F(u)}{(\beta^2-(\alpha+u)^2)^2} \tag{2.4}$$

By picking suitable orthonormal frame and using (2.2) again

$$u_{ji}^2 \geq u_{11}^2 \geq \frac{|\nabla u|^4(\alpha+u)^2}{(\beta^2-(\alpha+u)^2)^2} \tag{2.5}$$

Hence

$$0 \geq g^2\frac{(\alpha+u)^2}{\beta^2-(\alpha+u)^2} + (F_u + (n-1)K)g + g^2 + \frac{(\alpha+u)F(u)}{\beta^2-(\alpha+u)^2}\, g \tag{2.6}$$

Therefore

$$-(F_u + (n-1)K)(\beta^2 - (\alpha + u)^2) - (\alpha + u)F(u) \geq \beta^2 g \tag{2.7}$$

and the theorem follows.

THEOREM 5. Let M be a compact manifold with boundary. Suppose u is a positive solution of (1.1) and (1.2). If $H \leq 0$ is the lower bound of the mean curvature of ∂M with respect to the outward normal $\frac{\partial}{\partial \nu}$, then either

$$|\nabla u|^2 \leq \sup_{x \in M} \left[\frac{-(F_u + (n-1)K)(\beta^2 - (\alpha + u)^2) - (\alpha + u)F(u)}{\beta^2} \right] \times (\beta^2 - (\alpha + u)^2)$$

or

$$|\nabla u|^2 \leq (n-1)^2 H^2 \left(\frac{\beta^2 - \alpha^2}{\alpha^2} \right) (\beta^2 - (\alpha + u)^2)$$

for any $\alpha \geq 0$ and $\beta^2 \geq \sup(\alpha + u)^2$.

PROOF. Again we consider the function g defined by (2.1). If the supremum point x_o of g is in the interior of M, the estimates of theorem 4 can be applied. If $x_o \in \partial M$, then

$$0 < \frac{1}{2} g_\nu (x_o) \tag{2.8}$$

Hence at x_o

$$0 < u_{\nu\nu} u_\nu + \frac{|\nabla u|^2 \alpha u_\nu}{\beta^2 - \alpha^2} \tag{2.9}$$

By (1.17)

$$0 < -(n-1)Hu_\nu^2 + \frac{|\nabla u|^2 \alpha u_\nu}{\beta^2 - \alpha^2} \tag{2.10}$$

Since u is a positive solution, this implies

$$-(n-1)H > \frac{\alpha |\nabla u|}{\beta^2 - \alpha^2} \tag{2.11}$$

Hence

$$(n-1)^2 H^2 > \frac{\alpha^2 |\nabla u|^2}{(\beta^2 - \alpha^2)^2} \tag{2.12}$$

Therefore

$$(n-1)^2 H^2 \left(\frac{\beta^2 - \alpha^2}{\alpha^2} \right) > \frac{|\nabla u|^2 (x_o)}{\beta^2 - \alpha^2} \geq g \tag{2.13}$$

which proves the theorem.

THEOREM 6. Let M be a compact manifold with boundary. Suppose u is a solution of (1.1) and (1.3). If ∂M is convex with respect to the outward normal, then

$$|\nabla u|^2 \leq \sup_{x \in M} \left[\frac{-(F_u + (n-1)K)(\beta^2 - (\alpha+u)^2) - (\alpha+u)F(u)}{\beta^2} \right] \times (\beta^2 - (\alpha+u)^2)$$

for any constant α and $\beta^2 \geq \sup(\alpha + u)^2$.

PROOF. Similar to theorem 3, we only need to show that the supremum of the function g occurs in the interior of M.

If not, say $x_0 \in \partial M$ is the supremum point of g. Then

$$(2.14) \qquad 0 < \frac{u_j u_{j\nu}}{\beta^2 - (\alpha+u)^2} + \frac{|\nabla u|^2 (\alpha+u) u_\nu}{(\beta^2 - (\alpha+u)^2)^2}$$

Since $u_\nu \equiv 0$ on ∂M, this gives

$$(2.15) \qquad \begin{aligned} 0 &< \sum_j u_j u_{j\nu} \\ &= - h_{\alpha\beta} u_\alpha u_\beta \qquad 1 \leq \alpha, \beta < n \end{aligned}$$

which is a contradiction to the convexity assumption of ∂M. Hence the estimate of theorem 4 can be applied.

3. APPLICATIONS AND EIGENVALUE ESTIMATES. We will give applications of the previous theorems to obtain lower bounds for the first non-zero eigenvalues of the Laplacian.

For manifold without boundary, we consider the eigenvalues of the equation

$$(3.1) \qquad \Delta u = - \lambda u$$

In the case if M has boundary ∂M, again we impose either the Dirichlet or Neumann condition:

$$(3.2) \qquad u \equiv 0 \quad \text{on} \quad \partial M$$

$$(3.3) \qquad \frac{\partial u}{\partial \nu} \equiv 0 \quad \text{on} \quad \partial M .$$

THEOREM 7. Let M be a compact manifold. Suppose d denotes the diameter of M. If λ_1 is the first non-zero eigenvalue of (3.1), then

$$\lambda_1 \geq \frac{\exp - [1 + (1 - 4(n-1)^2 d^2 K)^{1/2}]}{2(n-1)d^2} .$$

PROOF. If u is the first eigenfunction, by theorem 1

$$(3.4) \qquad |\nabla u|^2 \leq \max \left\{ 4(n-1) \left[-(n-1)K + \lambda_1 + \frac{\lambda_1}{\mu - 1} \right], \sqrt{8} \frac{\lambda_1}{\mu - 1} \right\} (\mu \sup u - u)^2$$

for any $\mu > 1$. Therefore

$$\frac{|\nabla u|}{\mu \sup u - u} \leq \left[4(n-1)\left(\frac{\mu\lambda_1}{\mu - 1} - (n-1)K\right)\right]^{1/2} \tag{3.5}$$

However since u satisfied

$$\int_M u = 0 \tag{3.6}$$

This implies that the nodal set N of u divides M into two parts. If $x \in M$ is the point where u achieves its supremum and γ be the shortest geodesic joining x and N, then γ has length at most d. Integrating (3.5) along γ, we have

$$\begin{aligned} \log \frac{\mu}{\mu - 1} &\leq \int_\gamma \frac{|\nabla u|}{\mu \sup u - u} \\ &\leq \left[4(n-1)\left(\frac{\mu\lambda_1}{\mu - 1} - (n-1)K\right)\right]^{1/2} d \end{aligned} \tag{3.7}$$

Hence

$$\frac{\mu - 1}{\mu}\left[\frac{1}{4(n-1)d^2}\left(\log \frac{\mu}{\mu - 1}\right)^2 + (n-1)K\right] \leq \lambda_1 \tag{3.8}$$

Clearly the left hand side can be made to be positive by choosing μ closed enough to 1. The theorem is then proved by maximizing (3.8) with

$$\frac{\mu}{\mu - 1} = \exp[1 + (1 - 4(n-1)^2 d^2 K)^{1/2}] .$$

THEOREM 8. Let M be a compact manifold with boundary. Suppose i denotes the inscribed radius of M, i.e. the radius of the biggest geodesic ball than can be fitted into M. If μ_1 is the first eigenvalue of (3.1) and (3.2), then

$$\mu_1 \geq \frac{1}{\gamma}\left[\frac{1}{4(n-1)i^2}(\log \gamma)^2 + (n-1)K\right]$$

where

$$\gamma = \max\left\{\exp[1 + (1 - 4(n-1)^2 i^2 K)^{1/2}],\ \exp[-2(n-1) \times Hi]\right\}$$

PROOF. It is well known that the first eigenfunction u of the Dirichlet boundary problem does not change sign. We may then assume u is non-negative and apply theorem 2. We get

$$|\nabla u| < -2(n-1)H(\mu \sup u - u) \tag{3.9}$$

or

$$|\nabla u|^2 \leq 4(n-1)\left(\frac{\mu\lambda}{\mu - 1} - (n-1)K\right)(\mu \sup u - u)^2 \tag{3.10}$$

If (3.9) holds, we have

$$\frac{|\nabla u|}{\mu \sup u - u} < -2(n-1)H \tag{3.11}$$

By integrating along the shortest geodesic γ joining $\sup u$ to ∂M, we have

$$\log \frac{\mu}{\mu - 1} < -2(n-1)Hi \tag{3.12}$$

Hence if we take μ such that

$$\frac{\mu}{\mu - 1} \geq e^{-2(n-1)Hi} \tag{3.13}$$

we get a contradiction. Hence (3.10) holds, and the estimate of theorem 7 follows for μ satisfying (3.13).

By applying theorem 3 instead, one can easily obtain lower estimate for λ_1 of the Neumann problem.

THEOREM 9. Let M be a compact manifold with convex boundary. (i.e. the principal curvatures of ∂M are non-negative). If λ_1 is the first non-zero eigenvalue of (3.1) and (3.3), then

$$\lambda_1 \geq \frac{\exp - [1 + (1 - 4(n-1)^2 d^2 K)^{1/2}]}{2(n-1)d^2}$$

PROOF. Similar to theorem 7.

For compact manifolds with non-negative Ricci curvature we apply the gradient estimates of section 2 to obtain the following sharp estimates for the first non-zero eigenvalues of (3.1), (3.2) and (3.3).

THEOREM 10. Let M be a compact manifold. Suppose λ_1 is the first non-zero eigenvalue of (3.1). Then

$$\lambda_1 + \max\{-(n-1)K, 0\} \geq \frac{\pi^2}{4d^2}$$

In particular, if the Ricci curvature of M is non-negative

$$\lambda_1 \geq \frac{\pi^2}{4d^2}$$

PROOF. Applying theorem 4, we have

$$\begin{aligned}|\nabla u|^2 &\leq \sup\left[\frac{(\lambda_1 - (n-1)K)(\beta^2 - (\alpha + u)^2)}{\beta^2} + \frac{\lambda_1 u(\alpha + u)}{\beta^2}\right](\beta^2 - (\alpha + u)^2) \\ &= \sup\left[\frac{\lambda_1(\beta^2 - \alpha(\alpha + u)) - (n-1)K(\beta^2 - (\alpha + u)^2)}{\beta^2}\right] \\ &\quad \times (\beta^2 - (\alpha + u)^2)\end{aligned} \tag{3.14}$$

By setting $\alpha = 0$ and $\beta^2 = (\sup u)^2$ (we may assume that $\sup u \geqslant |\inf u|$) gives

(3.15)
$$|\nabla u|^2 \leqslant [\lambda_1 + \max\{-(n-1)K,0\}] \times ((\sup u)^2 - u^2)$$

Hence

(3.16)
$$\frac{|\nabla u|}{\sqrt{(\sup u)^2 - u^2}} \leqslant [\lambda_1 + \max\{-(n-1)K,0\}]^{1/2}$$

Integrating along the shortest geodesic joining $\sup u$ and the nodal set of u yields

(3.17)
$$\frac{\pi}{2} \leqslant [\lambda_1 + \max\{-(n-1)K,0\}]^{1/2} d$$

This proves the theorem.

THEOREM 11. Let M be a manifold with boundary. Suppose μ_1 is the first eigenvalue of (3.1) and (3.2), then

$$\mu_1 + \max\{-(n-1)K,0\} \geqslant \frac{1}{i^2}\left[\frac{\pi}{2} - \sin^{-1}\left(\frac{-(n-1)H}{(\mu_1 + \max\{-(n-1)K,0\} + (n-1)^2H^2)^{1/2}}\right)\right]^2$$

where i = inscribe radius and $H \leqslant 0$ is the lower bound of the mean curvature of ∂M.

In particular, if M has non-negative Ricci curvature then

$$\mu_1 \geqslant \frac{1}{i^2}\left[\frac{\pi}{2} - \sin^{-1}\left(\frac{-(n-1)H}{\mu_1 + (n-1)^2H^2}\right)\right]^2$$

PROOF. By theorem 5, if we set $\alpha = \gamma \sup u$ and $\beta = (\gamma + 1)\sup u$, then either

(3.18)
$$g = \frac{|\nabla u|^2}{\beta^2 - (\alpha + u)^2} \leqslant (\mu_1 + \max\{-(n-1)K,0\}) \times \frac{2\gamma + 1}{(\gamma + 1)^2}$$

or

(3.19)
$$g \leqslant (n-1)^2H^2\left(\frac{2\gamma + 1}{\gamma^2}\right)$$

where u is the non-negative first eigenfunction. We claim that for appropriate choice of γ

$$g \leq \mu_1 + \max\{-(n-1)K, 0\}. \tag{3.20}$$

If not, since $\frac{2\gamma+1}{(\gamma+1)^2} \leq 1$, g must satisfied (3.19). Therefore, we have

$$(n-1)^2H^2\left(\frac{2\gamma+1}{\gamma^2}\right) \geq g > \mu_1 + \max\{-(n-1)K, 0\} \tag{3.21}$$

However if $\gamma = \left[\frac{(n-1)^2H^2}{\mu_1 + \max\{-(n-1)K,0\}}\left(\frac{(n-1)^2H^2}{\mu_1 + \max\{-(n-1)K,0\}} + 1\right)\right]^{1/2}$ $+ \frac{(n-1)^2H^2}{\mu_1 + \max\{-(n-1)K,0\}}$, it can be easily be checked that

$$(n-1)^2H^2(2\gamma+1) = \gamma^2(\mu_1 + \max\{-(n-1)K,0\})$$

which is a contradiction to (3.21). Hence for the above choice of γ, we have

$$\frac{|\nabla u|}{\sqrt{(\gamma+1)^2(\sup u)^2 - (\gamma \sup u - u)^2}} \leq (\mu_1 + \max\{-(n-1)K,0\})^{1/2} \tag{3.22}$$

Integrating from $\sup u$ to ∂M yields the theorem.

REMARK. If M is a compact domain in $\mathbb{R}^n$ with non-negative mean curvature, theorem 11 generalizes the result obtained by Hersch [7].

THEOREM 12. Let M be a manifold with convex boundary. Suppose λ_1 is the first non-zero eigenvalue of (3.1) and (3.3), then

$$\lambda_1 + \max\{-(n-1)K,0\} \geq \frac{\pi^2}{4d^2}$$

PROOF. Follows from theorem 6.

The gradient estimate of theorem 4 is in some sense best possible for compact manifold with non-negative Ricci curvature, as the next theorem will show.

THEOREM 13. Let M be a compact manifold with non-negative Ricci curvature. If u is an eigenfunction on M then

$$\|u\|_\infty^2 \geq \frac{2\|u\|_2^2}{V}$$

where V = volume of M. Equality holds iff the universal covering $\tilde{M}$ isometrically splits into $\mathbb{R}^k \times N$ and u is an eigenfunction on $\mathbb{R}^k$ alone.

PROOF. By setting $\alpha = 0$, $\beta = \|u\|_\infty$ and applying theorem 4, we have

$$|\nabla u|^2 \leq \lambda \left[\|u\|_\infty^2 - u^2\right] \tag{3.23}$$

Integrating both sides yields

$$2\lambda \; \|u\|_2^2 \leq \|u\|_\infty^2 V\lambda \tag{3.24}$$

Hence equality holds iff

$$|\nabla u|^2 = \lambda(\|u\|_\infty^2 - u^2) \tag{3.25}$$

Differentiating equation (3.25) gives

$$u_j u_{ji} = -\lambda u u_i \tag{3.26}$$

If one chooses suitable orthonormal frame, this yields

$$\begin{aligned} u_{11} &= -\lambda u \\ &= \Delta u \end{aligned} \tag{3.27}$$

and

$$u_1 = \nabla u$$

Covariant differentiating (3.26) in the ith direction and sum

$$\begin{aligned}(3.28)\qquad 0 &= u_{ji}^2 + u_j u_{jii} + \lambda u_i^2 + \lambda u u_{ii} \\ &\geqslant u_{11}^2 + (n-1)K|\nabla u|^2 - \lambda^2 u^2 \\ &= (n-1)K|\nabla u|^2 \qquad \text{(by 3.27)} \\ &\geqslant 0\end{aligned}$$

This implies that equality holds on each step of (3.28). Hence the Ricci curvature vanishes along ∇u and $u_{ij} = 0$ unless $i = j = 1$.

We define a 1-form ω to be the dual of $(\text{sgn } u)\frac{\nabla u}{|\nabla u|}$. We claim that ω can be defined smoothly on M. In fact, since $u_{11} \equiv \Delta u$, this means $e_1 = (\text{sgn } u)\frac{\nabla u}{|\nabla u|}$ is well-defined up to a sign even at the critical points of u. Hence ω is a smooth 1-form.

Consider the equation

$$\begin{aligned}(3.29)\qquad (\omega,\Delta\omega) &= -\tfrac{1}{2}\Delta|\omega|^2 + |\nabla\omega|^2 \quad (\text{since } R_{11} \equiv 0) \\ &= |\nabla\omega|^2 \qquad (|\omega|^2 \equiv 1)\end{aligned}$$

Clearly $u_{ij} = 0$ when $i \neq 1$ or $j \neq 1$ implies

$$(3.30)\qquad |\nabla\omega|^2 = 0$$

Hence ω is a parallel harmonic 1-form on M. By the splitting theorem of Cheeger and Gromoll [4], $\tilde{M}$ splits into $\mathbb{R}^k \times N$. The rest of the theorem follows trivially.

4. UPPER BOUNDS FOR λ_m

Given any compact manifold of dimension n, the well-known Weyl formula gives the asymptotic relationship between the number of eigenvalues less than any given number T and the number itself. If we take the increasing ordering of eigenvalues $\{0 < \lambda_1 \leqslant \lambda_2 \leqslant \cdots\}$ then the Weyl formula can be written as

$$\lambda_m \sim \left(\frac{m}{V}\right)^{\frac{2}{n}} \times 4\pi\Gamma\left(\frac{n}{2}+1\right)^{\frac{2}{n}}$$

The purpose of this section is to give upper bounds of λ_m with the right order in $\frac{m}{V}$ which corresponds to the Weyl formula. For general Riemannian manifold, we obtain

$$\lambda_m \leqslant C_1 + C_2\left(\frac{m+1}{V}\right)^{\frac{2}{n}}$$

where C_1 and C_2 are constants depending only on n, d = diameter of M and K = lower bound of Ricci curvature of M. When M is a manifold of non-negative Ricci curvature, we have

$$\lambda_m \leqslant \overline{C}\left(\frac{m+1}{V}\right)^{\frac{2}{n}} \times 4\pi\Gamma\left(\frac{n}{2}+1\right)^{\frac{2}{n}}$$

where $\overline{C}$ = constant depending on n along. Moreover $\overline{C}(n)$ is bounded above for all $n \in \mathbb{Z}^+$.

COVERING LEMMA. Let M be a compact manifold. For each $0 < r \leqslant d$, there exists a collection of geodesic balls $\{B_i(r)\}_{i=1}^{k(r)}$ such that they satisfied the following properties

(i) $\bigcup_i B_i(r)$ covers M

(ii) $B_j(\frac{r}{2}) \cap B_i(\frac{r}{2})$ is a set of measure zero for $i \neq j$

(iii) for $y \in M$, there are at most

$$P = 3\left(2\,\frac{\sinh\sqrt{-K}\,r}{\sqrt{-K}\,r}\right)^{n}$$

balls which contained y, where $\kappa \leq 0$ is the lower bound of the sectional curvature of M.

REMARK. If M is non-negatively curved, $P \leq 3 \cdot 2^n$.

PROOF. For any $0 < r \leq d$, there exists a maximal set of points $\{x_i\}_{i=1}^{k(r)}$ such that they are mutually of distance at least r apart. Property (i) follows trivially from the maximality of $\{x_i\}_{i=1}^{k(r)}$ if we take $\{B(x_i),r\}_{i=1}^{k(r)}$ to be our collection. The fact that all the x_i's are of at least distance r apart gives property (ii). For $y \in M$, the number of balls which contained y is the same as the number of x_i's which are in $B(y,r)$. By the comparison theorem and lifting $B(y,r)$ to the tangent space at y, it reduces to counting the maximal number of points with euclidean distance

$$\frac{r\sqrt{-\kappa}}{\sinh r\sqrt{-\kappa}}$$

apart that can be fitted into the euclidean ball of radius 1. A simple counting argument gives (iii).

REMARK. By (i) and (ii) one obtains the following estimates of $k(r)$

$$\begin{aligned} V(M) &\leq \sum_{i=1}^{k(r)} V(B_i(r)) \\ &\leq k(r) \max\{V(B_i(r))\} \\ &\leq k(r)\alpha(n-1)\int_0^r [(-K)^{-1/2}\sinh\sqrt{-K}\, t]^{n-1}dt \end{aligned} \tag{4.1}$$

and

$$\begin{aligned} V(M) &\geq \sum_{i=1}^{k(r)} V(B_i(\tfrac{r}{2})) \\ &\geq k(r) \min\{B_i(\tfrac{r}{2})\} \\ &\geq k(r)\alpha(n-1)(\sqrt{T})^{1-n}\int_0^{r/2} (\sin(\sqrt{T}\, t))^{n-1}dt \end{aligned} \tag{4.2}$$

where $\alpha(n-1)$ denotes the volume of the unit $n-1$ sphere in $\mathbb{R}^n$, and τ = upper bound of the sectional curvature of M.

Clearly $k(r)$ is a non-increasing function of r. In general, $k(r)$ is not a surjection onto $\mathbb{Z}^+$ = positive integers. However the following lemmas show that it is not far from being surjective.

LEMMA 14. For $\varepsilon > 0$, and $m \in \mathbb{Z}^+$, there exists a collection of geodesic balls $\{B_i(r_i)\}_{i=1}^m$ which satisfied:

(i) $\bigcup B_i(r_i) - M$ is a set of measure less than ε.

(ii) $B_j(\frac{r_j}{2}) \cap B_i(r_i/2)$ is a set of measure zero for $i \neq j$

(iii) there exists an r, such that $r \leqslant r_i \leqslant r + \varepsilon$ for all $1 \leqslant i \leqslant m$.

PROOF. In view of lemma 14, we know that for δ small enough, there exists $r > 0$ such that $k(r) = m_1 < m < m_2 = k(r - \delta)$. Consider $\{B_i(r)\}_{i=1}^m$ the collection which satisfied property (i-iii) of the covering lemma. We may assume that $B_1(r-\delta) \bigcup_{i=2}^{m} B_i(r)$ does not cover M. Let

$$x \in M - \{B_1(r-\delta) \bigcup_{i=2}^{m} B_i(r) \subset B_1(r) - B_1(r-\delta).$$

Clearly the collection $\{B(x,r-\delta), B_1(r-\delta)\} \cup \{B_i(r)\}_{i=2}^m$ satisfies the required properties. Inductively by shrinking the balls $B_i(r)$ by δ and adding balls of radius $r - \delta$, we can obtain m balls which satisfied the conclusion of the lemma.

LEMMA 15. For $\varepsilon > 0$, and $m \in \mathbb{Z}^+$, there exists a collection of geodesic balls $\{B_i(r_i)\}_{i=1}^m$ of which satisfied

(i) $\bigcup B_i(r_i)$ covers M

(ii) $B_j(\frac{r_j}{2}) \cap B_i({}^{r}i/2)$ is a set of measure less than ε for $i \neq j$

(iii) there exists an r, such that $r \leqslant r_i \leqslant r + \varepsilon$ for all $1 \leqslant i \leqslant m$

(iv) for $y \in M$, the number of $B_i(r_i)$'s which contained y is at most

$$P = 3\left(2\,\frac{\sinh((r+\varepsilon)\sqrt{-K})}{(r+\varepsilon)\sqrt{-K}}\right)^n$$

PROOF. We proceed the same way as in lemma 14. Instead we take $\{B(x,r)\} \cup \{B_i(r)\}_{i=1}^m$ to be our new collection. It is easy to check that properties (i-iv) are satisfied. Again by adding more balls inductively, the lemma follows.

We are now ready to give some estimates on the eigenvalues.

THEOREM 16. Let M be a compact manifold. If λ_m is the mth non-zero eigenvalue of the Laplacian on M, then

$$\lambda_m \leq \begin{cases} \dfrac{(2\beta+1)^2}{4}(-K) + 4(1+2^\beta)^2\pi^2\left(\dfrac{\sinh\sqrt{-K}\,d}{\sqrt{-K}\,d}\right)^{\frac{2n-2}{n}} \\ \qquad \times\left[(m+1)\,\dfrac{\alpha(n-1)}{n}\,\dfrac{1}{V(M)}\right]^{\frac{2}{n}} \\ \text{when } n = 2(\beta+1),\quad \beta = 0,1,2,\ldots \\ \dfrac{(2\beta+2)^2}{4}(-K) + 4(1+\pi^2)(1+2^{2\beta})^2\left(\dfrac{\sinh\sqrt{-K}\,d}{\sqrt{-K}\,d}\right)^{\frac{2n-2}{n}} \\ \qquad \times\left[(m+1)\,\dfrac{\alpha(n-1)}{n}\,\dfrac{1}{V(M)}\right]^{\frac{2}{n}} \\ \text{when } n = 2\beta+3,\quad \beta = 0,1,2,\ldots \end{cases}$$

where $K \leq 0$ is the lower bound of the Ricci curvature of M.

PROOF. Let $\{B_i(r_i)\}_{i=1}^{m+1}$ be the collection of balls which satisfied lemma 14. Consider φ_i, the first eigenfunctions of the Dirichlet boundary problem on the $B_i(r_i)$. If u_α, $0 \leq \alpha \leq m-1$, are the first m eigenfunctions on M (including the constant function), then by the variational principle

$$\lambda_m = \inf_{f \perp u_\alpha} \frac{\int|\nabla f|^2}{\int f^2} \tag{4.3}$$

where inf is taken over all f orthogonal to u_α, $0 \leq \alpha \leq m-1$.

By the essential disjointness of $B_i(\frac{r_i}{2})$, the set of functions φ_i

on M are linearly independent. Hence there exists $0 \neq (a_i)_{i=1}^{m+1} \in \mathbb{R}^{m+1}$, such that

$$\varphi = \Sigma a_i \varphi_i \tag{4.4}$$

is orthogonal to u_α, for all α.

Therefore

$$\lambda_m \leqslant \frac{\int |\nabla\varphi|^2}{\int \varphi^2} \tag{4.5}$$

But

$$\begin{aligned} \int |\nabla\varphi|^2 &= \Sigma \int a_i^2 |\nabla\varphi_i|^2 \\ &= \Sigma\, \mu_1(B_i)\, a_i \int \varphi_i^2 \\ &\leqslant \max \mu_1(B_i) \int \varphi^2 \end{aligned}$$

where $\mu_1(B_i)$ denotes the first eigenvalue of the Dirichlet problem on $B_i(\frac{r_i}{2})$. This gives

$$\lambda_m \leqslant \max \mu_1(B_i) \tag{4.6}$$

However by the monotonuity of μ_1 and a theorem of Cheng [5]

$$\lambda_m \leqslant \mu_1(B(K,\tfrac{r}{2})) \tag{4.7}$$

$$\leqslant \begin{cases} \dfrac{(2\beta+1)^2}{4}(-K) + \dfrac{4(1+2^\beta)^2\pi^2}{r^2} \\ \text{when } n = 2(\beta+1), \quad \beta = 0,1,2,\ldots \\ \dfrac{(2\beta+2)^2}{4}(-K) + \dfrac{4(1+\pi^2)(1+2^{2\beta})^2}{r^2} \\ \text{when } n = 2\beta+3, \quad \beta = 0,1,2,\ldots \end{cases}$$

On the other hand, by (4.1)

$$
\begin{aligned}
(4.8)\qquad V(M) &\leqslant (m+1)\alpha(n-1)(-K)^{\frac{1-n}{2}} \int_0^{r+\varepsilon} (\sinh\sqrt{-K}\,t)^{n-1}dt \\
&\leqslant (m+1)\alpha(n-1)(-K)^{\frac{-n}{2}} \int_0^{\sqrt{K}\,(r+\varepsilon)} \left(\frac{\sinh\sqrt{-K}\,(r+\varepsilon)}{\sqrt{-K}\,(r+\varepsilon)}\,t\right)^{n-1}dt \\
&= (m+1)\alpha(n-1)(-K)^{\frac{-2n+1}{2}} \left(\frac{\sinh\sqrt{-K}\,(r+\varepsilon)}{r+\varepsilon}\right)^{n-1} \times \frac{[\sqrt{-K}\,(r+\varepsilon)]^n}{n} \\
&= (m+1)\frac{\alpha(n-1)}{n}(r+\varepsilon)(\sinh\sqrt{-K}\,(r+\varepsilon))^{n-1}(-K)^{\frac{-n+1}{2}} \\
&\leqslant (m+1)\frac{\alpha(n-1)}{n}(r+\varepsilon)^n \left(\frac{\sinh\sqrt{-K}\,d}{\sqrt{-K}\,d}\right)^{n-1}
\end{aligned}
$$

Hence

$$
(4.9)\qquad \frac{1}{r^2} \leqslant \frac{(r+\varepsilon)^2}{r^2}\left[(m+1)\,\frac{\alpha(n-1)}{n}\,\frac{1}{V(M)}\right]^{\frac{2}{n}} \left(\frac{\sinh\sqrt{-K}\,d}{\sqrt{-K}\,d}\right)^{\frac{2n-2}{n}}
$$

Together with (4.6)

$$
\lambda_m \leqslant \begin{cases}
\leqslant \dfrac{(2\beta+1)^2}{4}(-K) + 4(1+2^{\beta})^2\pi^2\,\dfrac{(r+\varepsilon)^2}{r^2}\left(\dfrac{\sinh\sqrt{-K}\,d}{\sqrt{-K}\,d}\right)^{\frac{2n-2}{n}} \\
\qquad \times\left[(m+1)\,\dfrac{\alpha(n+1)}{n}\,\dfrac{1}{V(M)}\right]^{\frac{2}{n}} \\
\text{when } n = 2(\beta+1),\quad \beta = 0,1,2,\ldots \\
\text{and} \\
\dfrac{(2\beta+2)^2}{4}(-K) + \dfrac{4(1+\pi^2)(1+2^{2\beta})^2(r+\varepsilon)^2}{r^2} \\
\qquad \times\left(\dfrac{\sinh\sqrt{-K}\,d}{\sqrt{-K}\,d}\right)^{\frac{2n-2}{n}}\left[(m+1)\dfrac{\alpha(n-1)}{n}\,\dfrac{1}{V(M)}\right]^{\frac{2}{n}}. \\
\text{when } n = 2\beta+3,\quad \beta = 0,1,2,\ldots
\end{cases}
$$

Letting $\varepsilon \to 0$, we obtain the desired result.

When $K = 0$, we can obtain a sharper estimate on λ_m.

THEOREM 17. Let M be a compact manifold of non-negative Ricci curvature. Then

$$\lambda_m \leqslant (n+4)n^{1-\frac{2}{n}} \left[\frac{m+1}{V}\alpha(n-1)\right]^{\frac{2}{n}}$$

PROOF. As in theorem 16, we have

$$\lambda_m \leqslant \mu_1(B(0,\tfrac{r}{2}))$$

However, it is known [5] that

$$\mu_1(B(0,\tfrac{r}{2})) \leqslant \frac{n(n+4)}{r^2}$$

Similarly to (4.7)

$$V(M) \leqslant (m+1)\max\{V(B(x_i, r+\varepsilon)\}$$

$$\leqslant \frac{(m+1)\alpha(n-1)}{n}(r+\varepsilon)^n$$

Hence

$$\lambda_m \leqslant n(n+4)\left[\frac{(m+1)\alpha(n-1)}{nV}\right]^{\frac{2}{n}}$$

5. LOWER BOUNDS FOR λ_m

In this section we will obtain lower bounds for λ_m with the right order in $\frac{m}{V}$. The first part is devoted to estimate higher eigenvalues. For m big enough, we have

$$\lambda_m \geqslant C_3\left(\frac{m}{V}\right)^{\frac{2}{n}}$$

where C_3 depends only on K, d, upper bound of δ = radius of convexity, and κ = lower bound of the sectional curvature of M. When M has non-negative sectional curvature we show that

$$\lambda_m \geq \bar{\bar{C}} \left(\frac{m}{V}\right)^{\frac{2}{n}}$$

where $\bar{\bar{C}}$ is a constant depending on n alone.

For general $m \geq 2$, we employ a method of the second author [15] to obtain lower bounds for λ_m, namely

$$\lambda_m \geq C_4 \left(\frac{m}{V}\right)^{\frac{2}{n}}$$

However C_4 now depends on n, δ, d, k and V.

THEOREM 18. Let M be a compact manifold of dimension n. Suppose λ_m is the mth non-zero eigenvalue of M. If

$$m \geq \frac{V(\sqrt{\tau})^{n-1}}{\alpha(n - 1)\int_0^{\delta/2} (\sin(\tau t))^{n-1} dt}$$

where τ = upper bound of the sectional curvature of M and δ = radius of convexity. Then

$$\lambda_m \geq \frac{\exp -[1 + (1 - 16(n - 1)^2 d^2 \kappa)^{1/2}]}{32(n - 1)p} \left(\frac{\alpha(n - 1)}{n}\right)^{\frac{2}{n}} \left(\frac{m}{V}\right)^{\frac{2}{n}}$$

where

$$P = 3\left(2 \frac{\sinh \delta\sqrt{-\kappa}}{\delta\sqrt{-\kappa}}\right)^n$$

with $0 \geq \kappa$ = lower bound of the sectional curvature of M.

PROOF. By (4.2) and lemma 15, it is clear that if

$$(5.1)\qquad m \geqslant \frac{V}{\alpha(n-1)(\sqrt{\tau})^{1-n}\int_0^{\delta/2}(\sin(\sqrt{\tau}\, t))^{n-1}\, dt}$$

and for any $\varepsilon > 0$, there exists a collection of geodesic balls $\{B_i(r_i)\}_{i=1}^m$ which satisfied properties (i-iv) of lemma 15. Moreover $r_i \leqslant r + \varepsilon \leqslant \delta$ for all $i = 1,\ldots, m$.

The maximum principle of λ_m gives

$$(5.2)\qquad \lambda_m = \max_{\{\varphi_i\}_{i=1}^m} \min_{f \perp \varphi_i} \frac{\int |\nabla f|^2}{\int f^2}$$

where we maximize over any set of m functions on M and minimize over all function f which are perpendicular to φ_i, $1 \leqslant i \leqslant m$. In particular, if we take φ_i to be the charateristic functions on $B_i(r_i)$, then the conditions $f \perp \varphi_i$ means

$$(5.3)\qquad \int_{B_i(r_i)} f = 0$$

Hence

$$(5.4)\qquad \lambda_m \geqslant \min_{f \perp \varphi_i} \frac{\int |\nabla f|^2}{\int f^2}$$

However if $\lambda_1(B_i)$ denotes the first non-zero eigenvalue of the Neumann problem on $B_i(r_i)$ then (5.3) implies

$$(5.5)\qquad \int_{B_i(r_i)} |\nabla f|^2 \geqslant \lambda_1(B_i) \int_{B_i(r_i)} f^2$$

Summing up both sides of (5.5) over i, we have

$$(5.6)\qquad \sum_{i=1}^{m} \int_{B_i(r_i)} |\nabla f|^2 \geq \min\{\lambda_1(B_i)\} \sum_{i=1}^{m} \int_{B_i(r_i)} f^2$$

$$\geq \min\{\lambda_1(B_i)\} \int_M f^2$$

by property (i) of lemma 15. On the other hand, property (iv) gives

$$(5.7)\qquad P\int_M |\nabla f|^2 \geq \sum_{i=1}^{m} \int_{B_i(r_i)} |\nabla f|^2$$

Therefore

$$(5.8)\qquad \lambda_m \geq \frac{\min\{\lambda_1(B_i)\}}{P}$$

Applying theorem 9, we have

$$(5.9)\qquad \lambda_1(B_i) \geq \frac{\exp - [1 + (1 - 16(n-1)^2 r_i^2 K)^{1/2}]}{8(n-1)r_i^2}$$

$$\geq \frac{\exp - [1 + (1 - 16(n-1)^2 (r+\varepsilon)^2 K)^{1/2}]}{8(n-1)(r+\varepsilon)^2}$$

by property (iii). However (4.2) and property (ii) gives

$$(5.10)\qquad V \geq m\alpha(n-1)\frac{1}{(\sqrt{\tau})^{n-1}} \int_0^{\frac{r}{2}} (\sin(\sqrt{\tau}\, t))^{n-1}\, dt - m\varepsilon$$

$$\geq m\left[\alpha(n-1)\frac{1}{(\sqrt{\tau})^{n-1}} \int_0^{\frac{r}{2}} \left(\frac{2\sin\left(\frac{r\sqrt{\tau}}{2}\right)}{t}\, t\right)^{n-1} dt - \varepsilon\right]$$

since $r \leqslant \delta$ = radius of convexity

$$= m\left[\alpha(n-1)\left(\frac{2\sin\left(\frac{r\sqrt{\tau}}{2}\right)}{r\sqrt{\tau}}\right)^{n-1}\left(\frac{r}{2}\right)^n\frac{1}{n} - \varepsilon\right]$$

$$\geqslant m\left[\frac{\alpha(n-1)}{n2^n}\,r^n - \varepsilon\right]$$

Combining (5.8), (5.9) and (5.10) yields

$$\lambda_m \geqslant \frac{\exp - [1 + (1 - 16(n-1)^2(r+\varepsilon)^2K)^{1/2}]}{8(n-1)p} \tag{5.11}$$

$$\times \left(\frac{r}{r+\varepsilon}\right)^2 \frac{1}{4}\left(\frac{m\alpha(n-1)}{n}\right)^{\frac{2}{n}} \frac{1}{(V + m\varepsilon)^{\frac{2}{n}}}$$

Theorem 18 is proved by letting $\varepsilon \to 0$.

THEOREM 19. Let M be a compact manifold of non-negative Ricci curvature. If

$$m \geqslant \frac{V(\sqrt{\tau}\,)^{n-1}}{\alpha(n-1)\int_0^{\delta/2}(\sin(\sqrt{\tau}\,t))^{n-1}\,dt}$$

then

$$\lambda_m \geqslant \frac{\pi^2}{3\cdot 2^{n+6}}\left(\frac{\alpha(n-1)}{n}\right)^{\frac{2}{n}}\left(\frac{m}{V}\right)^{\frac{2}{n}}$$

PROOF. We follow the idea of the proof of theorem 18. The non-negative Ricci curvature assumption allows us to apply theorem 12, hence we can replace (5.9) by

$$\lambda_1(B_i) \geqslant \left(\frac{\pi}{4(r+\varepsilon)}\right)^2 \tag{5.12}$$

and the theorem follows.

COROLLARY 20. Let M be a compact manifold of non-negative sectional curvature. If

$$m \geqslant \frac{V(\sqrt{\tau})^{n-1}}{\alpha(n-1)\int_0^{\delta/2}(\sin(\tau t))^{n-1}dt}$$

then

$$\lambda_m \geqslant \frac{\pi^2}{64 \cdot 3 \cdot 2^n}\left(\frac{\alpha(n-1)}{n}\right)^{\frac{2}{n}}\left(\frac{m}{V}\right)^{\frac{2}{n}}$$

For lower eigenvalues λ_m, the task of getting a lower bound becomes more difficult since theorem 9 and theorem 12 applied only to manifolds with convex boundaries. However utilizing the method of Yau [15], we can avoid the difficulty of obtaining a lower bound for the first eigenvalue of the Neumann problem.

Let M be a compact manifold, consider the ball of radius $r \leqslant d$, $B_p(r)$, centered at a given point $p \in M$. We define the isoperimetric constant $\Pi(B_p(r))$ by

$$\Pi(B_p(r)) = \inf\left[\frac{A(\partial M_1 \cap \partial M_2 \cap B_p(r))}{\min\{V(M_1 \cap B_p(\frac{r}{2})), V(M_2 \cap B_p(\frac{r}{2}))\}}\right]$$

where inf is taken over all decompositions $B_p(\frac{r}{2}) = M_1 \cup M_2$ with $V(M_1 \cap M_2) = 0$. This constant $\Pi(B_p(r))$ is similar to the constant $I(M)$ defined by Yau [15, p. 499]. Moreover, theorem 4 and corollary 1 in [15] can also be generalized to $\Pi(B_p(r))$ with the appropriate modification in the proofs and the statements. We will give their analogue for $\Pi(B_p(r))$ in the next two propositions, however since their proofs follow directly as in [15] and will be omitted.

PROPOSITION 21. $\Pi(B_p(r)) = \inf\left[\dfrac{\int_{B_p(r)}|\nabla f|}{\inf_{\beta\in\mathbb{R}}\int_{B_p(\frac{r}{2})}|f-\beta|}\right]$

where inf is taken over all functions $f \in H_{1,1}(B_p(r))$

PROPOSITION 22. For any $f \in H_{1,2}(B_p(r))$, we have

$$\int_{B_p(r)} |\nabla f|^2 \geq \frac{\Pi\,(B_p(r))^2}{4} \int_{B_p(\frac{r}{2})} (f - k)^2$$

for any $k \in \mathbb{R}$ satisfying

$$V(B_p(\tfrac{r}{2}) \cap \{x | f(x) \geq k\}) \geq \tfrac{1}{2} V(B_p(\tfrac{r}{2}))$$

and

$$V(B_p(\tfrac{r}{2}) \cap \{x | f(x) \leq k\}) \geq \tfrac{1}{2} V(B_p(\tfrac{r}{2}))$$

In particular, if $\int_{B_p(\frac{r}{2})} f = 0$ then

$$\int_{B_p(r)} |\nabla f|^2 \geq \frac{\Pi\,(B_p(r))^2}{4} \int_{B_p(\frac{r}{2})} f^2$$

The following proposition is essentially theorem 8 of [15] together with proposition 21.

PROPOSITION 23. Let M be a compact manifold. Consider the ball $B_p(r) \subset M$ with $r \leq d$. Define $G = \sup_\sigma \exp \left[\int_0^\ell \left(- \frac{1}{t^2} \int_0^t K(\sigma^1(s)) s^2 ds\right) dt\right]$ where σ ranges over all minimal geodesic segments with length $\ell \leq r$ and $\sigma(0) \in B_p(\frac{r}{2})$. Then for all $f \in H_{1,1}(B_p(r))$,

$$\Pi\,(B_p(r)) \geq \frac{nV(B_p(\frac{r}{2}))}{2r^{n+1}\alpha(n - 1)G}$$

Combining the above propositions, we can show that for any

$f \in H_{1,2}(B_p(r))$ satisfying $\int_{B_p(\frac{r}{2})} f = 0$ then

$$\int_{B_p(r)} |\nabla f|^2 \geq \left(\frac{nV(B_p(\frac{r}{2}))}{4r^{n+1}\alpha(n-1)G}\right)^2 \int_{B_p(\frac{r}{2})} f^2 . \tag{5.13}$$

The next lemma enables us to estimate the volume of $B_p(\frac{r}{2})$ from below.

LEMMA 24. Let M be a compact manifold. The ball $B_p(\frac{r}{2})$ has volume bounded below by either

$$V(B_p(\tfrac{r}{2})) \geq \frac{V(M)}{2}$$

or

$$V(B_p(\tfrac{r}{2})) \geq \frac{r^n C_o}{(2n)^n}$$

where C_o is the isoperimetric constant defined by

$$C_o = \inf \frac{A(N)^n}{\min\{V(M_1), V(M_2)\}^{n-1}}$$

with $\inf$ taken over all codimension-1 submanifold N which divides M into M_1 and M_2.

PROOF. Suppose $V(B_p(\frac{r}{2})) \leq \frac{V(M)}{2}$. In particular $V(B_p(t)) \leq \frac{V(M)}{2}$ for all $t \leq \frac{r}{2}$. Hence by the definition of C_o

$$\frac{A(\partial B_p(t))}{V(B_p(t))^{\frac{n-1}{n}}} \geq C_o^{\frac{1}{n}} \tag{5.14}$$

However it is well known that $A(\partial B_p(t)) = \frac{d}{dt}V(B_p(t))$. Therefore integrating (5.14) gives

$$nV(B_p(\tfrac{r}{2}))^{\frac{1}{n}} = \int_o^{\frac{r}{2}} \frac{\frac{d}{dt}V(B_p(t))}{V(B_p(t))^{\frac{n-1}{n}}}\, dt \geq \frac{rC_o^{\frac{1}{n}}}{2} \tag{5.15}$$

The lemma follows.

THEOREM 25. Let M be a compact manifold. Suppose λ_m is the mth non-zero eigenvalue of M with $m \geqslant 2$. Then

$$\lambda_m \geqslant C_5 \left(\frac{m}{V}\right)^{\frac{2}{n}}$$

where C_5 is a constant bounded below by

$$C_5 \geqslant \frac{\min\{\frac{2\delta}{d}, 1\}^{2n+2}\, \alpha(n-1)^{\frac{2}{n}} (d\sqrt{-k})^n}{3 \cdot 2^{6n+10} (\sinh(d\sqrt{-k}))^n n^{\frac{2}{n}} G^2}$$

or

$$C_5 \geqslant \frac{\min\{\frac{2\delta}{d}, 1\}^2 (d\sqrt{-k})^n C_o^2}{3 \cdot 2^{4n+8} (\sinh(d\sqrt{-k}))^n n^{2n-2+\frac{2}{n}} \alpha(n-1)^{2-\frac{2}{n}} G^2}$$

where C_o is the isoperimetric constant defined in lemma 24 and

$$G = \sup_\sigma \exp\left[\int_o^d \left(-\frac{1}{t^2}\int_o^t K(\sigma'(s))s^2 ds\right) dt\right]$$

PROOF. For $m \in \mathbb{Z}^+$, let $\{B_i(r_i)\}_{i=1}^m$ be a collection of geodesic balls which satisfied properties (i-iv) of lemma 15. In view of the proof of theorem 18 and 19 where we obtain the estimates for λ_m with $r \leqslant r_i \leqslant r + \varepsilon$ and by letting $\varepsilon \to 0$ in the final process, we may hence assume $r_i = r$ for $1 \leqslant i \leqslant m$.

If we take φ_i to be charateristic functions on $B_i(r)$, equation (5.2) gives

$$\lambda_m \geqslant \min_{f \perp \varphi_i} \frac{\int |\nabla f|^2}{\int f^2} \tag{5.16}$$

However proposition 23 and lemma 24 imply either

$$(5.17)\qquad \int_{B_i(2r)} |\nabla f|^2 \geq \left(\frac{nV}{8(2r)^{n+1}\alpha(n-1)G}\right)^2 \int_{B_i(r)} f^2$$

or

$$(5.18)\qquad \int_{B_i(2r)} |\nabla f|^2 \geq \left(\frac{C_o}{2^{n+3}n^{n-1}\alpha(n-1)rG}\right)^2 \int_{B_i(r)} f^2$$

Summing over i on both sides gives

$$(5.19)\qquad \bar{p}\int_M |\nabla f|^2 \geq \xi \int_M f^2$$

where

$$\xi = \min\left\{\left(\frac{nV}{2^{n+4}r^{n+1}\,\alpha(n-1)G}\right)^2, \left(\frac{C_o}{2^{n+3}n^{n-1}\,\alpha(n-1)rG}\right)^2\right\}$$

and

$$\bar{p} = 3\left(4\frac{\sinh d\sqrt{-K}}{d\sqrt{-K}}\right)^n$$

$\geq$ maximal number of $B_i(2r)$ that contained any given point in M

On the other hand by (4.2) and property (ii) of lemma 15

$$(5.20)\qquad V \geq m\alpha(n-1)\frac{1}{(\sqrt{\tau})^{n-1}}\int_o^{\frac{r}{2}} (\sin(\sqrt{\tau}\, t))^{n-1}\, dt$$

Since $r \leq d$, in the case when $2\delta \geq d$ we have $\delta \geq \frac{r}{2}$. If $2\delta < d$, then $\frac{r}{2} \geq \frac{r\delta}{d}$.

Therefore

$$(5.21)\qquad V \geq m\alpha(n-1)\frac{1}{(\sqrt{\tau})^{n-1}}\int_0^{\frac{r}{2}\times\min\{2\frac{\delta}{d},1\}}(\sin(\sqrt{\tau}))^{n-1}\,dt$$

$$\geq m\alpha(n-1)\frac{(\min\{\frac{2\delta}{d},1\})^n}{n2^n}r^n$$

$$\left(\text{since } \frac{r}{2}\times\min\{\frac{2\delta}{d},1\}\leq\delta\right)$$

Combining with (5.16) and (5.19), we have

$$\lambda_m \geq \frac{\xi}{\bar{p}}$$

where $\bar{p} = 3\left(4\frac{\sinh d\sqrt{-\kappa}}{d\sqrt{-\kappa}}\right)^n$

and

$$\xi \geq \min\left\{\frac{m^{2+\frac{2}{n}}\min\{\frac{2\delta}{d},1\}^{2n+2}\,\alpha(n-1)^{\frac{2}{n}}}{V^{\frac{2}{n}}n^{\frac{2}{n}}\,2^{4n+10}G^2},\ \frac{M^{\frac{2}{n}}\min\{\frac{2\delta}{d},1\}^2C_o^2}{V^{\frac{2}{n}}n^{2n-2+\frac{2}{n}}\,2^{2n+8}\,\alpha(n-1)^{2-\frac{2}{n}}G^2}\right\}$$

The theorem follows

COROLLARY 26. Let M be a compact manifold. Then

$$\lambda_m \geq C_4\left(\frac{m}{V}\right)^{\frac{2}{n}}$$

where C_4 = constant depending on n, δ, d, k, and V.

PROOF. This follows from theorem 25, plus the fact that $K \geq \kappa$, and a result of Croke [6] which gives a lower bound for C_o, namely

$$C_o \geqslant \frac{V^{n+1}}{4\alpha(n)^{n-1}\,\alpha(n-1)\left(\int_o^d (\sqrt{-K}^{-1}\sinh\sqrt{-K}\,t)^{n-1}\,dt\right)^{n+1}}$$

References

[1] T. Aubin, Fonction de Green et valeurs propres du laplacian (J. Math. Pure et Appl., 1974, 347-371).

[2] J. Cheeger, The Relation Between the Laplacian and the Diameter for Manifolds of Non-negative Curvature (Arch. der Math. Vol. 19, 1968, 558-560).

[3] J. Cheeger, A Lower Bound for the Smallest Eigenvalue of the Laplacian, in "Problems in Analysis, a symposium in honor of S. Bochner" (Princeton University Press, 1970).

[4] J. Cheeger and D. Gromoll, The Splitting Theorem for Manifolds of Non-negative Ricci Curvature (J. Diff. Geom., Vol. 6, 1971, 119-128).

[5] S. Y. Cheng, Eigenvalue Comparison Theorems and its Geometric Applications (Math. Z., 143, 1975, 289-297).

[6] C. Croke, Some Isoperimetric Inequalities and Consequences (to appear).

[7] J. Hersch, Sur la fréquence fondamentale d'une membrane vibrante: évaluations par défaut et principé de maximum (J. Math. Phy. Appl., Vol. XI, Fasc. 5, 1960, 387-413).

[8] P. Levy, Problèmes concrets d'analyse fonctionelles. (Paris, Gauthier-Villars, 1951).

[9] P. Li, A Lower Bound for the First Eigenvalue of the Laplacian on a Compact Riemannian Manifold (to appear in Indiana J. Math.).

[10] A. Lichnerowicz, Geometrie des groupes de transformations (Dunod, 1958).

[11] M. Obata, Certain Conditions for a Riemannian Manifold to be Isometric with a Sphere (J. Math. Soc. Japan, Vol. 14, 1962, 333-340).

[12] L. E. Payne and I. Stakgold, On the mean value of the fundamental mode

in the fixed membrane problem (Applicable Analysis, Vol. 3, 1973, 295-303).

[13] L. E. Payne and H. Weinberger, The optional Poincaré inequality for convex domains.

[14] R. C. Reilly, Applications of the Integral of an Invariant of the Hessian (Bulletin A.M.S., Vol. 82, No. 4, 1974).

[15] S. T. Yau, Isoperimetric Constants and the First Eigenvalue of a Compact Riemannian Manifold (Ann. scient. Éc. Norm. Sup. 4^c série, t. 8, 1975, 487-507).

Department of Mathematics
University of California, Berkeley
Berkeley, California 94720

Department of Mathematics
Stanford University
Stanford, California 94305

Proceedings of Symposia in Pure Mathematics
Volume 36, 1980

THE NUMBER OF BOUND STATES OF ONE-BODY SCHROEDINGER OPERATORS AND THE WEYL PROBLEM

Elliott H. Lieb[1]

ABSTRACT. If $\tilde{N}(\Omega,\lambda)$ is the number of eigenvalues of $-\Delta$ in a domain Ω in a suitable Riemannian manifold of dimension n, we derive bounds of the form $\tilde{N}(\Omega,\lambda) \leq D_n \lambda^{n/2}|\Omega|$ for *all* Ω, λ, n, Likewise, if $N_\alpha(V)$ is the number of nonpositive eigenvalues of $-\Delta + V(x)$ which are $\leq \alpha \leq 0$, then $N_\alpha(V) \leq L_n \int_M [V - \alpha]_-^{n/2}$ for *all* α and V and $n \geq 3$.

I. INTRODUCTION AND BACKGROUND.

Two closely related problems will concern us here : One is to bound the nonpositive eigenvalues of the one-body Schroedinger operator

$$H = -\Delta + V(x). \tag{1.1}$$

The other is to find an upper bound for

$$\tilde{N}(\Omega,\lambda) = \text{number of eigenvalues of } -\Delta \leq \lambda \tag{1.2}$$

in a domain Ω with Dirichlet boundary conditions.

In both cases the setting is a Riemannian manifold, M, and Δ is the Laplace-Beltrami operator. The only way in which the properties of the manifold will appear in our results will be through the fundamental solution of the heat equation

$$G(x, y; t) = [\exp(t\Delta)](x, y) \tag{1.3}$$

1980 Mathematics Subject Classification 35P15.
[1] Work supported by U.S.National Foundation grants PHYS-7825390 and INT 78-01160.

evaluated on the diagonal x = y, or else through the Green function

$$(-\Delta + e)^{-1}(x, y) = \int_0^\infty dt\, e^{-et}\, G(x, y; t) \tag{1.4}$$

for $e \geq 0$. (Note: G is defined for the *whole* manifold, not the subdomain Ω in the case of (1.2). One could of course, use G defined for the domain Ω in all the formulas, but then the dependence of the result on Ω will be complicated. It is precisely to avoid this complication that we use the G for the whole of M).

Let us begin with the problem defined by (1.2), which we may term the *Weyl problem.* Weyl [1] proved the asymptotic formula (for suitable domains) :

$$\tilde{N}(\Omega,\lambda) \approx C_n\, \lambda^{n/2}|\Omega| \tag{1.5}$$

where $|\Omega|$ is the Riemannian volume of Ω, n is the dimension of M and

$$\begin{aligned} C_n &= (4\pi)^{-n/2}\, \Gamma(1 + n/2)^{-1} \\ &= (2\pi)^{-n}\, \tau_n , \end{aligned} \tag{1.6}$$

where τ_n is the volume of the unit ball in $\mathbb{R}^n$. The constant C_n is called the *classical constant* for reasons which will become clear later. This result (1.5) is discussed in [2] and [3], §XIII.15, vol.4. The proof uses Dirichlet-Neumann bracketing.

Polya's conjecture is that (1.5) holds in $\mathbb{R}^n$ for all λ and Ω, not just asymptotically. Here we will prove

THEOREM 1. *For all λ and Ω there exist constants D_n and E_n (depending on the manifold M) such that*

$$\tilde{N}(\Omega,\lambda) \leq D_n\, \lambda^{n/2}|\Omega| \tag{1.7}$$

if

$$G(x, x; t) \leq A_n\, t^{-n/2}, \ \forall x \in M, \ \forall t > 0 \tag{1.8}$$

for some $A_n < \infty$, while

$$\tilde{N}(\Omega,\lambda) \leq (D_n\, \lambda^{n/2} + E_n)|\Omega| \tag{1.9}$$

if

$$G(x, x; t) \leq A_n t^{-n/2} + B_n, \ \forall x \in M, \ \forall t > 0, \tag{1.10}$$

D_n and E_n are proportional to A_n and B_n respectively.

In particular (1.8) and (1.7) hold for $\mathbb{R}^n$ (with $A_n = (4\pi)^{-n/2}$) and for many noncompact M, e.g. homogeneous spaces with curvature ≤ 0. (1.10) and (1.9) hold for compact M.

Next we turn to the Schroedinger problem (1.1). Let $E_1(V) \leq E_2(V) \leq \ldots \leq 0$ be the nonpositive eigenvalues of H on $L^2(M)$. If we write

$$V = V_+ - V_- \tag{1.11}$$

with

$$\begin{aligned} V_\pm(x) &= |V(x)|, \quad \text{when } V(x) \gtrless 0 \\ &= 0, \text{ otherwise} \end{aligned} \tag{1.12}$$

then the negative spectrum of H is discrete if $V_- \in L^{n/2}$, for example.

DEFINITION. $N_\alpha(V)$ *is the number of eigenvalues of* H *which are* $\leq \alpha \leq 0$.

Our main result is

THEOREM 2. *Suppose* (1.8) *holds and suppose* $n \geq 3$. *Then*

$$N_0(V) \leq L_n \int_M V_-(x)^{n/2}\, dx \tag{1.13}$$

for some constant L_n *depending on* M.

There are many remarks to be made about Theorem 2 and its connection with Theorem 1. First, the history of (1.13). Rosenbljum [4] first announced Theorem 2 in 1972. Unaware of this, Simon [5] proved an inequality of the form

$$N_0(V) \leq S_{n,\varepsilon}[\, \|V_-\|_{n/2+\varepsilon} + \|V_+\|_{n/2+\varepsilon}]^{n/2}$$

with $S_{n,\varepsilon} \to \infty$ as $\varepsilon \downarrow 0$. Also unaware of [4], Cwikel and myself [6,7] simultaneously found a proof of Theorem 2. Reed and Simon [3] call Theorem 2 the Cwikel - Lieb - Rosenbljum bound. Cwikel's method exploits some ideas in [5]. All three methods are different, Cwikel's and Rosenbljum's are applicable to a wider class of operators than the Schroedinger operator, but my method [7], based on Wiener integrals, which is the one presented here, gives the best constant by far. This result was announced in [7] and the proof was written up in [3], Theorem XIII.12 and in [8]. Because all the technical details can easily be found in [8], the presentation given here will ignore technicalities . Not only am I indebted to B.Simon for his help with the technical details, as just mentioned, but I also wish to acknowledge his role in stimulating my interest in the problem, and for his constant encouragement while the ideas were taking shape.

The connection between the two theorems is

PROPOSITION 3. *Let* $\alpha \leq 0$. *Then for all* M

$$\tilde{N}(\Omega,\lambda) \leq N_\alpha(\,(\alpha-\lambda)\,\chi_\Omega) \tag{1.14}$$

where χ_Ω *is the characteristic function of* Ω.

PROOF. Let ψ_j be the j th eigenfunction of $-\Delta$ in Ω and let $\tilde{\psi}_j$, defined on all of M, be ψ_j in Ω and zero outside. (1.14) is obtained by using the $\tilde{\psi}_j$ as variational functions for $-\Delta + (\alpha - \lambda)\,\chi_\Omega$ in the Rayleigh - Ritz variational principle. QED.

Similarly, we have

PROPOSITION 4. *if* $0 \leq \beta \leq -\alpha$, *then for all* M

$$N_\alpha(V) \leq N_{\alpha+\beta}(-[V+\beta]_-). \tag{1.15}$$

PROOF. Same as for proposition 3. Alternatively, one can remark that $V(x) \geq -[V+\beta]_-(x) - \beta$, and adding a positive operator can never decrease any eigenvalue. QED.

In particular, $N_\alpha(V) \le N_0(-[V-\alpha]_-)$ and $N_0(V) \le N_0(-V_-)$, whence we have

COROLLARY 5. *If* $n \ge 3$ *and* (1.13) *holds, then*

$$N_\alpha(V) \le L_n \int_M [V-\alpha]_-(x)^{n/2}\,dx. \tag{1.16}$$

Moreover to prove (1.13) *it is sufficient to consider* V *satisfying* $V(X) \le 0$, $\forall x$.

Proposition 3 will be used to derive Theorem 1 from Theorem 2 (actually from a generalization of Theorem 2, namely Theorem 8 and (4.3), which holds for all n). However, at this point we can, under the assumption that (1.8) holds, deduce (1.7) of Theorem 2 from proposition 3 and Theorem 2. Choosing $\alpha = 0$, we have

$$\begin{aligned}\tilde{N}(\Omega,\lambda) &\le L_n\,\lambda^{n/2} \int_M \chi_\Omega{}^{n/2}\,dx \\ &= L_n\,\lambda^{n/2}|\Omega|\end{aligned} \tag{1.17}$$

for $n \ge 3$.

It is to be emphasized that Theorem 2 is more delicate than Theorem 1. For one thing Theorem 1 holds for all n, whereas Theorem 2 holds only for $n \ge 3$. The analogue of Theorem 2 is definitely false for $n = 1$ and 2. In $\mathbb{R}^n$, at least, an arbitrarily small, nonpositive potential always has a negative eigenvalue when $n < 3$; cf. [3], Theorem XIII.11. For another thing, the best constant in (1.7) is , according to the Polya conjecture, the classical constant C_n (in $\mathbb{R}^n$ at least). However, the best constant L_n is bigger than C_n in general. It is easy to prove [9] (by considering a very "large" V and using Dirichlet-Neumann bracketing) that $L_n \ge C_n$. In [9] it was shown that $L_n > C_n$ for $3 \le n \le 7$ and recently [16] it was shown that $L_n > C_n$ for $n \ge 7$; in both cases explicit examples were constructed.

It is somewhat ironic that although Theorem 2 starts to hold only for $n = 3$, Theorem 1 is easiest to prove for $\mathbb{R}^n$ with $n = 1$. In that case the only domain that need be considered is a finite interval, and there the eigenvalues of $-\Delta$ can be computed explicitly. The Polya conjecture is easily seen to be true.

The intuition behind Theorem 2, and thereby the reason for calling C_n the classical constant is important. In the semiclassical picture of quantum mechanics in $\mathbb{R}^n$, which is similar to WKBJ theory, one has the mystical postulate that "each nice set in phase space $\Gamma = \{(p,x) \mid p \in \mathbb{R}^n,\ x \in \mathbb{R}^n\}$ of volume $(2\pi)^n$ can accomodate one eigenstate of H". This postulate can be made more precise by mean of the Dirichlet-Neumann bracketing method mentioned before. In any event, since the "eigenvalues" of $-\Delta$ are p^2 and the "eigenvalues" of V are $V(x)$, the postulate implies that

$$N_\alpha(V) \approx (2\pi)^{-n} \int_{\mathbb{R}^n\times\mathbb{R}^n} dpdx\ \Theta(\alpha - (p^2+V(x))) \tag{1.18}$$

with $\Theta(a) = 1$, for $a \geq 0$ and $= 0$ for $a < 0$. The p integration in (1.18) for fixed x, is easy to do, namely

$$\int_{p^2 \leq \alpha - V(x)} dp = \tau_n\ [V - \alpha]_-(x)^{n/2} \quad . \tag{1.19}$$

Thus, (1.18) yields

$$N_\alpha(V) \approx C_n \int [V - \alpha]_-(x)^{n/2}\ dx \quad . \tag{1.20}$$

While the chief purpose of this paper is to prove Theorems 1 and 2, quantities of no less interest are the moments of the nonpositive eigenvalues of H.

DEFINITION. *For* $\gamma > 0$

$$\begin{aligned} I_\gamma(V) &= \sum_i |E_i(V)|^\gamma \\ &= \gamma \int_{-\infty}^0 |\alpha|^{\gamma-1} N_\alpha d\alpha \end{aligned} \tag{1.21}$$

$I_0(V)$ *is defined to be* $N_0(V)$.

For $n \geq 3$ we can use Corollary 5 and Fubini's theorem to obtain

$$\begin{aligned} I_\gamma(V) &\leq \gamma L_n \int_M dx \int_{-\infty}^0 |\alpha|^{\gamma-1} [V - \alpha\ ]_-(x)^{n/2}\ d\alpha \\ &= \gamma L_n \int_M dx \int_{-V_-(x)}^0 d\alpha\ |\alpha|^{\gamma-1} (V_-(x) + \alpha)^{n/2} \\ &= L_{\gamma,n} \int_M V_-(x)^{\gamma+n/2}\ dx \end{aligned} \tag{1.22}$$

with

$$L_{\gamma,n} = L_n\ \Gamma(\gamma+1)\ \Gamma(1+n/2)\ \Gamma(1+\gamma+n/2)^{-1} \tag{1.23}$$

There are several things to be said about (1.22). Although it was derived from Corollary 5 under the assumptions (1.13) and $n \geq 3$, it holds much more generally. For example it holds in $\mathbb{R}^n$ for $n = 2$, $\gamma > 0$ and for $n = 1$, $\gamma > 1/2$ provided $\alpha > 0$. This was first given in [9]. In [9] it was stated that it holds for $n = 1$ and $\gamma = 1/2$. That was an error ; it is not known if (1.22) holds for $n = 1$, $\gamma = 1/2$ but it is known [9] that (1.22) does not hold for $n = 1$, $\gamma < 1/2$. In section II we shall briefly mention how to deduce (1.22).

The best constant $L_{\gamma,n}$ in (1.22) is not given by (1.23), as the foregoing remark already indicates. If we use C_n in place of $L_{\gamma,n}$ in (1.23) we have the classical value of $L_{\gamma,n}$ namely,

$$L^C_{\gamma,n} = (4\pi)^{-n/2}\ \Gamma(\gamma+1)\ \Gamma(1+\gamma+n/2)^{-1} \tag{1.24}$$

As in the case of L_n, it is easy to prove that $L_{\gamma,n} \geq L^C_{\gamma,n}$. The classical constant $L^C_{\gamma,n}$ can also be "derived" from the semiclassical assumption as in (1.18), (1.19), namely

$$\sum_i |E_i|^\gamma \approx (2\pi)^{-n} \int_{\mathbb{R}^n\times\mathbb{R}^n} dpdx\ |p^2 + V(x)|^\gamma\ \Theta(-p^2-V(x)). \tag{1.25}$$

If the p integration is done in (1.25), the result is (1.22) with $L^C_{\gamma,n}$.

An important question is :

When is $L_{\gamma,n} = L^C_{\gamma,n}$?

It seems to be true that $L_{\gamma,n} = L^C_{\gamma,n}$ for γ large enough, depending on n. This is known to be true [9, 10] for n = 1 and $\gamma \geq 3/2$. In fact [10], if $L_{\gamma_0,n} = L^C_{\gamma_0,n}$ for some γ_0, then equality holds for all $\gamma \geq \gamma_0$.

The case of primary *physical interest* is $\gamma = 1$, $n = 3$, where it is conjectured [9] that equality holds. If this were so, it would have important consequences for physics and it is hoped that someone will be motivated to solve the problem.

We now turn to the proof of Theorems 1 and 2 in the next three sections.

II. THE BIRMAN-SCHWINGER KERNEL

As stated in Corollary 5 we can assume $V(x) \leq 0$. Therefore write $V(x) = -U(x)$, $U \geq 0$.

A useful device for studying the nonpositive eigenvalues of $-\Delta-U$ was discovered by Birman [11] and Schwinger [12]. If $(-\Delta-U)\psi = E\psi$, $E \leq 0$, then

$$\psi = (-\Delta + |E|)^{-1}\ U\psi\ . \tag{2.1}$$

Defining $U^{1/2}\psi = \phi$, and multiplying (2.1) by $U^{1/2}$, we have

$$\phi = K_{|E|}(U)\phi \tag{2.2}$$

where $K_e(U)$, for $e \geq 0$, is the positive *Birman-Schwinger Kernel* given explicitly by

$$K_e(x, y; U) = U(x)^{1/2}(-\Delta + e)^{-1}(x, y)U^{1/2}(y)\ . \tag{2.3}$$

What (2.2) says is that for every nonpositive eigenvalue, E, of $-\Delta-U$, $K_{|E|}(U)$ has an eigenvalue 1. The converse is also easily seen to hold (see [3,8] for more details). $K_e(U)$ is to be thought of as an operator on L^2 ; we will see that it is compact, when $e > 0$ at least, and U is in a suitable L^p space.

In addition to the advantage that the study of the E's reduces to the study of a compact operator, there is the following important fact: Since $(-\Delta + e)^{-1}$ is operator monotone decreasing as a function of e, so is K_e. Hence (with $V = -U$),

$$N_\alpha(V) = k_{|\alpha|}(U) \equiv \textit{number of eigenvalues of } K_{|\alpha|}(U) \geq 1. \tag{2.4}$$

(2.4) will be exploited in the following way.

PROPOSITION 6. *Let* $F : \mathbb{R}^+ \to \mathbb{R}^+$ *be any function such that* $F(x) \geq 1$ *for* $x \geq 1$.

Then

$$\operatorname{Tr} F(K_e(U)) = \sum_i F(l_e^i(U))$$
$$\geq k_e(U) = N_{-e}(V) \tag{2.5}$$

where Tr *means trace and the* $l_e^i(U)$ *are the eigenvalues of* $K_e(U)$.

For example, consider $\mathbb{R}^n$, $n \leq 3$, and $F(x) = x^2$. Then

$$\operatorname{Tr} K_e(U)^2 = \iint U(x)\, U(y)\, [(-\Delta + e)^{-1}(x - y)]^2\, dxdy$$
$$\leq \|U\|_2^2\, \|(-\Delta + e)^{-1}(x)\|_2^2 \tag{2.6}$$

by Young's inequality. The last factor is of the form

$$\|(-\Delta + e)^{-1}(x)\|_2^2 = h_n e^{-2+n/2} \tag{2.7}$$

This shows that $K_e(U)$ is Hilbert-Schmidt when $U \in L^2$ and $e > 0$. When $n > 3$, $h_n = \infty$ but one can show that $K_e(U)$ is compact by considering the trace of a higher power of $K_e(U)$, cf. [9].

At this point we can derive the aforementioned bound, (1.22), on $I_\gamma(V)$. If we use (2.5) and (2.6) and insert the latter in (1.21), the α integration will diverge. The trick [9] is to use Proposition 4 with $2\beta = e = -\alpha$. Thus

$$N_{-e}(V) \leq \operatorname{Tr} K_{e/2}([V + e/2]_-)$$
$$\leq \|[U - e/2]_+\|_2^2\, h_n (e/2)^{-2+n/2} \tag{2.8}$$

Inserting (2.8) into (1.21), and doing the α integration first, we obtain

$$I_\gamma(V) \leq 2^{2-n/2}\, \gamma h_n \int dx \int_{-2U(x)}^{0} d\alpha\, |\alpha|^{\gamma-3+n/2} [U(x) + \alpha/2]^2$$
$$= L_{\gamma,n} \int U(x)^{\gamma+n/2} dx \tag{2.9}$$

if $\gamma > 2 - n/2$ and $n \leq 3$.

(2.9) can be extended to other values of n and γ (but with $\gamma > 1/2$ for $n = 1$), cf. [9] .

There is no way, however, to make this method work with $F(x) = x^a$ when $\gamma = 0$ and $n \geq 3$. Quiet reflection shows that if theorem 2 is to be provable by this method then we need $x^{-n/2} F(x) \to 0$ as $x \to 0$ but $x^{-1}F(x)$ remains bounded as $x \to \infty$. The tool we will use to bound $\operatorname{Tr}F(K_e(U))$ for such F's is the Wiener integral. That is the subject of the next section, which is really the main point of this paper.

III. THE WIENER INTEGRAL REPRESENTATION OF $F(K_e(U))$

Let $d\mu_{x,y;t}$ be *conditional* Wiener measure on paths $\omega(t)$ with $\omega(0) = x$ and $\omega(t) = y$. This measure gives a representation for G (cf. (3)) by

$$\int d\mu_{x,y;t}(\omega) = G(x,y;t) = e^{t\Delta}(x, y). \tag{3.1}$$

G itself has the *semigroup property*

$$\int_M G(x, y; t)\, G(y, z; s)\, dy = G(x, z; t+s). \tag{3.2}$$

The well known Feynman-Kac formula [13] is

$$\int d\mu_{x,y;t}(\omega) \exp\left[-\lambda\int_0^t U(\omega(s))ds\right] = e^{-t(-\Delta + \lambda U)}(x, y). \tag{3.3}$$

(Note the signs in (3.3).). Take $e \ge 0$, $\lambda \ge 0$, and $U \ge 0$, multiply (3.3) by $U(x)^{1/2}U(y)^{1/2}$ exp $(-et)$ and integrate with respect to t. The result is

$$\begin{aligned} A &= U(x)^{1/2}U(y)^{1/2}\int_0^\infty dt\, e^{-et}\int d\mu_{x,y;t}(\omega) \exp\left[-\lambda\int_0^t U(\omega(s))ds\right] \\ &= \{U^{1/2}(-\Delta + \lambda U + e)^{-1}\, U^{1/2}\}\,(x, y). \end{aligned} \tag{3.4}$$

Now

$$(-\Delta + e)^{-1} = (-\Delta + \lambda U + e)^{-1} + \lambda(-\Delta + \lambda U + e)^{-1}\, U(-\Delta + e)^{-1} \tag{3.5}$$

Multiplying (3.5) on both sides by $U^{1/2}$ we obtain

$$A = \{K_e(U)\ [1 + \lambda K_e(U)]^{-1}\}(x, y). \tag{3.6}$$

(3.6) can be cast in a more general form. If $g(x) = \exp(-\lambda x)$ and $F(x) = x(1 + \lambda x)^{-1}$ then

$$F(x) = x\int_0^\infty dy\, e^{-y}\, g(xy) \tag{3.7}$$

and

$$\begin{aligned} &U(x)^{1/2}U(y)^{1/2}\int_0^\infty dt\, e^{-et}\int d\mu_{x,y;t}(\omega)\, g\left(\int_0^t U(\omega(s))ds\right) \\ &\qquad = F(K_e(U))\,(x, y). \end{aligned} \tag{3.8}$$

Next we want to take the trace of both sides of (3.8). This is obtained by setting $y = x$ and integrating. (To be precise, one must first take $U \in C^\infty$ and then use a limiting argument.) The point to notice is that the x dependence occurs through

$$\int_M dx\, U(x)\, d\mu_{x,x;t}(\omega) = \int_M dx\, d\mu_{x,x;t}(\omega)\, U(\omega(0)). \tag{3.9}$$

By the semigroup property of $\exp[-t(-\Delta + \lambda U)]$, however, the same result is obtained (after the $d\mu$ integration) if $U(\omega(0))$ in (3.9) is replaced by $U(\omega(s))$ for any $0 \le s \le t$. Thus,

$$\int_0^\infty dt\, t^{-1}\, e^{-et}\int_M dx \int d\mu_{x,x;t}(\omega)\, f\left(\int_0^t U(\omega(s))ds\right) = \mathrm{Tr}\, F(K_e(U)). \tag{3.10}$$

with

$$f(x) = xg(x), \tag{3.11}$$

i.e.

$$F(x) = \int_0^\infty dy\, y^{-1}\, e^{-y}\, f(xy) \quad . \tag{3.12}$$

Now the relations between F, f and g are linear, as are the relations (3.8) and (3.10). The latter therefore extend to a large class of f's. Since we are here not particularly interested in the operator version, (3.8), we will concentrate on the trace version (3.10). The g's of the form $\exp(-\lambda x)$ are norm dense in the continuous functions which vanish at infinity. Furthermore, using the semigroup property of $\exp[-t(-\Delta + U)]$, one can see explicitly that (3.8) and (3.10) hold for f's of the form $x^k \exp(-\lambda x)$, with k a positive integer. By a monotone convergence arguement one arrives at

THEOREM 7. *Let* f *be a nonnegative lower semicontinuous function on* $[0, \infty)$ *satisfying:*

(i) $f(0) = 0$

(ii) $x^p f(x) \to 0$ *as* $x \to \infty$ *for some* $p < \infty$.

Let $U \geq 0$ *and* $U \in L^p + L^q$ *with* $p = n/2$ $(n \geq 3)$, $p > 1$ $(n = 2)$, $p = 1$ $(n = 1)$ *and* $p < q < \infty$ *Then* (3.10) *holds, with* F *given by* (3.12), *in the sense that both sides may be* $+\infty$

The reader is referred to [8] for details. Obviously the class of f's in Theorem 7 is not the largest possible, but it is more than adequate for our intended application.

The remark that allows us to bound the left side of (3.10) is the following.

THEOREM 8. *Suppose that* f *satisfies the conditions of Theorem 7 and that* f *is also convex. Then*

$$\operatorname{Tr} F(K_e(U)) \leq \int_0^\infty dt\, t^{-1} e^{-et} \int_M dx\, G(x, x; t)\, f(tU(x)). \tag{3.13}$$

PROOF. By Jensen's inequality, for any fixed path $t \to \omega(t)$,

$$f\left(\int_0^t U(\omega(s))\, ds\right) = f\left(\int_0^t (t^{-1}ds)\, t\, U(\omega(s))\right) \leq \int_0^t (t^{-1}ds)\, f(tU(\omega(s)))\,.$$

By the same remark as that preceding (3.10), $\int_M dx \int d\mu_{x,x;t}(\omega)\, h(\omega(s))$ is independent of s for any fixed function h. Inserting this in (3.10) gives (3.11). QED.

IV. APPLICATIONS OF THEOREM 8

PROOF OF THEOREM 1: We use Proposition 3 and Proposition 6. f is chosen to be of the following form for some $0 < a < b$

$$\begin{aligned} f(x) &= 0, && 0 \leq x \leq a \\ &= b(x - a), && a \leq x. \end{aligned} \tag{4.1}$$

F given by (3.12) is monotone, and the condition that $F(x) = 1$ is that a and b are related by

$$1 = h(a, b) = b\int_a^\infty dy(1 - a/y)e^{-y} = be^{-a} - abE_1(a) \tag{4.2}$$

where E_1 is the exponential integral. Assuming (1.10), (1.9) will be proved; (1.7) follows from the special assumption $B_n = 0$ in what follows.

In Proposition 3 write $\alpha = -e$ so that $\tilde{N}(\Omega, \lambda) <$ (3.13) with $U = (e + \lambda)\chi_\Omega$. The x integration can be done last in (3.13) in which case $f(tU(x)) = 0$ if $x \notin \Omega$. For $x \in \Omega$ we change variables to $t \to t/(e + \lambda)$. Thus,

$$b|\Omega| \int_a^\infty dt(1 - a/t)(A_n(e + \lambda)^{n/2} t^{-n/2} + B_n)\exp[-et/(e + \lambda)]$$
$$\geq \tilde{N}(\Omega, \lambda) \ . \tag{4.3}$$

The simplest choice for e, which is arbitrary, is $e = 0$. This will work only when $n \geq 3$ and $B_n = 0$. But for all cases we can choose $e = c\lambda$, $c > 0$ and thereby prove Theorem 1. QED.

PROOF OF THEOREM 2: Proposition 6 is used again and the proof parallels that of Theorem 1. (4.1) and (4.2) are assumed. We change variables in (3.13) to $t \to tU(x)^{-1}$ if $U(x) \neq 0$. If $e = 0$ the result is (1.13) with

$$L_n = A_n \int_0^\infty dt\ t^{-1-n/2} f(t)$$
$$= 4A_n b a^{1-n/2}[n(n-2)]^{-1} \ . \qquad \text{QED.} \tag{4.4}$$

REMARKS: (i) If $\alpha = -e \neq 0$ then the only estimate we have for $N_\alpha(V)$ is contained in Corollary 5, which is valid only for $n \geq 3$. Alternatively, one could try to estimate (3.13) directly with $e \neq 0$, but this is messy. As stated earlier, no inequality of the form (1.16) holds for all α, V when $n = 1$ or 2. But recently Itō [14] has bounded (3.13) when $e \neq 0$ and $n = 2$. He uses the fact that $f(x) \leq bx$ for $x \geq a$ and obtains a complicated upper bound for $N_\alpha(V)$ in terms of $\|V_-\|_2$ and $\|V_-|\ln V_-|^{1/2}\|_2$.

(ii) If $n \geq 3$ and $B_n = 0$ we can choose $e = 0$ in (4.3). This estimate for D_n is, of course, the same as L_n given by (4.4).

As an illustration of how good our bound is let us consider the case of $\mathbb{R}^3$, where $A_3 = (4\pi)^{-3/2}$. We choose $a = 0.25$ in (4.1) and find that $E_1(.25) = 1.0443$ and $b = 1.9315$ according to (4.2). Using (4.4),

$$D_3 = L_3 = 0.1156 \ . \tag{4.5}$$

This value of D_3 can be used in (1.7). When compared with $C_3 = 0.0169$, which is supposed to be the sharp constant, it is not very good. The estimate for D_3 can be improved by using (4.3) with $e = c\lambda$, $c > 0$.

If, however, the same number, L_3, is used in (1.13) the result is quite good. As already stated, the best $L_3 > C_3$. In fact, by an explicit example,

$$L_3 \geq (3\pi)^{-3/2}\Gamma(3)/\Gamma(3/2) = 0.0780, \tag{4.6}$$

cf. [9], eqn. (4.24). It is conjectured that the right side of (4.6) is, in fact, the sharp constant in (1.13) for $\mathbb{R}^3$. In any case our result, (4.5), is off by at most 49%.

As stated in Section 1, a quantity of physical interest is I_1, the sum of the absolute values of the eigenvalues, in $\mathbb{R}^3$. Using the bound (1.22), (1.23), together with (4.5), we have

$$L_{1,3} \leq (2/5)L_3 = .04624 \tag{4.7}$$

This result was announced in [15].

BIBLIOGRAPHY

1. H. Weyl, "Das asymptotische Verteilungsgesetz der Eigenwerte Linearer partieller Differentialgleichungen", Math. Ann. 71 (1911), 441-469.

2. M. Kac, "Can one hear the shape of a drum?", Slaught Memorial Papers, no. 11, Amer. Math. Monthly 73 (1966), no. 4, part II, 1-23.

3. M. Reed and B. Simon, Methods of Modern Mathematical Physics, Acad. Press, N. Y., 1978.

4. G. V. Rosenbljum, "Distribution of the discrete spectrum of singular differential operators", Dokl. Aka. Nauk SSSR, 202 (1972), 1012-1015 (MR 45 #4216). The details are given in "Distribution of the discrete spectrum of singular differential operators", Izv. Vyss. Ucebn. Zaved. Matematika 164 (1976), 75-86. [English trans. Sov. Math. (Iz. VUZ) 20 (1976), 63-71.]

5. B. Simon, "Weak trace ideals and the number of bound states of Schroedinger operators", Trans. Amer. Math. Soc. 224 (1976), 367-380.

6. M. Cwikel, "Weak type estimates for singular values and the number of bound states of Schroedinger operators", Ann. Math. 106 (1977), 93-100.

7. E. Lieb, "Bounds on the eigenvalues of the Laplace and Schroedinger operators", Bull. Amer. Math. Soc. 82 (1976), 751-753.

8. B. Simon, Functional Integration and Quantum Physics, Academic Press, N. Y., to appear 1979.

9. E. Lieb and W. Thirring, "Inequalities for the moments of the eigenvalues of the Schroedinger equation and their relation to Sobolev inequalities", in Studies in Mathematical Physics: Essays in Honor of Valentine Bargmann (E. Lieb, B. Simon and A. Wightman eds.), Princeton Univ. Press, Princeton, N. J., 1976. These ideas were first announced in "Bound for the kinetic energy of fermions which proves the stability of matter", Phys. Rev. Lett. 35 (1975), 687-689, Errata 35 (1975), 1116.

10. M. Aizenman and E. Lieb, "On semi-classical bounds for eigenvalues of Schroedinger operators", Phys. Lett. 66A (1978), 427-429.

11. M. Birman, "The spectrum of singular boundary problems", Math. Sb. 55 (1961), 124-174. (Amer. Math. Soc. Trans. 53 (1966), 23-80).

12. J. Schwinger, "On the bound states of a given potential", Proc. Nat. Acad. Sci. U.S.A. 47 (1961), 122-129.

13. M. Kac, "On some connections between probability theory and differential and integral equations", Proceedings of the Second Berkeley Symposium on Mathematical Statistics and Probability, Univ. of Calif. Press, Berkeley, 1951, 189-215.

14. K. R. Ito, "Estimation of the functional determinants in quantum field theories", Res. Inst. for Math. Sci., Kyoto Univ. (1979), preprint.

15. E. Lieb, "The stability of matter", Rev. Mod. Phys. 48 (1976), 553-569.

16. V. Glaser, H. Grosse and A. Martin, "Bounds on the number of eigenvalues of the Schroedinger operator", Commun. Math. Phys. 59 (1978), 197-212.

DEPARTMENTS OF MATHEMATICS AND PHYSICS
PRINCETON UNIVERSITY
JADWIN HALL
P.O.BOX 708
PRINCETON, N. J. 08544

Proceedings of Symposia in Pure Mathematics
Volume 36, 1980

GEOMETRY IN THE SCATTERING PHASE

A. Majda and J. Ralston[1]

Let M be a Riemannian manifold, and let Δ denote the Friedrichs extension in $L^2(M)$ of the Laplace-Beltrami operator acting on $C_0^\infty(M)$. If M is not compact, the heat kernel, $\exp(t\Delta)$, need not belong to trace class. However, one can consider the differences

$$D(t) = e^{t\Delta} - e^{t\Delta'}$$

where Δ' is the operator corresponding to Δ when

A) the metric on M is changed on K, a compact subset of M, or

B) a compact subset K is removed from M. In this case one defines $\exp(t\Delta')$ to be the zero operator on functions supported in K.

If $D(t)$ belongs to trace class for $t > 0$, then a trace theorem of M.G. Krein implies that there is a unique $s(\lambda)$, such that

$$\int_0^\infty e^{-\lambda t} s(\lambda)d\lambda = \frac{1}{t} \operatorname{tr}\left(e^{t\Delta} - e^{t\Delta'}\right) . \tag{1}$$

One thinks of (1) as a generalization of the familiar formula

$$\int_0^\infty e^{-\lambda t} n(\lambda)d\lambda = \frac{1}{t} \operatorname{tr}(e^{t\Delta})$$

$$n(\lambda) = \{ \# \mu : -\mu \le \lambda, \mu \in \operatorname{spec}(\Delta)\}$$

from the compact case.

The relation (1) has been studied in the following sinple cases:

a) $M = \mathbb{R}^n$ with the Euclidean metric and the metric is perturbed as in A).

1980 Mathematics Subject Classification 35P25, 58G11, 58G17, 58G16, 58G25.

[1]Authors supported by National Science Foundation, Grant MCS 76-10227

b) $M = \mathbb{R}^n$ with the Euclidean metric and a compact subset K with smooth boundary is removed as in B).

In case b) Jensen and Kato showed in [1] that Kato's invariance principle for wave operators combined with further work of Birman and Krein imply that

$$e^{2\pi i s(\lambda)} = \det S(\lambda) \tag{2}$$

where $S(\lambda)$ is the scattering matrix relating Δ and Δ', and the same argument yields this result in case a) as well. Because of (2) $s(\lambda)$ is known as "the scattering phase".

Formula (1) suggests that in examples a) and b) the scattering phase is the natural analogue of the counting function $n(\lambda)$, and at the same time shows that if $s(\lambda)$ is well behaved at $\lambda \to \infty$, its asymptotics will contain precisely the same geometric information – in a "relative" form – as those of $n(\lambda)$. Thus, one has motivation for studying the asymptotic behavior of $s(\lambda)$. However, as things never seem to proceed in the order they ought to, the authors [2, I], [2, II] actually began studying the asymptotics of $s(\lambda)$ without knowing (1), and we found the resemblance of the asymptotics to those of $n(\lambda)$ quite surprising.

The detailed behavior of $s(\lambda)$ for λ large depends directly on the asymptotics of certain generalized eigenfunctions $\Phi(x,\lambda,\omega)$, $\omega \in S^{n-1}$, of Δ' known as "distorted plane waves", defined by the conditions

(i) $(\Delta' + \lambda)\Phi = 0$, (eigenfunction)

(ii) $\Phi = e^{i\sqrt{\lambda}\omega\cdot x} + v$ (distorted plane wave)

(iii) $v(x) = |x|^{\frac{1-n}{2}} e^{i\sqrt{\lambda}|x|}\left(f\left(\frac{x}{|x|}\right) + O\left(\frac{1}{|x|}\right)\right)$ (radiation condition)

As their name suggests, one thinks of the distorted plane waves as the spatial parts of time-harmonic solutions to the wave equation $u_{tt} = \Delta u$, and makes use of techniques from the theory of hyperbolic PDE to study their asymptotic behavior. This leads to the following result (cf [2, III]:

THEOREM: Assume in case a) that the perturbed metric has no trapped geodesics, i.e. for all R there is a T(R) such that any geodesic $x(s)$ with $x(0) \in K$ will satisfy $|x(s)| > R$ for $s \geq T(R)$. In case b) assume that K is strictly convex. Then $s(\lambda)$ has a complete asymptotic expansion in descending powers of λ for λ large, and using (1) one can read off the coefficients in this expansion, e.g., in case b) one has a formula given by V.S. Buslaev

$$s(\lambda) = \frac{(4\pi)^{-\frac{n}{2}}}{\Gamma\left(1+\frac{n}{2}\right)} \lambda^{\frac{n}{2}} (\text{Volume } K) - \frac{(4\pi)^{\frac{1-n}{2}}}{\Gamma\left(1+\frac{n-1}{2}\right)} \lambda^{\frac{1-n}{2}} (\text{Area } K) + o\left(\lambda^{\frac{n}{2}-1}\right)$$

The proof of the Theorem in case a) depends on asymptotic expressions for the distorted plane waves in terms of oscillatory integrals. The case b) is deeper and makes essential use of recent work of R.B. Melrose.

For compact manifolds one knows that the presence of periodic geodesics on M tends to make $n(\lambda)$ more irregular as $\lambda \to \infty$. In cases a) and b) we suspect that the effect of periodic geodesics — in b) one is allowed to reflect geodesics off K — will be significant. It is conceivable that even estimates like

$$(3) \qquad s(\lambda) = c_n \lambda^{\frac{n}{2}} + o\left(\lambda^{\frac{n}{2}}\right) \qquad \lambda \text{ large}$$

can fail to hold when there are periodic geodesics, but we have no examples. However, one notices that the hypothesis of the Theorem in case b) is much stronger than that of no periodic geodesics. We believe that at least (3) should hold in case b) as long as there are no trapped relected geodesics. Progress on this problem has been made by Jensen and Kato who show in [1] that, combining (1), a Tauberian theorem of G. Freud and results of J. Helton and Ralston, one can prove (3) holds when K is starlike.

1. A. Jensen and T. Kato, "Asymptotic behavior of the scattering phase for exterior domains", Commun. P.D.E., 3 (1978), 1165-1195.

2. A. Majda and J. Ralston, "An analogue of Weyl's theorem for unbounded domains, I, II, III", Duke Math. J. 45 (1978), 183-196, 513-536. III is available in preprint.

DEPARTMENT OF MATHEMATICS
UNIVERSITY OF CALIFORNIA, BERKELEY
BERKELEY, CALIFORNIA 94720

and

DEPARTMENT OF MATHEMATICS
UNIVERSITY OF CALIFORNIA, LOS ANGELES
LOS ANGELES, CALIFORNIA 90024

Proceedings of Symposia in Pure Mathematics
Volume 36, 1980

WEYL'S CONJECTURE FOR MANIFOLDS WITH CONCAVE BOUNDARY

R.B. Melrose*

ABSTRACT. Using results from the theory of hyperbolic equations and methods of microlocal analysis it is shown that for a generic class of compact Riemann manifolds with geodesically concave boundary the counting functions for the eigenvalues of the Dirichlet and Neumann problems for the Laplacian have two asymptotic terms. This resolves in these cases, in the affirmative, conjectures of Weyl and Polya.

Let M be a compact Riemann manifold with boundary and let Δ be its Laplace-Beltrami operator. We shall denote by $sp(M) = sp_N(M) = \{0 = \lambda_1 < \lambda_2 \leq \ldots\}$ the eigenvalue sequence of the square root of the self-adjoint extension of $-\Delta$ with Neumann boundary condition, $\partial_\nu u|_{\partial M} = 0$. (The results discussed here all extend to the Dirichlet boundary condition, $u|_{\partial M} = 0$, with the proofs being slightly simpler). The number of such eigenvalues less than λ is then given by:

$$N(\lambda) = \min \{j; \lambda_{j+1} \geq \lambda\}.$$

In order to discuss the behavior of $N(\lambda)$ for large λ we shall impose certain geometric conditions on M. First we require

(1.1) ∂M is everywhere strictly geodesically concave.

Under this condition (and indeed more generally, see [20]) if we denote by S^*M the unit cosphere bundle and by B^*M the unit coball bundle on M the geodesic flow of M is well-defined on the topological space $BS^*M = S^*(M\backslash\partial M) \cup B^*(\partial M) = (S^*M\backslash N^*(\partial M)/\iota^*$ where $\iota : \partial M \hookrightarrow M$ and $N^*\partial M$ is the conormal bundle. Moreover, the flow $\Phi^t : BS^*M \longrightarrow BS^*M$ is C^∞ in the complement of the closed set

$$A(t) = \{\rho \in BS^*M;\ \Phi^s(\rho) \in S^*(\partial M) \quad \text{for some} \quad |s| \leq t\},$$

1970 Mathematics Subject Classification 65N25.
*This work was supported in part by NSF Grant #7704148.

with respect to the natural C^∞ structure in $BS^*M\setminus S^*\partial M$. Note that BS^*M carries a natural measure preserved by Φ and with respect to which $S^*\partial M$ has measure zero, so meas.$(A(\infty)) = 0$. Let $B \subset BS^*M\setminus A(\infty)$ be the set of points at which $\Phi^t = \mathrm{Id}$ to infinite order for some $t \neq 0$. We impose a condition introduced first by Duistermaat & Guillemin:

$$\text{(1.2)} \qquad \text{meas.}(B) = 0.$$

Finally, for technical reasons discussed briefly below we shall demand that

(1.3) For $t \neq 0$, Φ^t fixes no points in $S^*\partial M$.

It can be shown that (1.2), (1.3) hold generically amongst manifolds satisfying (1.1).

(1.4) THEOREM. If M is a compact Riemann manifold with boundary satisfying (1.1), (1.2), and (1.3) then as $\lambda \to \infty$

$$\text{(1.5)} \qquad N(\lambda) = (2\pi)^{-n} c_n \mathrm{vol}(M)\lambda^n + c_{n,N}\mathrm{vol}(\partial M)\lambda^{n-1} + o(\lambda^{n-1}),$$

with $c_{n,N} > 0$ an absolute constant depending only on the dimension of M, c_n the volume of the unit ball in $\mathbb{R}^n$.

Under conditions (1.1) - (1.3) the counting function for the Dirichlet problem has a similar asymptotic expansion:

$$N_D(\lambda) = (2\pi)^{-n} c_n \mathrm{vol}(M)\lambda^n - c_{n,D}\mathrm{vol}(\partial M)\lambda^{n-1} + o(\lambda^{n-1})$$

with $c_{n,D} > 0$.

(1.6) COROLLARY. If M satisfies (1.1) - (1.3) then the Polya conjecture is asymptotically valid: $\exists \bar{\lambda} = \bar{\lambda}(M)$ such that

$$N_N(\lambda) \geq (2\pi)^{-n} c_n \mathrm{vol}(M)\lambda^n \qquad \lambda \geq \bar{\lambda}$$

$$N_D(\lambda) \leq (2\pi)^{-n} c_n \mathrm{vol}(M)\lambda^n \qquad \lambda \geq \bar{\lambda}.$$

When $\partial M = \phi$ and $\mathrm{sp}(M) = \{0 < \lambda_1 \leq \lambda_2 \ldots\}$ is taken to be the eigenvalue sequence of the square root of the self-adjoint extension of $-\Delta$, Avakumovic ([2]) and Lewitan ([12], [13]) proved the Weyl estimate with optimal remainder term: $N(\lambda) = (2\pi)^{-n}\mathrm{vol}(M)\lambda^n + O(\lambda^{n-1})$. This result was extended to more general, and higher order, operators by Hörmander ([8]) and using his simpler approach Duistermaat & Guillemin ([4]) showed that condition (1.2), limiting the size of the set of closed geodesics, implies the improved remainder estimate $N(\lambda) = (2\pi)^{-n}\mathrm{vol}(M)\lambda^n + o(\lambda^{n-1})$ in that case. For a general class of manifolds with boundary Seeley ([21], [22]) has recently proved the original Weyl estimate without any restrictive assumption, such as (1.1), on the boundary curvature. However, it seems difficult to extend his direct estimation to prove the refined bounds (1.5) which show that the volume of the boundary can be seen concretely in the eigenvalue asumptotics.

The proof of Theorem 1.4 is based on the method, used by Lewitan, Hörmander, Duistermaat & Guillemin and Seeley, of the analyis of the cosine transform, $\hat{f}(t)$, or Fourier transform, $\hat{e}(t)$, of the spectral measure

$$e(\lambda) = \sum_{\lambda_j \in \mathrm{sp}(M)} \delta(\lambda-\lambda_j) \in \mathcal{S}'(\mathbb{R})$$

$$\hat{e}(t) = \int e^{-i\lambda t} e(\lambda)d\lambda$$

$$\hat{f}(t) = \tfrac{1}{2}(\hat{e}(t) + \hat{e}(-t)).$$

The basis of this method is that in recovering $e(\lambda)$, and hence $N(\lambda)$, from $\hat{e}(t)$ by inverse Fourier transformation the leading contribution comes from the neighborhood of $t = 0$.

(1.7) THEOREM. If M is a compact Riemann manifold with strictly geodesically concave boundary there exists $T > 0$ such that if $\rho \in C_0^\infty(\mathbb{R})$ has support in $(-T,T)$, then $\rho(t)\hat{e}(t)$ is a classical Lagrangian distribution associated to $T_0^*\mathbb{R}$. Thus

$$\rho(t)\hat{e}(t) = \int e^{i\lambda t} a(\lambda)d\lambda \tag{1.8}$$

where $a(\lambda) \in S_{cl}^{n-1}(\mathbb{R})$, i.e., $a(\lambda) \sim \sum_{j\leq n-1} a_j\lambda^j$ as $\lambda \to \infty$.

Here, as in the boundaryless case T is the length of the shortest closed geodesic on M. Using a simple Tauberian argument to exploit the positivity of the measure e (cf. Hörmander [8, pp. 209-211]) the standard Weyl estimate, as obtained by Seeley, follows directly from Theorem 1.7 once the leading term a_{n-1} in the expansion of $a(\lambda)$ is identified with a multiple of

$\mathrm{vol}(M)$. In fact, the asymptotic expansion of a can be directly related to the known asymptotic expansion of the trace of the heat kernel so that the first two terms are readily identified with multiples of $\mathrm{vol}(M)$ and $\mathrm{vol}(\partial M)$ respectively. To improve the remainder estimates sufficiently to give (1.5) extra estimates on $\hat{e}(t)$ for large values of t are needed. It is in obtaining these estimates, following Duistermaat & Guillemin ([4]) that the conditions (1.2), (1.3) are used. As originally pointed out by Avakumovic, (1.2) is necessary if $N(\lambda)$ is to have a simple asymptotic expansion down to $o(\lambda^{n-1})$; on the other hand (1.3) is imposed for purely technical reasons and the result, (1.6), is probably true without it.

In outline the contents of this paper are as follows. The basis for the computation of $\hat{e}(t)$ is laid in Section 2 where the parametrix for the initial value problem, with homogeneous boundary condition:

$$\begin{cases} (\partial_t^2 - \Delta)u = 0 & \text{in } \mathbb{R}_t \times M \\ \partial_\nu u|_{\partial M} = 0 & \\ u|_{t=0} = u_0, \quad \partial_t u|_{t=0} = 0 & \end{cases}$$

is written down in terms of oscillatory integrals for $|t|$ small. In doing so, the results of [16], [19], Taylor [23] on the form of the parametrix for the associated boundary value problem are used freely, as are the results on symplectic geometry contained in [17]. The notion of almost-analytic extension (see Hörmander [9]) and transmission properties of oscillatory integrals (see Boutet de Monvel [3], Gårding [5], Hischowitz & Piriou [7]) also enter. Theorem 1.7 is proved in Section 3. By applying stationary phase arguments, and analogous methods for integrals with complex phase (cf. Melin & Sjöstrand [15]), the asymptotic expansion of the cosine transform, $\hat{f}(t)$, is obtained from the parametrix. The properties of Airy integrals and some associated Airy functions are used here. Then, the modifications needed to discuss the sine transform $\hat{g}(t) = \frac{1}{2i}(\hat{e}(t) - \hat{e}(-t))$ and so obtain $\hat{e}(t)$ are briefly discussed. Theorem 1.4 itself is proved in Section 4 using results of Andersson & Melrose ([1]) and the hypotheses (1.2), (1.3) to extend the argument of Duistermaat & Guillemin [4, Theorem 3.5].

The author is grateful to Bob Seeley, Jim Ralston, Andy Majda and Victor Guillemin for helpful discussions on the matters considered here.

2. WAVE KERNEL FOR SHORT TIMES.

The methods of geometric optics were refined by Lax, [11], allowing the solution of the Cauchy problem for the wave operator over short times through the use of oscillatory integrals. Thus, the solution to

$$\partial_t^2 u - \Delta u = 0$$

$$u(0,z) = u_0(z), \quad \partial_t u(0,z) = 0$$

can be written locally in the form:

$$u(t,z) = \int e^{i\phi_+(t,z,\zeta) - i\phi_+(0,z',\zeta)} a_+(t,z,\zeta)u_0(z')dz'd\zeta + \int e^{i\phi_-(t,z,\zeta) - i\phi_-(0,z',\zeta)} a_-(t,z,\zeta)u_0(z')dz'd\zeta \tag{2.1}$$

for t small with $a_\pm$ classical symbols and $\phi_\pm(0,z',\zeta) = z'\cdot\zeta$. Hörmander, [7], further exploited this idea in order to estimate the spectral function of the Laplacian near an open point of the manifold; in so doing he made a special choice of the phase functions in (2.1), namely with the $\phi_\pm$ linear in the distinguished variable t. To allow such estimation near boundary points it is useful to make an even more special choice of phase functions, retaining linearity in t but demanding extra properties with respect to the boundary of the manifold.

To introduce the form of the phase functions, and for later reference, we first give the parametrix for the (uniquely solvable) boundary value problem (see [16], [19], Taylor [23])

$$\begin{aligned} &\partial_t^2 v - \Delta v = 0 \\ &\partial_\nu v|_{\partial X} = v_0 \in \mathcal{E}'(\mathbb{R} \times \partial X) \\ &v = 0 \quad t << 0 . \end{aligned} \tag{2.2}$$

Unlike the Cauchy problem, the boundary value problem (2.2) exhibits different behavior microlocally in various regions of $T^*(\mathbb{R} \times \partial X)$; only in the hyperbolic part of the cotangent bundle to the boundary can the solution v be written in the form of simple oscillatory integrals (2.1). First we introduce normal coordinates $z = (x,y)$ near ∂X so that

$$\Delta = \partial_x^2 + Q(x,y,\partial_y)$$

where the symbol $q(x,y,\eta)$ is the induced Riemannian structure on $x = \text{const.}$ The three natural regions of $T^*(\mathbb{R} \times \partial X)$, in canonically dual coordinates (t,y,τ,η), are then the hyperbolic region(s) $H_\pm$ where $\tau^2 > q(0,y,\eta)$, $\pm\tau > 0$, the elliptic region E where $\tau^2 < q(0,y,\eta)$ and the glancing surface(s) $G_\pm$ where $\tau^2 = q(0,y,\eta)$, $\pm\tau > 0$.

Now the solution to (2.2) can be expressed as

$$v = v_H^\pm + v_G^\pm + v_E \tag{2.3}$$

near $x = 0$, where

$$v_H^\pm = \int e^{i\psi_\pm(x,t,y,\tau,\eta) - i\psi_\pm(0,t',y',\tau,\eta)} a_H^\pm(x,t,y,\tau,\eta)v(t',y')dt'dy'd\tau d\eta \tag{2.4}$$

$$\begin{aligned} v_G^\pm = \int e^{i\alpha(x,t,y,\tau,\eta) - i\alpha(0,t',y',\tau,\eta)} & (A_\pm(\beta(0,\tau,\eta))^{-1} \\ \times (a_G^\pm(x,t,y,\tau,\eta)A_\pm'(\beta(x,y,\tau,\eta)) & + b_G^\pm(x,t,y,\tau,\eta)A_\pm'(\beta(x,y,\tau,\eta))) \\ & (N_+^{-1}v)(t',y')dt'dy'd\tau d\eta \end{aligned} \tag{2.5}$$

$$v_E = \int e^{i\psi_E(x,t,y,\tau,\eta) - i\psi_E(0,t',y',\tau,\eta)} a_E(x,t,y,\tau,\eta)v(t',y')dt'dy'd\tau d\eta \tag{2.6}$$

In (2.4) $a_H^\pm \in S_{cl}^{-1}$ are supported in $H_\pm$ and the phase functions $\psi_\pm$ are the analogues of those used by Hörmander in [7] - in this case they are easy to construct since the solutions to the characteristic equations

$$\partial_x\psi_\pm = \pm((\partial_t\psi_\pm)^2 - q(x,y,\partial_y\psi_\pm))^{\frac{1}{2}} \tag{2.7}$$

with initial data $\psi_\pm(0,t,y,\tau,\eta) = t\tau + y\cdot\eta$, obtained by Hamilton-Jacobi theory, are linear in t. The $a_H^\pm$ are determined by transport equations in the usual way. Note that both phase functions are defined throughout $H = H_+ \cup H_-$, the use of ψ_+ in the region H_- and ψ_- in the region H_+ in (2.4) corresponds to the solution of the backward analogue of (2.2).

The more complicated oscillatory integrals, (2.5), near the glancing surfaces $G_\pm$, are based on the properties of the Airy functions $A_\pm(s) = Ai(e^{\pm 2\pi i/3}s)$, where Ai is the solution of $Ai''(s) = sAi(s)$ which is exponentially decreasing as $s \to \infty$. The symbols $a_G^\pm \in S_{cl}^{-2/3}$, $b_G^\pm \in S_{cl}^{-1}$ have supports in small conic neighborhoods of the images $\tau^2 = |\eta|^2$, $\pm\tau > 0$ of the glancing surfaces. Both α and β are real C^∞

functions, homogeneous of degrees 1 and $\frac{2}{3}$ respectively, in (τ,η). To gain simplicity in (2.5) we impose the following conditions on α,β:

(2.8) $\alpha(x,t,y,\tau,\eta)$ and $\beta(x,y,\tau,\eta)$ are respectively linear in and independent of t and both are invariant under the reflection $(t,\tau) \mapsto (-t,-\tau)$,

(2.9) $\beta(0,y,\tau,\eta) = \tau^2 - |\eta|^2, \quad \partial_x\beta < 0,$

(2.10) $\det(\partial_{(t,y)}\partial_{(\tau,\eta)}\alpha(0,t,y,\tau,\eta)) \neq 0 \quad \text{on} \quad \tau^2 = |\eta|^2.$

The main property of the phase functions α, β is that they satisfy the modified characteristic equations: with p the symbol of $P = \partial_t^2 - \Delta$,

$$p(x,y,d_{(x,t,y)}\alpha) = p(x,y,d_{(x,t,y)}\beta)\cdot\beta \tag{2.11}$$

$$\partial_x\alpha\partial_x\beta + q(x,y,d_{(t,y)}\alpha,d_{(t,y)}\beta) = 0 \tag{2.12}$$

where $q(x,y,\tau,\eta,\tau',\eta')$ is the polarized form of q and equations (2.11), (2.12) are only required to hold in the region $x \geq 0$, $\beta \leq 0$ and in the sense of formal power series on $x = 0$ in the region $\beta \geq 0$.

The existence of such phase functions α,β follows from the basic assumption (1.1), of geodesic concavity, using the results of [17] and the t-translation invariance. The symbols $a_G^{\pm}$, $b_G^{\pm}$ are obtained by solving transport equations in the same region $\beta \leq 0$, $x \geq 0$ and as formal power series equations at $x = 0$ in $\beta \geq 0$. The important property of a, b is that the Airy operator (see [18]) $v_G|_{x=0} = Av$ is classical and elliptic near $G_{\pm}$. The operator N_+^{-1} is then an Airy operator so that $\partial_\nu v_G|_{x=0} \equiv v$ microlocally near $G_{\pm}$. The invertibility of N_+ follows from the ellipticity of A, N_+^{-1} is an Airy operator of negative type (see [17]) - a nonclassical pseudodifferential operator:

$$(N_+^{-1})v(t',\eta') = \int e^{i\alpha(0,t',y',\tau',\eta') - i\alpha(0,t'',y'',\tau',\eta')}[\tilde{a}(t',y',t'',y'',\tau',\eta')$$

$$+ \tilde{b}(t',y',t'',y'',\tau',\eta')(A_{\pm}/A'_{\pm})(\beta(0,\tau',\eta'))]$$

$$A'_{\pm}(\beta(0,\tau',\eta'))^{-1}\lambda y''dt''d\tau'\lambda y'.$$

where $\tilde{a}$, $\tilde{b}$ are classical of orders -1 and $-2/3$ respectively and $\tilde{a} = 0$ on $\beta_0 = 0$ $((\tau')^2 = |\eta'|^2)$. Inserting this into (2.5) the integral over (t',y'), (τ',η') can be carried out by stationary phase, giving the following form

$$(2.13) \quad v_G^{\pm} = \int e^{i\alpha(x,t,y,\tau,\eta) - i\alpha(0,t',y',\tau,\eta)} \times \left[\tilde{a}_G \frac{A_{\pm}(\beta(x,y,\tau,\eta))}{A_{\pm}(\beta(0,\tau,\eta))} + \tilde{b}_G \frac{A'_{\pm}(\beta(x,y,\tau,\eta))}{A_{\pm}(\beta(0,\tau,\eta))} \right] v(t',y')dt'dy'd\tau\lambda y$$

where a, b are no longer classical symbols but are of the form:

$$(2.14) \quad \begin{cases} \tilde{a}_G \sim \sum_{j\geq 0} a_{j+1}(x,t,y,t',y',\tau,\eta)\Phi_{\pm}^{-1,j}(\beta(0,\tau,\eta)) + a_0(x,t,y,t',y',\tau,\eta) \\ \tilde{b}_G \sim \sum_{j\geq 0} b_{j+1}(x,t,y,t',y',\tau,\eta)\Phi_{\pm}^{-1,j}(\beta(0,\tau,\eta)) + b_0(x,t,y,t',y',\tau,\eta) \end{cases}$$

with the a_j, b_j classical and $\Phi_{\pm}^{-1}(z) = \frac{A_{\pm}(z)}{A'_{\pm}(z)}$, $\Phi^{-1,j}(z) = \frac{d^j}{dz^i}\Phi(z)$. The orders of a_j, b_j are $\frac{-1}{3} - \frac{j}{3}$, $\frac{-2}{3} - \frac{j}{3}$ respectively.

The third element v_E, the elliptic part of the solution to (2.2), is similar to the hyperbolic part (2.5) except that it is given by a Fourier integral operator with complex phase function. Indeed, in the region $\tau^2 < q(0,y,\eta)$ the characteristic equation $p(x,y,d_{(x,t,y)}\psi_E) = 0$, $\psi_E(0,t,y,\tau,\eta) = t\tau + y\cdot\eta$ has no real solutions, but has a unique solution in the sense of formal power series at $x = 0$ with $\operatorname{Im}\psi_E \geq 0$; ψ_E is linear in t and can be extended to be C^∞ and homogeneous of degree 1 so that these characteristic equations hold to infinite order at $x = 0$. The Taylor series of the symbol a_E is then obtained as the solution of a transport equation.

The phase functions α,β not only allow the solution to the problem (2.2) but can also be used to give the solution to the initial-boundary value problem

$$(2.15) \quad \begin{aligned} &\partial_t^2 w - \Delta w = 0 \\ &\partial_\nu w|_{\partial X} = 0 \\ &w(0,z) = w_0(z),\ \partial_t w(0,z) = 0 \end{aligned}$$

in the form of a sum of oscillatory integrals. Working in the same normal coordinates (x,y) we shall show how to solve (2.15) for $w_0 = \delta(x-x')\delta(y-y')$ for $x' > 0$, but uniformly as $x' \to 0$. The first step is to extend the quadratic form $q(x,y,\eta)$ smoothly into $x < 0$ and express the solution to the pure initial value problem in a form adapted to the boundary $x = 0$.

Observe that the characteristic equations (2.11), (2.12) can be written

$$p(x,y,d_{(x,t,y)}(\alpha \pm \tfrac{2}{3}(-\beta)^{3/2}) = 0 \qquad \beta \le 0, x \ge 0$$

so that the phase function

$$(2.16) \qquad \Phi(x,t,y,s,\tau,\eta) = \alpha(x,t,y,\tau,\eta) + s|\eta|^{\frac{1}{3}}\beta(x,y,\tau,\eta) + \tfrac{1}{3}s^3|\eta|$$

(defined near $\tau = \pm|\eta|$) is a non-degenerate phase function which can be used in place of $\phi_\pm$ in (2.1) at least in the region $|\xi| \le \varepsilon|\zeta|$, ξ being the dual variable to x. Thus, the solution to the initial value problem can be written $u = u_T^\pm + u_G$ (transversal and glancing terms):

$$(2.17) \qquad u_G = \int e^{i\Phi(x,t,y,s,\tau,\eta)-i\Phi(x',0,y',r,\tau,\eta)} c_G(x,t,y,s,r,\tau,\eta)\,d\tau d\eta dr ds,$$

$$(2.18) \qquad u_T^\pm = \int e^{i\phi_\pm(x,t,y,\zeta) - i\phi_\pm(x',0,y',\zeta)} c_T^\pm(x,t,y,\zeta)\,d\zeta$$

where in the integrals (2.18) we can assume that $c_T^\pm$ have support in the region where $|\partial_x\phi_\pm| > \delta|\partial_y\phi_\pm|$ for some $\delta > 0$, and the support of c_G can be taken to be in $|s|$, $|r| \le \varepsilon$, $|\tau^2 - |\eta|^2| \le \varepsilon\tau^2$ for some $\varepsilon > 0$; $c_G \in S^1_{cl}$, $c_T^\pm \in S^0_{cl}$.

To obtain the solution to (2.15) we add terms to u so that the boundary condition is satisfied. Since $WF(u)$ is contained in the characteristic variety of P the product $u' = H(t)u(x,t,y;x',y')$ is well-defined where H is the Heaviside function; it satisfies

$$(\partial_t^2 - \Delta)u' = \partial_t\delta(t)\delta(x-x')\delta(y-y')$$

$$u' = 0 \quad \text{in} \quad t < 0.$$

In view of Peetre's theorem on partial hypoellipticity with respect to a non-characteristic boundary the limits from $x < 0$, $b_1(u) = \partial_x u'|_{x=0}$, $b_0(u) = u'|_{x=0}$ are well-defined uniformly for $x' \ge 0$ and are supported in $t \ge 0$. Similarly the distribution u_c, obtained by cutting off u' in $x < 0$ is defined and satisfies

$$(\partial_t^2 - \Delta)u_c = \partial_t\delta(t)\delta(x-x')\delta(y-y') - 2b_1(t,y;x',y') \otimes \delta(x)$$
$$- b_0(t,y;x',y') \otimes \delta'(x).$$

Thus, if v is the solution to (2.2) when $v_0 = b_1(u)$ then $w_c = u_c - v_c$

satisfies

$$(2.20)\qquad (\partial_t^2 - \Delta)w_c = \partial_t\delta(t)\delta(x-x')\delta(y-y') - \delta'(x) \otimes w'(t,y;x',y')$$

for some w' supported in $t \geq 0$. Since $w_c(x,t,y;x',y')$ is a C^∞ function of $x' \geq 0$ and y', with distributional values, and (2.20) is the weak form of (2.2), w_c is the kernel of the wave operator that we are seeking.

Observe that these operations on u to produce w_c can also be carried out on each of the terms u_G, $u_T^{\pm}$, since the data on $t = 0$ for these solutions to the initial value problem is C^∞ in the variable x in $x \leq 0$ uniformly for $x' \geq 0$. Thus, to find w_c we need only carry out these operations explicitly for $x' > 0$, obtaining w_1, $w_2^{\pm}$ such that $w = w_1 + w_2^+ + w_2^-$. From (2.17) we see that

$$b_1(u_G) = H(t) \int e^{i\Phi(0,t,y,s,\tau,\eta) - i\Phi(x',0,y',r,\tau,\eta)} d_G \, d\tau d\eta dr\, ds$$

with $d_G \in S^2_{cl}$ supported near $s = r = 0$, $\tau^2 = |\eta|^2$. The operations, on u_G, of multiplication by $H(t)$ and restriction to $x = 0$ are sanctioned by wavefront set considerations. Thus

$$(2.21)\qquad w_1 = (u_G)_c + w_{1,G} + w_{1,H}^+ + w_{1,H}^- + w_{1,E}$$

is given by the application of (2.4), (2.5), (2.6) to $b_1(u_G)$:

$$(2.22)\qquad w_{1,G} = \iint_0^\infty e^{i\alpha(x,t,y,\tau',\eta') - i\alpha(0,t',y',\tau',\eta')}[a_G A_{\pm}'(\beta) + b_G A_{\pm}(\beta)]/A_{\pm}(\beta_0)$$
$$e^{i\Phi(0,t',y',s,\tau,\eta) - i\Phi(x'',0,y'',r,\tau,\eta)} d_G dt'dy'dsdrd\tau d\tau' d\eta d\eta'$$

$$(2.23)\qquad w_{1,H} = \iint_0^\infty e^{i\psi_{\pm}(x,t,y,\tau',\eta') - i\psi_{\pm}(0,t',y',\tau',\eta')} a_H^{\pm}$$
$$e^{i\Phi(0,t',y',s,\tau,\eta) - i\Phi(x'',0,y'',r,\tau,\eta)} d_G dt'dy'dsdrd\tau d\tau' d\eta d\eta'$$

$$(2.24)\qquad w_{1,E} = \iint_0^\infty e^{i\psi_E(x,t,y,\tau',\eta') - i\psi_E(0,t',y',\tau',\eta')} a_E$$
$$e^{i\Phi(0,t',y',s,\tau,\eta) - i\Phi(x'',0,y'',r,\tau,\eta)} d_G dt'dy'dsdrd\tau d\tau' d\eta d\eta'$$

In (2.22), a_G, b_G have expansions in terms of Airy quotients of the form (2.14).

Notice that we cannot, quite, use standard techniques to simplify these integrals since the distribution $b_1(u_G)$ is not simply Lagrangian. In fact, the main difficulty in the proof of the asymptotic expansion (1.7) comes from the term (2.22) which is 'doubly glancing.'

From (2.18) we obtain the other terms in w:

$$w_2^{\pm} = (u_H)_c^{\pm} + w_{2,G}^{\pm} + w_{2,H}^{\pm,+} + w_{2,H}^{\pm,-} + w_{2,E}^{\pm} \tag{2.25}$$

$$\begin{aligned} w_{2,G}^{\pm} = \iint_0^\infty & e^{i\alpha(x,t,y,\tau',\eta') - i\alpha(0,t',y',\tau',\eta')}[a_G A_\pm'(\beta) \\ & + b_G A_\pm(\beta)]/A_\pm(\beta_0) \\ & e^{i\phi_\pm(0,t',y',\tau,\eta) - i\phi_\pm(x'',0,y'',\tau,\eta)} d_T^{\pm} dt'dy'd\tau d\tau'd\eta d\eta' \end{aligned} \tag{2.26}$$

$$\begin{aligned} w_{2,H}^{\pm\,\pm} = \iint_0^\infty & e^{i\psi_\pm(x,t,y,\tau',\eta') - i\psi_\pm(0,t',y',\tau',\eta')} a_H^{\pm} \\ & e^{i\phi_\pm(0,t',y',\tau,\eta) - i\psi_\pm(x'',0,y'',\tau,\eta)} d_H^{\pm}\, dt'dy'd\tau d\tau'd\eta d\eta' \end{aligned} \tag{2.27}$$

$$\begin{aligned} w_{2,E}^{\pm} = \iint_0^\infty & e^{i\psi_E(x,t,y,\tau',\eta') - i\psi_E(0,t',y',\tau',\eta')} a_E \\ & e^{i\phi_\pm(0,t',y',\tau,\eta) - i\phi_\pm(x'',0,y'',\tau,\eta)} d_T^{\pm}\, dt'dy'd\tau d\tau'd\eta d\eta'. \end{aligned} \tag{2.28}$$

From wave front set considerations it is only clear that these distributions are defined as smooth functions of x' in $x' > 0$. However, from the use of Peetre's theorem above this C^∞ dependence extends down to $x' = 0$. To see this explicitly in (2.26), (2.27) and (2.28) note that the only possible source of difficulty is the neighborhood of $t' = 0$, and the fact that the initial data for $u_T^{\pm}$ is C^∞ in x in $x \leq 0$ (even in $x \leq x'$) is reflected in appropriate transmission conditions on the Taylor series in t of $d_T^{\pm}$.

3. COSINE TRANSFORM OF THE SPECTRAL MEASURE.

In this section we show that the cosine transform

$$\hat{f}(t) = \int_{\mathbb{R}} \cos(t\lambda)e(\lambda)d\lambda = \int_{\mathbb{R}} w(t,z,z)dz$$

of the spectral measure $e(\lambda)$ has a classical asymptotic expansion as $t \downarrow 0$. To do this we first examine each of the terms (2.22), (2.23), (2.24), (2.26), (2.27), (2.28) integrated over the diagonal of $M\times M$.

We start with (2.22) in which $\beta_0 = (\tau')^2 - |\eta'|^2$ and $\beta = \beta(x,y,\tau',\eta')$. The independence of β_0 of the t', y' variables allows the (y',η)-integrals to be carried out by stationary phase. In fact, from (2.10),

$$-\alpha(0,t',y',\tau',\eta') + \alpha(0,t',y',\tau,\eta)$$

$$= (y' + t'g)(\eta-\eta') + t'(\tau-\tau')\mu$$

g,μ being C^∞ and homogeneous of degree zero, with $\mu\tau > 0$ so

$$(3.1)\quad w_{1,G}(t,z,z") = \iint_0^\infty e^{i\alpha(x,y,y,\tau',\eta') - it'(\tau'-\tau)\mu + is\beta_0'|\eta'|^{\frac{1}{3}} + \frac{is^3}{3}|\eta'|} \times e^{-i\alpha(x",0,y",\tau,\eta') - ir\beta"|\eta'|^{\frac{1}{3}} - i\frac{r^3}{3}|\eta'|} As(\beta,\beta_0)dt'd\tau'd\eta'd\tau dsdr$$

where from now on we write

$$As(\beta,\beta_0) = (aA(\beta) + bA'(\beta))/A(\beta_0)$$

for a general 'Airy term,' with a, b of the form (2.14) and

$$\beta_0' = (\tau^2 - |\eta'|^2)|\eta'|^{-4/3}, \quad \beta" = \beta(x",0,y",\tau,\eta').$$

Next we carry out the (t',τ)-integrals. The stationary points of the phase are at $\tau = \tau'$, $\bar\mu t' = -2s\tau' + 2r\tau'(1 + x"\gamma)$ where $\bar\mu > 0$; since $t' \geq 0$ we get two terms from (3.1)

$$(3.2)\quad w_{1,G} = w_{1,G}^{(1)} + w_{1,G}^{(2)}$$

$$(3.3)\quad w_{1,G}^{(1)}(t,z,z",$$

$$= \int_{\tau'(s-r(1+x"\gamma))\leq 0} e^{i\alpha(x,t,y,\tau',\eta') - i\alpha(x",0,y",\tau',\eta')} \times e^{is\beta_0|\eta'|^{\frac{1}{3}} + is^3\frac{|\eta'|}{3}} \times e^{ir\bar\beta - ir^2|\eta'|/3} As(\beta,\beta_0)dsdrd\tau'd\eta'$$

$$(3.4)\qquad w^{(2)}_{1,G}(t,z,z'') = \int e^{i\alpha(x,t,y,\tau',\eta') - i\alpha(x'',0,y'',\tau',\eta') + irx''\bar{\gamma}}\, As(\beta,\beta_0)\,dr\,d\tau'\,d\eta'$$

In (3.3) the region of integration, $\tau'(s-r(1+x''\gamma)) \leq 0$, depends on the sign of τ', changing with the Airy function, and $\bar{\beta} = \beta_0 - x''\gamma(x'',y'',r,s,\tau',\eta')$ with $\gamma > 0$ homogeneous of degree $2/3$. Similarly, in (3.4) $\bar{\gamma} > 0$ is homogeneous of degree 2/3. We proceed to examine the main term (3.3), restricted to $x = x''$, $y = y''$ and integrated. We can introduce $\theta = \partial_t\alpha(x,t,y,\tau',\eta')$, $\lambda = \beta_0|\theta|^{-2/3}$ as independent phase variables and then, setting $S = s|\eta'|^{-1/3}|\theta|^{1/3}$, $R = r|\eta'|^{-1/3}|\theta|^{1/3}$,

$$(3.5)\qquad \int w^{(1)}_{1,G}(t,z,z)\,dz = \int e^{it\theta} e(t,\theta)\,d\theta$$

with

$$(3.6)\qquad e(t,\theta) = \iint_{\substack{0\\ \theta(s-r(1+x\gamma))\leq 0}}^{\infty} e^{i(s\lambda+is^3/3 - R\lambda(1+x\gamma) - R^3/3)|\theta|} As(\lambda|\theta|^{2/3}(1-x\gamma),\lambda|\theta|^{2/3})\, dx\, dy\, dR\, dS\, d\lambda .$$

Our objective is to show that e is classical. Since $x,\gamma \geq 0$ the Airy quotients in (3.6) extend holomorphically into the region $\pm \operatorname{Im}\lambda \leq 0$ and in that region have uniform asymptotic expansions. For example

$$(3.7)\qquad \frac{A_\pm(\lambda|\theta|^{2/3}(1-x\gamma))}{A_\pm(\lambda|\theta|^{2/3})} = \exp\left(\tfrac{2i}{3}(-\lambda+x\gamma)^{3/2}|\theta| - \tfrac{2i}{3}(-\lambda)^{2/3}|\theta|\right)g(\lambda,\theta),$$

$$g \sim \sum c_j(\lambda,x\gamma)(\theta)^{\frac{1}{3}-j} \qquad \operatorname{Im}\lambda \leq 0,\quad |\lambda\theta^{\frac{2}{3}}| \geq 1$$

This, together with the limitation on the regions of integration in (3.6) allows us to apply Cauchy's (Stokes') theorem to replace the real integral over λ by a complex integral over $\lambda + i\sigma$, $\sigma\theta \leq 0$. Extending

$$\tilde{A}s(\beta,\beta_0) = \tilde{a}\frac{A_\pm(\beta)}{A'_\pm(\beta_0)} + \tilde{b}\frac{A'_\pm(\beta)}{A'_\pm(\beta_0)}$$

by choosing appropriate almost-analytic extensions $\tilde{a}_j$, $\tilde{b}_j$ of the components of a, b we obtain

(3.8) $$e(t,\tau) = \int_{\theta(S-R(1+x\gamma))\leq 0} \int_{\sigma\theta\leq 0} \int_0^\infty e^{i(S-R(1+x\gamma))(\lambda + i\sigma) + \frac{i}{3}(S^3 - R^3)|\theta|}$$

$$\bar{\partial}_{(\lambda+i\sigma)} \tilde{A}s((\lambda+i\sigma)|\theta|^{\frac{2}{3}}+x\gamma|\theta|^{\frac{2}{3}}, (\lambda+i\sigma)|\theta|^{\frac{2}{3}})$$

Now, by assumption $\bar{\partial}_{(\lambda+i\sigma)} \tilde{A}s$ can be rewritten as an Airy symbol

$$a'\frac{A_\pm(\beta)}{A'_\pm(\beta_0)} + b'\frac{A'_\pm(\beta)}{A'_\pm(\beta_0)}$$

where a', b' vanish to all orders at $\sigma = 0$. Observe that the expansion (3.7) becomes uniform in $\sigma\theta \geq 0, \lambda$ after multiplication by a function vanishing to all orders at $\sigma = 0$. Then the x-integral in (3.8) can be carried out by integration by parts since the function

$$m = \partial_x(-Rx\gamma(\lambda+i\sigma) + \tfrac{2}{3}(-(\lambda+i\sigma)+a\bar{\gamma})^{\frac{3}{2}} - \tfrac{2}{3}(-(\lambda+i\sigma))^{\frac{3}{2}})$$

whilst not smooth at $\sigma = 0$ is such that a/m is C^∞ at $\sigma = 0$ if a is C^∞ and vanishes to all orders at $\sigma = 0$. The new symbol so obtained depends only on the Taylor series of $\tilde{A}s$ at $x = 0$. Now, the coefficients $\tilde{a}_j$, $\tilde{b}_j$ in the Airy symbol $\tilde{A}s((\lambda+i\sigma)|\theta|^{2/3}, (\lambda+i\sigma)|\theta|^{2/3})$ vanish to all orders at $\sigma = 0$, so the Airy symbol itself is classical in $\sigma\theta \leq 0$. Thus, we see that

(3.9) $$e(t,\theta) = \int_{\sigma\theta\leq 0} \int_{\theta(S-R)\leq 0} e^{i[(S-R)(\lambda+i\sigma) + \frac{1}{3}(S^3-R^3)]|\theta|} \, a d\sigma d\lambda dS dR$$

with a classical, so e itself is classical and we conclude that $w_{1,G}^{(1)}$ makes a classical contribution to $k(t)$.

Proceeding to the analysis of $\int w_{1,G}^{(2)}(t,z,z)dz$ we see from (3.3) that similar arguments apply. After almost analytic extension into $\sigma\theta \leq 0$ with θ,σ as above, the x-integral can be carried out giving a classical expansion. The two terms (2.23) both come from transversal compositions of Lagrangian manifolds, i.e., can be evaluated by stationary phase after integration over $z = z'$, whilst $w_{1,E}$ in (2.24) involves an analogous computation for a complex phase function. Since $(u_G)_c$ integrated over $z = z'$ is also classical, as in the open manifold case, we conclude that w_1 in (2.21) makes a classical contribution to $k(t)$.

Of the remaining terms in w, (2.26) is easily shown to give a classical term in $k(t)$ by the methods used above whereas the term (2.27), (2.28) and $(W_T)_c$ are all intrinsically classical in form. Thus, the fact that $\hat{f}(t)$ has a classical expansion at $t = 0$ is verified.

To complete the proof of Theorem 1.7 we need to carry out a similar analysis of the sine transform $\hat{g}(t)$, near $t = 0$. Writing this as a trace

$$\hat{g}(t) = \int w'(t,z,z)\,dz$$

where w' is expressed in terms of its eigenfunction expansion:

$$w'(t,z,z') = \sum_j \sin(t\lambda_j)e_j(z)e_j(z')$$

we observe that w', and hence $\hat{g}$, can be analysed by solution of the wave equation:

$$\text{(3.10)} \qquad \begin{cases} (\partial_t^2 - \Delta_q)w' = 0 \\ \partial_\nu w'|_{\partial M} = 0 \\ w'(0,z,z') = 0,\ \partial_t w'(0,z,z') = \sqrt{-\Delta}\,\delta(z,z') . \end{cases}$$

Note that here, $(-\Delta)^{\frac{1}{2}}$ is the square-root of the self-adjoint extension of $-\Delta$ with boundary condition, extended to act on distributions. To carry through the program of section 2 on w' we only need show that $\sqrt{-\Delta}\delta(z,z')$ can be written as an oscillatory integral with the same formal properties as $\delta(z,z')$, itself-including the transmission condition. In fact this is not quite true. However, $(-\Delta)^{\frac{1}{2}}$ can be expressed in terms of oscillatory integral operators within the framework developed by Boutet de Monvel [3]. If $-\Delta$ is considered as a pseudodifferential operator with transmission condition, acting on the manifold with boundary and S is the positive square-root in this sense then

$$R = (-\Delta)^{\frac{1}{2}} - S$$

is the boundary correction term, and can be written as a Fourier integral operator with complex phase. In the local, normal, coordinates used in Section 2:

$$\text{(3.11)} \qquad Rf(x,y) = \iint_0^\infty e^{i(y-y')\eta - x'\mu' - x\mu} c(x,y,x',y',\eta) f(x',y')\,dx'dy'd\eta$$

where c is a classical symbol of order 0 and $\mu, \mu' > 0$ are homogeneous of degree one in η.

Thus, the terms in w' coming from the initial data $S\delta(z,z')$ are handled precisely as in Section 2, the extra terms from the other part $R\delta(z,z')$ of the initial data are easily treated by similar methods since the complex phase in (3.11) is precisely of the type already encountered in, for example, (3.8). This completes the proof of Theorem 1.7.

4. PROOF OF THEOREM 1.4.

All the ingredients needed for the proof of Theorem 1.4 are now available. First we show that the part $(1-\rho)(t)\hat{e}(t)$ of the Fourier transform of e not estimated in Theorem 1.7 contributes a term $o(\lambda^{n-1})$ to

$$N(\lambda) = \int_{-\infty}^{\lambda} e(\lambda)d\lambda.$$

To do this we reduce the question to that handled by Duistermaat & Guillemin in the boundaryless case. Their argument (see [4, §3]) rests on two main observations. First that the contribution of $(1-\rho(t))\hat{e}(t)$ to $N(\lambda)$, to $o(\lambda^{n-1})$, depends only on the singularities of $\hat{e}(t)$ (for $t \neq 0$ of course) and secondly that these singularities can be represented as the 'traces' of Fourier integral operators of the form:

$$\hat{e}(t) = \int \left(\int e^{i\phi(t,z,z',\zeta)} a(t,z,z'\zeta)d\zeta \Big|_{z=z'} \right) dz \tag{4.1}$$

with a symbol of order 0. Now, the results of Andersson & Melrose ([1]) show that the Poisson relation is valid for $\hat{e}$, its singularities are located only at the lengths of closed geodesics. Thus the assumption (1.3) shows that these geodesics are all simply reflected at most - they have no points of tangency to ∂M. Since a representation such as (4.1) is derived from the global symbol calculus of Fourier integral operators (Hörmander [10]) and depends only on the behavior of Δ microlocally near the closed geodesics we conclude that, under the assumption (1.3) it holds as in the boundaryless case (cf. [6]). Thus, the calculations of Duistermaat & Guillemin apply without change and show that, to $o(\lambda^{n-1})$, $N(\lambda)$ is determined as $\lambda \to \infty$ from the asymptotic expansion of $\rho(t)\hat{e}(t)$ in (1.8). Since Theorem 1.7 asserts that $\rho(t)\hat{e}(t)$ has a classical expansion at $t = 0$ it only remains to evaluate the coefficients in order to obtain (1.5) (and the corresponding estimates for $N_D(\lambda)$).

To evaluate the coefficients in the expansion of

$$e_1(\lambda) = (2\pi)^{-1} \int e^{i\lambda t} \rho(t)\hat{e}(t)dt$$

as $\lambda \to \infty$, when $\rho = 1$, near $t = 0$, we refer to standard results on the expansion of the trace of the heat kernel as $s \downarrow 0$. Thus

$$e_1(\lambda) = (2\pi)^{-1} \operatorname{tr} \int \rho(t) e^{i\lambda t} e^{-i\lambda\sqrt{-\Delta}} dt$$

so,

$$\int e^{-s\lambda^2} e_1(\lambda) d\lambda = (2\pi)^{-1} \operatorname{tr} \int \rho(t) e^{i\lambda t - s\lambda^2} e^{-i\lambda\sqrt{-\Delta}} d\lambda dt \sim \operatorname{tr} e^{s\Delta} \quad \text{as} \quad s \downarrow 0$$

$$= (4\pi)^{-n/2} \operatorname{vol}(M) s^{-n/2} + \tfrac{1}{4}(4\pi)^{-n-1/2} \operatorname{vol}(\partial M) s^{-n-1/2} + O(s^{-n/2+1}).$$

(see McKean & Singer [14, p. 45].) Comparing coefficients we obtain (1.5) with $c_{n,N} > 0$ and so complete the proof of Theorem 1.4.

REFERENCES

1. Andersson, K.G., & R.B. Melrose, The propagation of singularities along gliding rays. Invent. Math. 41 (1977), 197-232.

2. Avakumovic, V.G., Über die Eigenfunktionen auf geschlossenen Riemannschen Mannigfaltigkeiten. Math. Z. 65 (1956), 327-344.

3. Boutet de Monvel, L., Boundary value problems for pseudodifferential operators. Acta Math. 126 (1971), 11-51.

4. Duistermaat, J.J., & V.W. Guillemin, The spectrum of positive elliptic operators and periodic bicharacteristics. Invent. Math. 29 (1975), 39-79.

5. Gårding, L., Sharp fronts of paired oscillatory integrals. R.I.M.S. Symposium April 1976, Kyoto.

6. Guillemin, V.W., & R.B. Melrose, Poisson summation formula on a manifold with boundary. Adv. in Math. (to appear).

7. Hirschowitz, A., & A. Piriou, Proprietes de transmission pour les distributions integrales de Fourier. Comm. in P.D.E. 4 (1979), 113-217.

8. Hörmander, L., The spectral function of an elliptic operator. Acta Math. 121 (1968), 193-218.

9. Hörmander, L., Lecture notes at the Nordic Summer School of Mathematics, 1969.

10. Hörmander, L., Fourier integral operators. I. Acta Math. 127, (1971), 79-183.

11. Lax. P.D., Asymptotic solutions of oscillatory initial value problems. Duke Math. J. 24 (1957), 627-646.

12. Lewitan, B.M., On the asymptotic behavior of the spectral function of a self-adjoint differential equation of the second order. Izv. Akad. Nauk. SSSR Sv. Mat. 16 (1952), 325-352.

13. Lewitan, B.M., On the asymptotic behavior of the spectral function and the eigenfunction expansion of self-adjoint differential equations of the second order II. Izv. Akad. Nauk. SSSR Sv. Mat. 19 (1955), 33-58.

14. McKean, H., & I. Singer, Curvature and the eigenvalues of the Laplacian. J. Diff. Geom. 1 (1967), 43-69.

15. Melin, A., & J. Sjöstrand, Fourier integral operators with complex valued phase functions. Springer Lecture Notes 459, 120-223.

16. Melrose, R.B., Microlocal parametrices for diffractive boundary value problems. 42 (1975), 605-635.

17. Melrose, R.B., Equivalence of glancing hypersurfaces. Invent. Math. 37 (1976), 165-191.

18. Melrose, R.B., Airy operators. Comm. in P.D.E. 3 (1978), 1-76.

19. Melrose, R.B., Transformation of boundary value problems, in preparation.

20. Melrose, R.B., & J. Sjöstrand, Singularities of boundary value problems I. Comm. Pure Appl. Math. 31, (1978), 593-617.

21. Seeley, R.T., A sharp asymptotic remainder estimate for the eigenvalues of the Laplacian in a domain of $\mathbb{R}^3$. Adv. in Math. 29 (1978), 144-269.

22. Seeley, R.T., An estimate near the boundary for the spectral function of the Laplacian. (to appear).

23. Taylor, M.E., Grazing rays and reflection of singularities of solutions to wave equations. Comm. Pure Appl. Math. 29 (1976), 1-38.

24. Weyl, H., Das asymptotische Verteilungsgesetz der Eigenwerte linearer partieller Differentialgleichungen. Math. Ann. 71 (1912), 441-479.

Proceedings of Symposia in Pure Mathematics
Volume 36, 1980

EXTRINSIC ESTIMATES FOR λ_1

Robert C. Reilly

ABSTRACT. In this note I compare certain extrinsic estimates of mine for λ_1, with those of Hersch (surfaces of genus 0) and Yang and Yau (surfaces of genus $g \geq 1$). Hersch's estimate turns out always better than mine, but sometimes that of Yang and Yau is not. In addition, I show that for certain hypersurfaces one of my estimates implies an intrinsic estimate.

In this note I'll make some observations on my paper [7].

The major result in [7] is the following series of estimates for the first positive eigenvalue λ_1 of the Laplace-Beltrami operator Δ of a compact, oriented, connected n-manifold M isometrically immersed in $\mathbb{R}^{n+p}$:

$(A)_r$ If $p > 1$, then

$$\lambda_1(\int_M \sigma_{r-1} dA)^2 \leq nA \int_M |\vec{\sigma}_r|^2 dA \quad (1 \leq r \leq n\ ,\ r \text{ odd})\ ;$$

if $|\vec{\sigma}_r| \neq 0$ then equality implies that M is a minimal submanifold of a hypersphere in $\mathbb{R}^{n+p}$.

$(B)_r$ If $p = 1$ then

$$\lambda_1(\int_M \sigma_{r-1} dA)^2 \leq nA \int_M \sigma_r^2\, dA \quad (0 \leq r \leq n)\ ;$$

equality implies M is a hypersphere.

The σ's are the mean curvatures on M ; when $p = 1$ they are all scalar-valued but when $p > 1$ the curvatures of odd order are normal-vector valued. In either case the even order curvatures are intrinsic scalar quantities; if $p = 1$, set σ_{-1} = - (support function) .

OBSERVATIONS. 1. Recently, Yang and Yau [10] showed that if M is a surface of genus g then $\lambda_1 \leq \frac{8\pi(g+1)}{A}$. The case $g = 0$ had already been proved by Hersch [4] . Let us compare

1980 Mathematics Subject Classification 53C42

their estimates with some of mine.

Case (a): If M is the boundary of a domain D in R^3 and V is the volume of D then inequality $(B)_0$ reduces to $\lambda_1 \leq \frac{2}{9}(A/V)^2$. (Issac Chavel [1] has independently obtained this inequality). When $g = 0$, the isoperimetric inequality $A^3 \geq 36\pi V^2$ shows that Hersch's estimate is more precise than mine. However when $g \geq 1$, my estimate is sometimes more accurate than that of Yang and Yau. For example, let M be a surface of genus $g \geq 1$ obtained by attaching g "handles" to a (round) sphere in R^3 and then smoothing out the corners. If the handles are sufficiently small, then the ratio $\frac{2A^2}{9V^2}$ for M will almost equal the corresponding ratio for the sphere, and thus nearly equal $\frac{8\pi}{A}$, while the ratio $\frac{8\pi(g+1)}{A}$ will be much greater.

Case (b): Let M be any Riemannian 2-manifold of genus g. According to Nash's theorem, M can be isometrically embedded in some Euclidean space. Such an embedding yields the inequality $(A)_1 : \lambda_1 \leq \frac{2}{A}\int_M |\vec{\sigma}_1|^2 dA$. When $g = 0$ the inequality of Chen [2] and Willmore [9], $4\pi \leq \int_M |\vec{\sigma}_1|^2 dA$, shows that Hersch's inequality is better than mine. However when $g \geq 1$, my estimate is sometimes more accurate than that of Yang and Yau. Indeed, it is easy to prove that if M is the stereographic image in R^3 of a minimal surface of area $\bar{A}$ in the unit sphere S^3 then $\int_M |\sigma_1|^2 dA = \bar{A}$. However, Blaine Lawson has pointed out to me that for each $g \geq 1$ the minimal surface $\xi_{g,1}$ (See [5]) of genus g in the unit sphere has $\bar{A} < 8\pi$. If M is the stereographic image of $\xi_{g,1}$ then

$$\frac{2}{A}\int_M |\vec{\sigma}_1|^2 dA = \frac{2}{A}\bar{A} < \frac{16\pi}{A} \leq \frac{8\pi(g+1)}{A} .$$

REMARK. If we combine the formula $\int_M |\vec{\sigma}_1|^2 dA = \bar{A}$ with the inequality of Chen and Willmore we obtain a simple proof of the well-known fact that among all the compact minimal surfaces in a Euclidean unit sphere, the totally geodesic 2-sphere has smallest area.

2. Although the estimates $(A)_r$, $(B)_r$ for λ_1 are extrinsic, they sometimes imply intrinsic estimates for hypersurfaces. Here are two examples.

(a) Suppose that M is a hypersurface in R^{n+1} and that

$\sigma_2 > 0$ on M. Newton's inequality tells us that $\sigma_1^2 \geq \sigma_2 > 0$, so σ_1 never changes sign. Without loss of generality we may assume that σ_1 is positive, so that $(B)_2$, combined with Newton's inequality, implies that $\lambda_1 (\int_M \sqrt{\sigma_2} dA)^2 \leq nA \int_M \sigma_2^2 dA$ (equality only for spheres). This is an intrinsic estimate of λ_1 because, by the Gauss curvature equation, σ_2 is the normalized Riemannian scalar curvature function on M and thus intrinsic.

(b) Suppose that M is a strictly convex hypersurface in R^{n+1}. Then Cheng and Yau [3] proved that

$$\sigma_1^2 \leq \sup_M \{\frac{-\Delta R}{2 \min_{i \neq j} R_{ijij}} + R\},$$

where R_{ijkl} is the Riemann tensor (computed relative to an orthonormal frame) and R is the scalar curvature function. Combining this fact with $(B)_1$ yields the intrinsic estimate

$$\lambda_1 \leq n \sup_M \{\frac{-\Delta R}{2 \min_{i \neq j} R_{ijij}} + R\}$$

(Equality only for spheres).

REMARKS. (1) The intrinsic estimates just given require the extrinsic hypothesis that M be a hypersurface in R^{n+1}. If $n = 2$ we can drop that hypothesis because the other intrinsic hypotheses ($\sigma_2 > 0$ in (a); convexity in (b)) already imply that M can be isometrically immersed in $\mathbb{R}^3$. This follows from Nirenberg's solution of the Weyl problem [6].

(2) Weyl [8] obtained an estimate quite like that of Cheng and Yau given above in the case $n = 2$.

BIBLIOGRAPHY

[1] Issac Chavel, "On A. Hurwitz' method in isoperimetric inequalities", Proc. American Math. Soc., 71 (1978), 275-279.

[2] Bang-Yen Chen, "On the total curvature of immersed manifolds I", American J. Math. 93 (1971), 148-162.

[3] Shiu-Yueng Cheng and Shing-Tung Yau, "Hypersurfaces with constant scalar curvature", Math. Ann. 225 (1977), No. 3, 195-204.

[4] Joseph Hersch, "Quatre properiétés isopérimétriques de membranes sphérique homogènes", CR Acad. Sci. 270 (1970), 1645-1648.

[5] H. Blaine Lawson, Jr., "Local rigidity theorems for minimal hypersurfaces", Ann. Math. 89 (1969), 187-197.

[6] Louis Nirenberg, "The Weyl and Minkowski problems in differential geometry in the large", Comm. Pure and Appl. Math., 6 (1953), 337-394.

[7] Robert C. Reilly, "On the first eigenvalue of the Laplacian for compact submanifolds of Euclidean space", Comm. Math. Helv. 52 (1977), 525-533.

[8] Hermann Weyl, "Über die Bestimmung einer geschlossenen Konvexen Fläche durch ihr Linienelement", Viertelzahrsschrift der naturforschender Gesellschaft, Zurich 61 (1916), 40-72.

[9] T. J. Willmore, "Mean curvature of immersed surfaces", An. Sti. Univ. "Al. I Cuza" Iasi Sect. I a Mat. 14 (1968), 99-103.

[10] Paul C. Yang and Shing-Tung Yau, "Eigenvalues of the Laplacian of compact Riemann surfaces and minimal submanifolds", Preprint (1978)

DEPARTMENT OF MATHEMATICS
UNIVERSITY OF CALIFORNIA, IRVINE
IRVINE, CALIFORNIA 92717

Proceedings of Symposia in Pure Mathematics
Volume 36, 1980

GEOMETRIC BOUNDS ON THE LOW EIGENVALUES OF A COMPACT SURFACE

R. Schoen, S. Wolpert and S. T. Yau

In this note we announce upper and lower bounds for the low eigenvalues of a negatively curved compact surface in terms of the lengths of certain families of geodesics (Main Theorem). Let M be a compact oriented surface of genus $g > 1$ endowed with a metric of Gauss curvature K satisfying $-1 \leqq K \leqq -k$ for some constant $k > 0$. Let $0 = \lambda_0 < \lambda_1 \leqq \lambda_2 \leqq \cdots$ be the eigenvalues of the Laplace operator on M acting on functions. Our main result says that for $1 \leqq n \leqq 2g-3$, λ_n is bounded above and below by positive constants (depending on g and k) times the length of the shortest subdivision of M into $n+1$ pieces by simple closed geodesics. Also, it follows that λ_{2g-2} has positive upper and lower bounds depending on g, k. Basically, our results say that λ_n can be small only for a surface which is nearly divided into $n+1$ pieces, each piece having negative Euler characteristic. Since one can construct surfaces having specified lengths for the disjoint simple closed Poincaré geodesics, it follows that for any n with $1 \leqq n \leqq 2g-3$, there exists a sequence of surfaces of genus g and $K \equiv -1$ so that λ_n tends to zero and λ_{n+1} has a positive lower bound. We only sketch a proof of our result in this note. Full details will appear elsewhere.

For $1 \leqq n \leqq 2g-3$, we consider the family of curves which consist of a disjoint union of simple closed geodesics dividing M into $n+1$ components. We let $\mathcal{C}_n$ denote the class of all such curves. Define a number ℓ_n by

$$\ell_n = \min\{L(C) : C \in \mathcal{C}_n\}$$

where $L(C)$ is the length of C. We now state our main result.

MAIN THEOREM. Let M be a compact oriented surface of genus $g > 1$ with a metric of Gauss curvature K. Suppose for some constant $k > 0$ we have $-1 \leqq K \leqq -k$. There exist positive constants α_1, α_2 depending only on g such that for $1 \leqq n \leqq 2g-3$, we have $\alpha_1 k^{3/2} \ell_n \leqq \lambda_n \leqq \alpha_2 \ell_n$ and $\alpha_1 k \leqq \lambda_{2g-2} \leqq \alpha_2$.

We outline the main steps in the proof of the theorem.

1980 Mathematics Subject Classification 53-XX.

1. We use a variant of the Ahlfors-Schwarz lemma to reduce the case of variable curvature to that of curvature identically equal to -1. If our given metric is $\sigma = \sigma(z)|dz|^2$, and $\mu = \mu(z)|dz|^2$ is the Poincaré metric, we use the curvature bound to show

$$\mu(z) \leqq \sigma(z) \leqq k^{-1}\mu(z).$$

From this it follows directly that

$$\ell_n(\mu) \leqq \ell_n(\sigma) \leqq k^{-1/2}\ell_n(\mu).$$

The mini-max characterization of the n th eigenvalue then implies

$$k\lambda_n(\mu) \leqq \lambda_n(\sigma) \leqq \lambda_n(\mu).$$

Putting these inequalities together, we see that a proof of the Main Theorem for the Poincaré metric μ (i.e. for $k = 1$) implies the theorem for any metric σ satisfying the hypotheses. From now on we work only with the Poincaré metric.

2. To prove the upper bound on λ_n for the Poincaré metric, there are two important ingredients. Lemma 1 is due to Bers [1].

LEMMA 1. There is a constant β_1 depending only on g such that $\ell_{2g-2} \leqq \beta_1$.

Computing the number β_1 of Lemma 1 is a very tedious combinatorial problem, and we do not know its value. The second ingredient for the upper bound is the collar theorem of L. Keen [6], Halpern [4], and Randol [8].

LEMMA 2. Let γ be a simple closed geodesic in M having length L_γ. Let K_γ be the largest symmetric embedded cylinder in M which is swept out by geodesics of a fixed length normal to γ. Let A_γ be the area of K_γ. We then have the following inequality,

$$A_\gamma \geqq 2L_\gamma \operatorname{csch}(L_\gamma/2).$$

Now suppose that C is a curve in $\mathcal{C}_n$ satisfying $L(C) = \ell_n$. Suppose $C = \gamma_1 \cup \cdots \cup \gamma_p$ where the γ_i are simple closed geodesics in M. Let $K_1, \ldots, K_p$ be the collars about $\gamma_1, \ldots, \gamma_p$ defined in Lemma 2. It is shown in [7] that these collars are disjoint. We let $M_1, \ldots, M_{n+1}$ be the connected components of $M \sim C$ and let $M_i^0 = M_i \sim (\cup_{j=1}^p K_j)$ for $1 \leqq i \leqq n+1$. For $1 \leqq i \leqq n+1$, we choose a function φ_i which takes the value 1 on M_i^0, 0 on M_j^0 for $j \neq i$, and which is extended continuously to M as a linear function of the distance across each collar K_ℓ. Let V denote the linear span of $\varphi_1, \ldots, \varphi_{n+1}$. Thus V is an $(n+1)$-dimensional subspace of $L^2(M)$, and an easy calculation, using Lemma 2 and the fact that $\ell_n \leqq \beta_1$, shows that any $\varphi \in V$ satisfies

$$\int_M |\nabla\varphi|^2\, dv \leqq \alpha_2 \ell_n \int_M \varphi^2\, dv$$

where $\nabla\varphi$ is the metric gradient of φ, dv is the volume element and α_2 is a constant depending only on g. Since V is $(n+1)$-dimensional, we can find a $\varphi \in V$ satisfying

$\int_M \varphi f_i \, dv = 0$ for $i = 0, 1, \ldots, n-1$ where $f_0, f_1, \ldots, f_{n-1}$ are the first n eigenfunctions of Δ on M. The variational characterization of λ_n,

$$\lambda_n = \inf \{\int_M |\nabla \varphi|^2 \, dv : \int_M \varphi^2 = 1, \int_M \varphi f_i = 0, \; i = 0, 1, \ldots, n-1\}$$

then shows that $\lambda_n \leqq \alpha_2 \ell_n$ which is the upper bound.

3. The first step in the proof of the lower bound is to prove the (weaker) inequality $\beta_2 \ell_n^2 \leqq \lambda_n$ for a positive β_2 depending only on g. The proof of this uses the following result of Cheeger [2].

LEMMA 3. Let Ω be a domain in M. Let $h(\Omega)$ be the following isoperimetric constant

$$h(\Omega) = \inf \frac{L(\Gamma)}{\min(A_1, A_2)}$$

where the infimum is taken over all smooth curves Γ dividing Ω into two pieces of area A_1 and A_2. Let $\mu_1(\Omega)$ be the first eigenvalue of Ω for the Neumann problem. Then it is true that $\mu_1(\Omega) \geqq \frac{1}{4} h(\Omega)^2$.

We also use the following classical result which follows from the mini-max principle for λ_n.

LEMMA 4. Suppose M is partitioned into $n+1$ disjoint regions $M_1, \ldots, M_{n+1}$. Then we have

$$\lambda_n(M) \geqq \min \{\mu_1(M_1), \ldots, \mu_1(M_{n+1})\}.$$

Let $M_1, \ldots, M_{n+1}$ be the subdivision of M obtained by removing a curve $C \in \mathcal{C}_n$ satisfying $L(C) = \ell_n$. Because of Lemmas 3 and 4, the inequality $\beta_2 \ell_n^2 \leqq \lambda_n$ will follow if we can show that, for each M_i, we have $h(M_i) \geqq \beta_3 \ell_n$ for a positive constant β_3. To prove this, we let Γ be a smooth embedded curve which divides M_i into two pieces. We enumerate the possibilities.

Case i: A component of Γ is a circle bounding a disk D in M_i. The well-known isoperimetric inequality for simply-connected domains in the hyperbolic plane (see [9]) then implies Area (D) $\leqq L(\Gamma)$, so we have $h(M_i) \geqq 1$ in this case.

Case ii: A component Γ_0 of Γ is a segment having endpoints p, q on a boundary circle γ_j of M_i, and lying entirely within the collar neighborhood K_j of γ_j provided by Lemma 2. Since K_j is topologically an annulus, it follows that Γ_0 is homotopic with fixed endpoints p, q to an arc τ contained in γ_j. Since τ is a geodesic arc and $K \leqq 0$, it follows that $L(\tau) \leqq L(\Gamma_0)$. Since $\Gamma_0 \cup \tau$ bounds a disc D in M, it follows from the isoperimetric inequality, as in Case i, that Area (D) $\leqq L(\tau) + L(\Gamma_0)$. Therefore Area (D) $\leqq 2L(\Gamma_0) \leqq 2L(\Gamma)$ and $h(M_i) \geqq 1/2$.

Case iii: A component Γ_0 of Γ is a segment with endpoints on the boundary of M and Γ_0 is not contained in any boundary collar neighborhood K_j. In this case, Γ_0 extends at least the length of a boundary collar K_j. Thus Lemmas 1 and 2 imply that $L(\Gamma_0) \geqq \beta_4$ for a constant $\beta_4 > 0$ depending only on g. Thus we get $h(M_i) \geqq$

$L(\Gamma)/\text{Area}(M) \geqq [4\pi(g-1)]^{-1}\beta_4$.

Case iv: A component Γ_0 of Γ is a circle which is freely homotopic to a boundary circle γ_j of M. In this case, $\gamma_j \cup \Gamma_0$ bounds an annulus Ω, and the isoperimetric inequality of Ionin [5] and Yau [9] implies Area $(\Omega) \leqq L(\gamma_j) + L(\Gamma_0)$. Since $K \leqq 0$ and γ_j is a geodesic, we have $L(\gamma_j) \leqq L(\Gamma_0)$. Thus Area $(\Omega) \leqq 2L(\Gamma_0) \leqq 2L(\Gamma)$ and $h(M_i) \geqq 1/2$.

Case v: Two components of Γ are circles which are freely homotopic to each other. This case follows like the previous case.

Case vi: All components of Γ are circles, no two components are freely homotopic, and no component is freely homotopic to a boundary circle of M_i. In this case we will give a lower bound on $L(\Gamma)$, so we note that we can replace each component Γ_0 of Γ by a geodesic freely homotopic to Γ_0 and lying in M_i. This can be done because the boundary of M_i consists of geodesic circles. Thus we assume that Γ is a disjoint union of simple closed geodesics in M_i. Suppose $C = \cup_{j=1}^{p}\gamma_j$ where each γ_j is a simple closed geodesic. If it were true that $L(\Gamma) < p^{-1}\sum_{j=1}^{p} L(\gamma_j)$, then there would be a component γ_j of C satisfying $L(\Gamma) < L(\gamma_j)$. This would imply that by reattaching along γ_j and cutting along Γ we would have a subdivision of M into $n+1$ pieces by disjoint simple closed geodesics of total length less than ℓ_n. This contradiction implies $L(\Gamma) \geqq p^{-1}\ell_n$. Since the maximum possible number of disjoint simple closed geodesics is $3g-3$, we have $L(\Gamma) \geqq (3g-3)^{-1}\ell_n$. Therefore, $h(M_i) \geqq L(\Gamma)/\text{Area}(M) \geqq [12\pi(g-1)^2]^{-1}\ell_n$.

The above cases include all possibilities for Γ, so we conclude $h(M_i) \geqq \min\{1/2, [4\pi(g-1)]^{-1}\beta_4, [12\pi(g-1)^2]^{-1}\ell_n\}$. Since $\ell_n \leqq \beta_1$, we have $h(M_i) \geqq \beta_3\ell_n$ where

$$\beta_3 = \min\{1/2\beta_1^{-1}, [4\pi(g-1)]^{-1}\beta_4\beta_1^{-1}, [12\pi(g-1)^2]^{-1}\}.$$

This concludes the proof that $\beta_2\ell_n^2 \leqq \lambda_n$ for a positive constant β_2 depending only on g.

4. The proof that $\alpha_1 \leqq \lambda_{2g-2}$ also follows from the above analysis. The upper bound on λ_{2g-2} follows from the work of S. Y. Cheng [3].

5. Finally we sketch a proof that $\alpha_1\ell_n \leqq \lambda_n$ for $1 \leqq n \leqq 2g-3$. We first note that if we set $\ell_{2g-1} = 1$ we have

$$\ell_1 = \prod_{i=1}^{2g-2}(\ell_i/\ell_{i+1}^2)^{2^{i-1}}.$$

We let $\beta_5 > 0$ be a small constant which will be chosen to depend only on g. If it were true that $\ell_i/\ell_{i+1}^2 > \beta_5$ for $1 \leqq i \leqq 2g-2$, we would have

$$\ell_1 > (\beta_5)^{2^{2g-2}-1}$$

and our previous lower bound would imply that λ_1 is bounded below. It would then easily follow that $\alpha_1\ell_n \leqq \lambda_n$ for a small $\alpha_1 > 0$. We may henceforth assume that

$\ell_i/\ell_{i+1}^2 \leqq \beta_5$ for some i with $1 \leqq i \leqq 2g-2$. Let m be the largest such i. It is easy to show that there is a constant $\beta_6 > 0$ depending on β_5 so that

$$\beta_6 \leqq \lambda_n \quad \text{for } n > m.$$

This implies $\alpha_1 \ell_n \leqq \lambda_n$ for $m < n \leqq 2g-2$. To prove the inequality for $1 \leqq n \leqq m$, we let $C = \gamma_1 \cup \cdots \cup \gamma_p$ be a curve in $\mathcal{C}_m$ with $L(C) = \ell_m$ where the γ_i are simple closed geodesics in M. Let $M_1, \ldots, M_{m+1}$ be the connected components of $M \sim C$ and let $M_i^0 = M_i \sim (\cup_{j=1}^p K_j)$. For $1 \leqq i \leqq m+1$, let φ_i be a function which takes the value 1 on M_i^0, 0 on M_j^0 for $j \neq i$ and which varies linearly with distance across the boundary collars of M_i. We are going to compare the φ_i with the first m eigenfunctions of Δ on M. Let V be the linear span of $\varphi_1, \ldots, \varphi_{m+1}$ in $L^2(M)$. Thus V is an $(m+1)$-dimensional vector space. Let $Q(\varphi, \psi)$ be the symmetric quadratic form on V defined by

$$Q(\varphi, \psi) = \int_M (\nabla\varphi \cdot \nabla\psi)\, dv.$$

Using the explicit nature of the collars and the definition of V we can bound the eigenvalues $\mu_0, \ldots, \mu_m$ of Q taken with respect to the $L^2(M)$ inner product as follows.

$$\text{(1)} \qquad \mu_m \leqq \beta_7 \ell_m, \quad \tilde{\beta}_7 \ell_n \leqq \mu_n \quad \text{for } 1 \leqq n \leqq m$$

where we are assuming $0 = \mu_0 < \mu_1 \leqq \mu_2 \leqq \cdots \leqq \mu_m$.

Let $f_0, f_1, f_2, \ldots$ be an orthonormal system of eigenfunctions and let W be the $(m+1)$-dimensional space spanned by $f_0, f_1, \ldots, f_m$. For any $\varphi \in V$, let

$$\varphi = \sum_{i=0}^{\infty} a_i f_i$$

be the series expansion of φ in terms of the f_i. We then have

$$Q(\varphi, \varphi) = \sum_{i=0}^{\infty} \lambda_i a_i^2.$$

From (1) we have $Q(\varphi, \varphi) \leqq \beta_7 \ell_m \int_M \varphi^2\, dv$. Thus

$$\lambda_{m+1} \sum_{i=m+1}^{\infty} a_i^2 \leqq Q(\varphi, \varphi) \leqq \beta_7 \ell_m \int_M \varphi^2\, dv = \beta_7 \ell_m \sum_{i=0}^{\infty} a_i^2.$$

Using our lower bound $\beta_2 \ell_n^2 \leqq \lambda_n$, we get

$$\beta_2 \ell_{m+1}^2 \sum_{i=m+1}^{\infty} a_i^2 \leqq \beta_7 \ell_m \sum_{i=0}^{\infty} a_i^2.$$

Let $F: V \to W$ be the linear map

$$F(\varphi) = \sum_{i=0}^{m} a_i f_i.$$

We have shown $\|\varphi - F(\varphi)\|_2^2 \leqq \beta_2^{-1}\beta_7(\ell_m/\ell_{m+1}^2)\|\varphi\|_2^2$ where $\|\cdot\|_2$ is the L^2-norm. By the choice of m, this gives

$$\|\varphi - F(\varphi)\|_2^2 \leqq \beta_2^{-1}\beta_7\beta_5 \|\varphi\|_2^2.$$

If β_5 is chosen sufficiently small, this implies that F is a linear isomorphism. Letting $G = F^{-1}$ we see that

$$\| G(f) - f \|_2^2 \leqq \beta_8 \beta_5 \| f \|_2^2 \tag{2}$$

for any $f \in W$. The proof will be finished if we can show

$$\beta_9 \mu_n \leqq \lambda_n \leqq \mu_n \quad \text{for } 1 \leqq n \leqq m, \tag{3}$$

since this in combination with (1) gives the desired result. The inequality $\lambda_n \leqq \mu_n$ is immediate since the minimax properties for μ_n and λ_n imply

$$\lambda_n \leqq \inf \{ Q(\varphi, \varphi) : \varphi \in V, \|\varphi\|_2 = 1, \int_m \varphi f_i = 0 \text{ for } i = 0, 1, \ldots, n-1 \} \leqq \mu_n .$$

To complete the proof of (3), we let W_0 be the linear span in W of $f_0, \ldots, f_n$ and let V_0 be the linear span in V of $G(f_0), \ldots, G(f_n)$. Because of (2), V_0 is $(n+1)$-dimensional (β_5 is small). We denote by σ_j the first Neumann eigenvalue for the collar K_j. One can show

$$\beta_{11} L(\gamma_j) \leqq \sigma_j \leqq \beta_{10} L(\gamma_j) .$$

For any $f \in W_0$, and for $1 \leqq j \leqq p$, we let δ_j denote the absolute value of the change of G(f) across the collar K_j. Denote by τ_j the average value of f on the collar K_j. By the minimal property of σ_j we have

$$\sigma_j \int_{K_j} (f - \tau_j)^2 \, dv \leqq \int_{K_j} |\nabla f|^2 \, dv .$$

Let ϵ be a positive number which will be chosen later. If β_5 is chosen small relative to ϵ and we assume $\|f\|_2 = 1$, it follows from (2) that

$$\beta_{12} \epsilon^2 \leqq \int_{K_j} (f - \tau_j)^2 \, dv \quad \text{if } \delta_j > \epsilon$$

for a positive constant β_{12}. Combining these we have

$$\sigma_j \leqq \beta_{12}^{-1} \epsilon^{-2} \int_{K_j} |\nabla f|^2 \, dv \quad \text{if } \delta_j > \epsilon .$$

Let $J_\epsilon = \{ j : \delta_j > \epsilon \}$ and sum on j,

$$\sum_{j \in J_\epsilon} \sigma_j \leqq \beta_{12}^{-1} \epsilon^{-2} \int_M |\nabla f|^2 \, dv .$$

Since $f \in W_0$ and $\| f \|_2 = 1$, it follows that

$$\sum_{j \in J_\epsilon} \sigma_j \leqq \beta_{12}^{-1} \epsilon^{-2} \lambda_n .$$

Finally we show that there is a $\bar{\varphi} \in V$ having jumps equal to $\bar{\delta}_j$ on K_j so that $\bar{\delta}_j = 0$ if $j \notin J_\epsilon$ and $\| \bar{\varphi} - G(f) \|_2 \leqq \beta_{13} \epsilon$. It then can be seen that $Q(\bar{\varphi}, \bar{\varphi}) \leqq \beta_{14} \sum_{j \in J_\epsilon} \sigma_j$. Therefore

$$Q(\bar{\varphi}, \bar{\varphi}) \leqq \beta_{14} \beta_{12}^{-1} \epsilon^{-2} \lambda_n .$$

We apply the above analysis with $f = f_i$ for $i = 0, 1, \ldots, n$ to get $\overline{\varphi}_0, \overline{\varphi}_1, \ldots, \overline{\varphi}_n$ in V satisfying

$$\| \overline{\varphi}_i - G(f_i) \|_2 \leqq \beta_{13} \epsilon \qquad Q(\overline{\varphi}_i, \overline{\varphi}_i) \leqq \beta_{14} \beta_{12}^{-1} \epsilon^{-2} \lambda_n$$

for $i = 0, 1, \ldots, n$. We fix ϵ small enough so that $\overline{\varphi}_0, \ldots, \overline{\varphi}_n$ are linearly independent and hence span an $(n+1)$-dimensional subspace $\overline{V}_0$ of V. It follows that

$$Q(\varphi, \varphi) \leqq \beta_9^{-1} \lambda_n \| \varphi \|_2^2 \quad \text{for} \quad \varphi \in \overline{V}_0$$

where β_9 is a positive constant. Since $\overline{V}_0$ is $(n+1)$-dimensional it follows that $\beta_9 \mu_n \leqq \lambda_n$, completing the proof of (3). This completes the proof of the Main Theorem.

BIBLIOGRAPHY

1. L. Bers, Spaces of degenerating Riemann surfaces, Ann. of Math. Studies 79 (1974), 43-55.

2. J. Cheeger, A lower bound for the smallest eigenvalue of the Laplacian, Problems in Analysis, a Symposium in Honor of S. Bochner, Princeton Univ. Press, Princeton, N.J., 1970.

3. S. Y. Cheng, Eigenvalue comparison theorems and its geometric applications, Math. Z. 143 (1975), 289-297.

4. N. Halpern, Collars on Riemann surfaces, preprint.

5. V. K. Ionin, Isoperimetric and various other inequalities on a manifold of bounded curvature, Sib. Math. J. 10 (1969), 329-342.

6. L. Keen, Collars on Riemann surfaces, Ann. of Math. Studies 79 (1974), 263-268.

7. J. P. Matelski, A compactness theorem for Fuchsian groups of the second kind, Duke Math. J. 43 (1976), 829-840.

8. B. Randol, Cylinders in Riemann surfaces, preprint.

9. S. T. Yau, Isoperimetric constants and the first eigenvalue of a compact Riemannian manifold, Ann. Scient. de l'Ecole Norm. Sup., 4e série, t. 8, fasc. 4, 1975.

Proceedings of Symposia in Pure Mathematics
Volume 36, 1980

COMBINATORIAL HODGE THEORY AND SIGNATURE THEOREM

Nicolae Teleman

ABSTRACT. In this paper we generalize the smooth Hodge theory, and the smooth signature operator into the category of all closed, oriented combinational manifolds.

1. INTRODUCTION. In this paper we present, briefly, the solution to some problems posed by I. M. Singer in [5], §4, concerning elliptical operators on PL-manifolds.

The problems posed by I. M. Singer and solved in this paper are the following, quote:

A. "If M is a PL-manifold, the L-polynomials are still well defined (Thom [8]) from which one can still define the *rational* Pontrjagin classes. The Hirzebruch signature theorem still holds. Is there an associated elliptic operator (as in the smooth case) whose index is the signature? On what spaces does it operate? Does it have a symbol and where does the symbol lie?"

B. "...Are present analytic techniques strong enough to prove the Hodge theorem for smooth manifolds by approximations via combinatorial Laplacians associated to smooth triangulations? If so, this might shed some light on the PL-problem".

In order to solve these problems, we treatise the exterior derivative and its adjoint from a PL-distributional point of view.

On smooth manifolds, the smooth forms are used as testing functions in the C^∞-distribution theory. In addition, the algebra of smooth forms contains all the information concerning the real singular cohomology of the manifold (de Rham's isomorphism).

Our leading idea in constructing a parallel theory on PL-manifolds consists of substituting the testing algebra of smooth differential forms by a new algebra which satisfies the following basic requirements: (i) in the case of PL-manifolds which are smoothable, it contains the algebra of smooth differentiable forms, (ii) it is invariant under PL-homeomorphisms, (iii) it

1980 Mathematics Subject Classification 58-A-14, 58-G-05, 58-G-99.

contains the cohomological information, (iv) it is the smallest in the class of algebras satisfying (i) - (iii).

Clearly, such an algebra is just the algebra of piecewise-differentiable forms.

This paper gives first a solution to Problem B. It provides the construction of a combinatorial Hodge isomorphism, strong Hodge decomposition and Poincaré duality. Those results, we guess, entitle us to speak of a veritable combinatorial Hodge theory.

The paper gives also an upper bound for the class of regularity of the combinatorial harmonic forms: that is, in general, the combinatorial harmonic forms do not have square integrable first-order derivatives.

The main tool in constructing our Hodge theory is the L_2-description of the real singular cohomology, which turns out to be the proper substitute (on PL-manifolds) of the de Rham isomorphism theorem (on smooth manifolds).

There are two essential facts which make our constructions work.

The first is the local acyclicity of the closure (in L_2) of the Sullivan complex of PL-differentiable forms.

The second essential fact is the equality of the weak and strong exterior derivatives.

Indeed, as expected in Problem B, performance of a PL-Hodge theory has given us enough indications for the attack of the signature Problem A.

In fact, as a first application of our Hodge theory, we construct a Fredholm operator $D^+ : W_1^+ \to W_0^-$ whose index is the signature of the manifold. This solves Problem A except for the question concerning the symbol of D^+. (By the way, a second application of the combinatorial Hodge theory, is the construction of a Fredholm operator whose index is the Euler characteristic.)

Notice that all our constructions and objects are very natural in the sense that they reduce to the classical ones when the manifold and the Riemannian metric are smooth. This is insured, in part, by the property i) of the testing algebra of piecewise-differentiable forms.

At this point, we express our high expectation that the entire program of I. M. Singer in [5] §4 could be realized along these lines.

As I. M. Singer explains in [5], this would give a direct proof of D. Sullivan's theorem about existence of an orientation class in K-theory, and could lead to an analytic proof of S. Novikov's theorem on the invariance of the rational Pontrjagin classes.

As I knew from Jeff Cheeger's lecture at this Meeting, our results have a considerable intersection, even if our considerations are very different in spirit.

The author acknowledges Professor I. M. Singer for useful conversations and for his constant interest shown all along in the preparation of this paper.

Details will appear elsewhere.

2. PRELIMINARIES. All throughout this paper M will denote any compact, oriented PL-manifold of dimension m, without boundary. Though, many of the results of this paper can be stated with no orientation requirement.

$T = \left\{\sigma_\alpha^{(r)}\right\}_{\alpha\varepsilon\Lambda}$ being any triangulation of M, compatible with the PL-structure of M, $\sigma_\alpha^{(r)}$ denotes an arbitrary simplex of dimension r.

For any r, let $S^r(M,T)$ denote the vector space of all piecewise-differentiable forms of degree r on M, with respect to the triangulation T, see D. Sullivan [6].

The exterior derivative d is simplexwise defined:

$$d : S^r(M,T) \to S^{r+1}(M,T), \quad d^2 = 0.$$

The complex:

$$S^*(M,T) \equiv \{S^r(M,T), d\}_r$$

depends, obviously, on T, but the natural inductive limit:

$$S^*(M) \equiv \varinjlim_{T} S^*(M,T)$$

is a PL-invariant of M.

With T, a fixed triangulation of M, let Γ denote a combinatorial Riemannian metric of class C^2 on M, see N. Teleman [7].

For simplicity, we can assume that Γ is flat on each simplex of T.

If $\omega = \{\omega_\alpha\}_{\alpha\varepsilon\Lambda}$ is any form in $S^*(M)$, we define its L_2-norm:

$$||\omega||^2 = \sum_{\alpha\varepsilon\Lambda} \int_{\sigma_\alpha^{(m)}} \omega_\alpha \wedge *\omega_\alpha ,$$

$*$ denoting the Hodge star operator of Γ on $\sigma_\alpha^{(m)}$.

Let L_2^r denote the completion of $S^r(M)$ in the norm $||\ ||$.

L_2^r, as a Hilbertian space, does not depend on T, nor on Γ.

3. WEAK AND STRONG EXTERIOR DERIVATIVE. Let $\omega \varepsilon L^r$. We say that ω has "PL-distributional exterior derivative", (or "weak exterior derivative") $d\omega \varepsilon L_2^{r+1}$, if and only if there exists a form $\theta \varepsilon L_2^{r+1}$ such that:

$$\int_M \omega \wedge d\psi = (-1)^{r+1} \int_M \theta \wedge \psi, \quad \text{for } \forall \psi \varepsilon S^{m-r-1}(M);$$

we denote $d\omega = \theta$.

We say, instead, that ω has "strong exterior derivative" $T\omega$ if and only if ω belongs to the domain $\mathcal{D}^r_T$ of the closure T of the exterior derivative d, which is defined on the dense subspace $S^r(M) \subset L^r$, i.e. if and only if there exists a sequence $\{\psi_n\} \subset S^r(M)$ such that:

$$\lim_{n\to\infty} \psi_n = \omega, \quad \text{and} \quad \{d\psi_n\}_n \text{ is a Cauchy sequence;}$$

one defines $T\omega = \lim_{n\to\infty} d\psi_n$.

THEOREM 3.1. *Let* ω *be any form in* L^r_2. *Then* ω *has weak exterior derivative, in* L^{r+1}_2 *if and only if* ω *has strong exterior derivative and, in that case* $d\omega = T\omega$.

Let d^* denote the Hilbert space adjoint of the operator d. It is defined on a dense domain $\mathcal{D}^r_{d^*} \subset L^r_2$.

We may introduce a formal adjoint δ for the operator d by the formula:

$$\delta\omega = (-1)^{mr+m+1} {*d*}\omega, \qquad {*\omega} \in \mathcal{D}^{m-r}_d,$$

defined on the domain:

$$\mathcal{D}^r_\delta = \{\omega \mid {*\omega} \in \mathcal{D}^{m-r}_d\}.$$

A consequence of Theorem 3.1, is:

THEOREM 3.2. (i) $\mathcal{D}^r_{d^*} = \mathcal{D}^r_\delta$,

(ii) $d^* = \delta$.

4. L_2-de RHAM ISOMORPHISM. We have, clearly, $d\mathcal{D}^r_d \subset \mathcal{D}^{r+1}_d$, and $d^2 = 0$. Let $\mathcal{D}^*_d$ denote the complex $\{\mathcal{D}^r_d, d\}_r$.

THEOREM 4.1. (L_2-*de Rham isomorphism*) *For any compact Pl-manifold* M, $\partial M = \phi$, (i) *the differential sheaf of germs of elements* $\omega \in \mathcal{D}^*_d$:

$$0 \to \underline{R} \to \underline{\mathcal{D}^0_d} \xrightarrow{d} \underline{\mathcal{D}^1_d} \to \cdots \to \underline{\mathcal{D}^r_d} \xrightarrow{d} \underline{\mathcal{D}^{r+1}_d} \to \cdots$$

is a fine resolution of the constant sheaf $\underline{R}$. *Therefore,* (ii) *the homology of the complex* $\mathcal{D}^*_d$ *is canonically isomorphic to* $H^*(M,R)$, *the real singular cohomology of* M.

The proof of Theorem 4.1., (i) is based on Theorem 4.2.5. from Hörmander [4] restated for the de Rham complex instead of the Dolbeault complex.

5. COMBINATORIAL HODGE THEORY. The space:

$$H^r = \{h \mid h \in \mathcal{D}^r_d \cap \mathcal{D}^r_\delta, \ dh = 0 = \delta h\}$$

is called the "space of combinatorial harmonic forms".

Any combinatorial harmonic form is a cycle in the complex $\mathcal{D}_d^*$; let

$$\chi^r : \mathcal{H}^r \to H_r(\mathcal{D}_d^*) \simeq H^r(M,R)$$

denote the "combinatorial Hodge homomorphism" which associates to any harmonic cocycle its cohomology class.

THEOREM 5.1. *(Combinatorial Hodge Isomorphism)*

For any compact, oriented, combinatorial manifold M, $\partial M = \phi$ *and combinatorial metric* Γ, *the Hodge homomorphisms* χ^r *are isomorphisms.*

PROOF. The theorem follows by proving that the space of cocycles in the complex $\mathcal{D}_d^*$ is closed, together with Theorem 4.1.

From the proof of Theorem 5.1., we deduce also that $d\mathcal{D}_d^r$ is closed in L_2^{r+1}, which combined with Theorem 3.2. gives us that $\delta\mathcal{D}_\delta^r$ is closed in L_2^{r-1}. From the combinatorial Hodge isomorphism theorem, we conclude that $\mathcal{H}^r$ is closed. Now, we can formulate:

THEOREM 5.2. *(Strong Hodge decomposition) For any compact, oriented, combinatorial manifold* M, $\partial M = \phi$, *we have the orthogonal decomposition:*

$$L_2^r = d\mathcal{D}_d^{r-1} \oplus \delta\mathcal{D}_\delta^{r+1} \oplus \mathcal{H}^r ,$$

each sumand being closed in L_2^r .

THEOREM 5.3. *If* M *is as in Theorem 5.2., has flat combinatorial metric, and in at least one dimension* r *it is:*

$$\dim_R H^r(M,R) > \binom{m}{r} ,$$

then there exists at least one harmonic form on M *which does not have all coefficients in the Sobolev space* H_1.

The proof of Theorem 5.3. is based on a Bochner-type identity argument.

6. COMBINATORIAL SIGNATURE OPERATOR. In this section we suppose that $\dim M = 4\mu$, $\mu \in N$. We define the operator $D = d + \delta$ on:

$$W_1^r = \mathcal{D}_d^r \cap \mathcal{D}_\delta^r , \qquad 0 \le r \le m .$$

As in the smooth case, see M. F. Atiyah and I. M. Singer [1], let's consider on $W_0^* = \bigoplus_{r=0}^{m} L_2^r$ the involution τ:

$$\tau(\alpha) = i^{r(r-1)+2\mu} * \alpha , \qquad \alpha \in L_2^r ,$$

$$i = \sqrt{-1} .$$

Notice that τ keeps $W_1^* = \bigoplus_{r=0}^{m} W_1^r$ invariant. Therefore, as τ is an involution, if we put $W_j^\pm = (1 \pm \tau)W_j^*$, $j = 0$ or 1, we have $W_j^* = W_j^+ \oplus W_j^-$. It can be shown that $DW_1^\pm \subset W_0^\mp$; let be $D^\pm = D|W_1^\pm$.

THEOREM 6.1. (*Combinatorial Signature Theorem*) *For any compact, oriented, combinatorial manifold* M, $\partial M = \phi$, *and combinatorial metric* Γ : (i) *the operator* D^+ *is a Fredholm operator, and* (ii) Index D^+ = Sign M.

PROOF. The proof goes along with the same lines as in the smooth case, see M. F. Atiyah and I. M. Singer [1]. In the combinatorial case, we have to consider, obviously, Theorem 5.1. and Theorem 5.2.

7. EULER CHARACTERISTIC. As a second application of our combinatorial Hodge theory, we can formulate

THEOREM 7.1. *For any combinatorial manifold as before,* (i) *the operator*

$$D : \oplus W_1^{even} \to \oplus W_0^{odd}$$

is a Fredholm operator, and (ii) *its index is the Euler characteristic of* M.

BIBLIOGRAPHY

1. M. F. Atiyah and I. M. Singer, The Index of Elliptic Operators, Part III, Annals of Mathematics, 87 (1968), 546-604.

2. J. J. Kohn, Harmonic Integrals on Strongly Pseudo-Convex Manifolds, Part I, Annals of Mathematics, 78 (1963), 113-148.

3. L. Hörmander, L^2-Estimates and existence theorems for the $\overline{\delta}$-operator. Acta Mathem., 113 (1965), 89-152.

4. L. Hörmander, An Introduction to Complex Analysis in Several Variables, North-Holland, (1973).

5. I. M. Singer, Future extensions of index theory and elliptic operators, in: Prospects in Mathematics, Ann. Math. Stud., 70, Princeton (1971), 171-185.

6. D. Sullivan, Differential Forms and the Topology of Manifolds, Proc. Tokyo Conf. on Manifolds, (1973).

7. N. Teleman, Global Analysis on PL-manifolds. Trans. A.M.S. (will appear).

8. R. Thom, Les classes caractéristiques de Pontrjagin des variétés triangulées, Symposium Internacional de Topologia Algebraica, (1958), 54-67.

DEPARTMENT OF MATHEMATICS
CALIFORNIA INSTITUTE OF TECHNOLOGY
PASADENA, CALIFORNIA 91125

Current Address:
California Institute of Technology
Pasadena, California 91125

Proceedings of Symposia in Pure Mathematics
Volume 36, 1980

SPECTRAL DISTRIBUTION OF LAPLACIANS AND KATO'S INEQUALITY

D.A.UHLENBROCK[1]

ABSTRACT. The scalar Beltrami Laplacian and the Bochner connection Laplacian on suitable vectorbundles are related to each other via a generalization of an inequality originally employed by T.Kato.On a compact Riemannian manifold these Laplacians generate traceclass contraction semigroups.As a consequence of the inequality the scalar semigroup dominates the vectorsemigroup in a specified sense.This yields an inequality for the semigroup traces,the first eigenvalues and the resolvents of these Laplacians.

1.INTRODUCTION. This is a report on joint work with Harald Hess and Robert Schrader at the Freie Universitaet Berlin.Details,technical as well as historical,can be found in [4],[5].The original motivation stemmed from an interest in gauge-field theory and Yang-Mills fields.Applications of some of our results in these directions are treated in papers of Schrader [8] , Combes-Schrader-Seiler [3],Avron-Herbst-Simon [2].Kato's use of the original inequality occurred in connection with proofs of essential selfadjointness for Laplacians with "magnetic terms" [7] .Simon [9],[10] related an abstract version of this inequality to positivity preserving semigroups (in a probabilistic setting) and to a comparison of semigroups.Our work stressed the vectorbundle aspects needed in the gauge field theory applications and was consequently characterized by a more general comparison setting. We refer to this setting as "domination" and note that we are in fact dealing with a special category,whose objects are certain pairs of related Hilbert spaces with morphisms related pairs of bounded linear maps between such pairs.

2.LAPLACIANS. Let M denote a compact (finitedimensional,real,smooth) manifold without boundary which is carrying a smooth Riemannian metric g.The smooth sections $C^\infty(M\times R)$ of the trivial bundle $M\times R$ are dense in the real Hilbert space $\mathcal{H}=L^2(M\times R)$ of square integrable (w.r.t. the Riemannian measure m on M)

1980 Mathematics Subject Classification 58G99.

[1]Supported by NSF-Grant #MCS77-00097

sections.Those elements of $\mathcal{K}$ which correspond to nonnegative functions naturally form a cone $\mathcal{K}^+$closed under addition and multiplication by nonnegative scalars.We let ∇ denote the covariant derivative of the Riemannian connection and write Δ for the Beltrami Laplacian on M.Another Laplacian,which we intend to relate to Δ ,is formulated in terms of a finitedimensional complex vectorbundle V on M carrying a hermitian structure $(\ |\ )$ on each fiber, which varies smoothly with the fiber.Again it is natural to consider the complex Hilbert space $\mathcal{H}=L^2(V)$ consisting of the square integrable sections f of V,which are normed by

$$||f||^2:=((f\ |f))=\int (f\ |f)\ dm.$$

In particular,we assume given a connection on V in terms of its covariant derivative ${}^V\nabla : C^\infty (V) \to C^\infty (T^*M\otimes V)$,which is hermitian in the sense that for any f_1 , f_2 in $C^\infty (V)$

$$\nabla(f_1\ |f_2\ ) = ({}^V\nabla\ f_1\ |f_2\ ) + (f_1\ |{}^V\nabla\ f_2\ ). \tag{2.1}$$

If ∇^*denotes the Riemannian connection on T^*M then there is a natural connection ${}^V\nabla^*$ on $T^*M\otimes V$ mapping $C^\infty(T^*M \otimes V)$ to C^∞ $(T^*M \otimes T^*M \otimes V)$ and which for a in $C^\infty (T^*M)$,f in $C^\infty (V)$ is defined by

$${}^V\nabla^*(a \otimes f):=(\nabla^* a) \otimes f\ +\ a \otimes {}^V\nabla f. \tag{2.2}$$

Finally,the metric g induces the contraction on tensors $g^* \otimes id_V : C^\infty (T^*M \otimes T^*M \otimes V) \to C^\infty (V)$ given by

$$(g^* \otimes id_V)\ a \otimes b \otimes f := g^*(a,b)\ f. \tag{2.3}$$

Here $g^*(a,b)$ is the obvious dual Riemannian structure on T*M.We can therefore make the following

DEFINITION 2.1. The Bochner Laplacian ${}^V\Delta$ is the second order differential operator on $C^\infty(V)$ defined by the composition

$${}^V\Delta := (g^* \times id_V) \cdot {}^V\nabla^* \cdot {}^V\nabla\ . \tag{2.4}$$

Note that for the trivial bundle $M \times C$ (or $M \times R$) and the Riemannian connection the Bochner Laplacian is none other than the Beltrami Laplacian.

In order to be able to relate Δ and ${}^V\Delta$ we have to relate $\mathcal{K}=L^2(M \times R)$ and $\mathcal{H}=L^2(V)$.Thus,for f in $\mathcal{H}$ we define a function $|f|$ in $\mathcal{K}^+$ by

$$|f|(x) := \sqrt{(f(x)\ |f(x))}\ ,\text{where x is in M.} \tag{2.5}$$

The following properties of the nonlinear map sending f to $|f|$

$$|((f\ |f'))| \le ((\ |f|\ |\ |f'|)),\ ||f||=||\ |f|\ || \tag{2.6}$$

are immediate consequences of the definitions.

For the following we assume fixed a measurable section c of V such that $c(x) = 1$ for all x in M. Nevertheless the essential results will prove to be wholly independent of this choice. For any section f of V we introduce the <u>signum</u> of f as the section

$$(2.7) \qquad (\mathrm{sgn}\, f)(x) := \begin{cases} f(x)/|f(x)| & \text{if } f(x) \neq 0 \\ c(x) & \text{if } f(x) = 0 \,. \end{cases}$$

Therefore $f = |f| \cdot \mathrm{sgn}\, f$ and $|\mathrm{sgn}\, f| = 1$. We will now formulate Kato's inequality for ${}^V\Delta$ and Δ.

PROPOSITION 2.2. For any f in $C^2(V)$ the inequality

$$(2.8) \qquad \mathrm{Re}\,({}^V\Delta f \,|\, \mathrm{sgn}\, f) \leq \Delta |f|$$

holds in the sense of distributions on M.

PROOF: It is convenient to regularize the squareroot singularity in equ.(2.5) by introducing the smooth section $|f|_\varepsilon := \sqrt{|f|^2 + \varepsilon^2}$ for $\varepsilon > 0$, so that we can perform the following computation. In view of equ.(2.1) we find

$$(2.9) \quad \tfrac{1}{2}\nabla |f|_\varepsilon^2 = |f|_\varepsilon . \nabla |f|_\varepsilon = \tfrac{1}{2}\,({}^V\nabla f \,|\, f) + \tfrac{1}{2}\,(f \,|\, {}^V\nabla f) = \mathrm{Re}\,({}^V\nabla f \,|\, f).$$

Applying $\nabla^* \otimes \nabla$ we find analogously after contracting with $g^* \otimes \mathrm{id}_V$ that

$$(2.10) \qquad \mathrm{Re}\,({}^V\Delta f \,|\, f) + |{}^V\nabla f|^2 = |f|_\varepsilon . \Delta |f|_\varepsilon + |\nabla |f|_\varepsilon|^2 .$$

According to equ.(2.9) $|f|_\varepsilon \cdot \nabla |f|_\varepsilon = \mathrm{Re}\,({}^V\nabla f | f)$ so that

$$|f|_\varepsilon^2 . |\nabla |f|_\varepsilon|^2 = |\mathrm{Re}\,({}^V\nabla f \,|\, f)|^2 \leq |f|^2 \cdot |\nabla |f|_\varepsilon|^2 \leq |f|_\varepsilon^2 |\nabla |f|_\varepsilon|^2 = |\mathrm{Re}({}^V\nabla f | f)|^2 \leq |{}^V\nabla f|^2 |f|^2$$

Hence we find (a priori only on supp f, but then by the locality of differential operators everywhere) that $|\nabla |f|_\varepsilon|^2 \leq |{}^V\nabla f|^2$. Thus we get by equ.(2.10) that $\mathrm{Re}\,({}^V\Delta f |\, f/|f|_\varepsilon) \leq \Delta |f|_\varepsilon$. On letting ε descend to zero the assertion follows. QED

In order to derive some interesting consequences from this inequality we prefer to rephrase it in an abstract Hilbert space setting. To this end we consider a smooth element g in $\mathcal{K}^+$ and define for given f in $C^2(V)$ $f' := g . \mathrm{sgn}\, f$. It is then apparent that f and f' are related in a very special way. Namely

$$(2.11) \quad ((f|f')) = \int (f|f') = \int g (f \,|\, \mathrm{sgn}\, f) = \int |f|\;|f'| = ((|f| \;|\; |f'|)).$$

Now the distribution inequality (2.8) is clearly equivalent to the Hilbert space inequality for such f and f':

$$(2.12) \qquad \mathrm{Re}\,((-{}^V\Delta f | f')) \geq ((\, |f| \;|\; -\Delta |f'|\,)).$$

This inequality is an instance of the abstract Kato inequality which we intend to exhibit in the context of a notion called <u>domination</u>.

3.DOMINATION. For the following we let $\mathcal{H}$ be a complex (separable) Hilbert space and $\mathcal{K}$ be a real (separable) Hilbert space which is equipped with a distinguished positive cone $\mathcal{K}^+$ (i.e. $\mathcal{K}^+ + \mathcal{K}^+ \subset \mathcal{K}^+$ and $R^+ \cdot \mathcal{K}^+ \subset \mathcal{K}^+$) such that g is in $\mathcal{K}^+$ if and only if $((g|\mathcal{K}^+)) \subset R^+$.

DEFINITION 3.1: $\mathcal{H}$ is dominated by $\mathcal{K}$ if there exists an absolute value map $|\ |: \mathcal{H} \to \mathcal{K}^+ \subset \mathcal{K}$ such that

(3.1) $\quad ||f|| = ||\ |f|\ ||.$

(3.2) $\quad |((f|f'))| \leq ((|f|\ |\ |f'|\ )).$

(3.3) For all f in $\mathcal{H}$, g in $\mathcal{K}^+$ there exists f' in $\mathcal{H}$ such that $g = |f'|$ and

$$((\ f\ |f'\ )) = ((\ |f|\ ||f'|\ )).$$

REMARK: We shall call a pair of vectors f and f' in $\mathcal{H}$ absolutely paired in case they satisfy the absolute pairing condition (3.3).

In view of eqs. (2.5),(2.6) and (2.11) we see that the $\mathcal{H}$ and $\mathcal{K}$ of section 2 furnish a prime example for this definition.

DEFINITION 3.2: Let $\mathcal{H}_i$ be dominated by $\mathcal{K}_i$ for i=1,2. A (bounded) linear operator $B: \mathcal{K}_1 \to \mathcal{K}_2$ dominates a (bounded) linear operator $A: \mathcal{H}_1 \to \mathcal{H}_2$ if

(3.4) $\quad |A\ f| \leq B\ |f| \quad$ for all f in $\mathcal{H}_1$.

REMARK: According to equ.(3.3) the absolute value map takes $\mathcal{H}$ onto $\mathcal{K}^+$. Hence, if A is dominated by B the latter is automatically positivity preserving in the sense that B maps $\mathcal{K}_1^+$ into $\mathcal{K}_2^+$.

This notion of domination has the following easily established features:

LEMMA 3.3:

(i) If B_i dominates A_i and c_i are complex numbers for i=1,2, then $|c_1|\ B_1 + |c_2|\ B_2$ dominates $c_1\ A_1 + c_2\ A_2$.

(ii) If B_i dominates A_i for i=1,2 and the operators are composable, then the composite $B_1 \cdot B_2$ dominates $A_1 \cdot A_2$.

(iii) If B_i dominates A_i for i in a directed set I and B_i converges strongly (weakly) to B while A_i converges strongly (weakly) to A, then B dominates A.

(iv) If B dominates A then the adjoint B* dominates the adjoint A*.

We now assume fixed a space $\mathcal{H}$ which is dominated by a space $\mathcal{K}$. Let S(t)=exp(-tH) and T(t)=exp(-tK) be any (strongly continuous) contraction semigroups on $\mathcal{H}$ and $\mathcal{K}$ respectively. Recall that the generators H and K, which are densely defined on domains D(H) and D(K), are (maximally) accretive in the sense that Re $((Hf|f)) \geq 0$ for f in D(H) and Re $((Kg|g)) \geq 0$ for g in D(K).

Using the wellknown relations between such semigroups and their resolvents

(3.5) $(\lambda+H)^{-1} = \int dt e^{-t\lambda} S(t)$ and $(\lambda+K)^{-1} = \int dt e^{-t\lambda} T(t)$ for $\lambda > 0$ and

(3.6) $S(t) = \lim_{n\to\infty} (\frac{n}{t}(\frac{t}{n}+H)^{-1})^n$ and $T(t) = \lim_{n\to\infty} (\frac{n}{t}(\frac{t}{n}+K)^{-1})^n$

it follows immediately from Lemma 3.3 that the domination of S(t) by T(t) for t >0 is equivalent to the domination of $(\lambda+H)^{-1}$ by $(\lambda+K)^{-1}$ for $\lambda>0$.

In order to see what the domination of S(t) by T(t) for t >0 implies for the generators H and K we consider f in D(H) and g' in $\mathcal{K}^+ \cap D(K^*)$.Condition (3.3) of Definition 3.1 guarantees the existence of f' in $\mathcal{H}$ with $g' = |f'|$ such that f and f' are absolutely paired.But then ,if S(t) is dominated by T(t),we find for t >0,since

$$\mathrm{Re}\,((\exp(-tH)f|f')) \le |((\exp(-tH)f|f'))| \le ((|\exp(-tH)f|\,||f'|)) \le ((\exp(-tK)|f|\;||f'|\,)) \quad \text{that}$$

$$\frac{1}{t}\{((f|f')) - \mathrm{Re}\,((\exp(-tH)f|f'))\} = \mathrm{Re}\,((t^{-1}.(1-\exp(-tH))f\;|f')) \ge$$
$$\frac{1}{t}\{((|f|\,||f'|)) - ((\exp(-tK)|f|\,||f'|))\} = ((|f|\;|t^{-1}(1-\exp(-tK^*))g')).$$

Letting t descend to 0 we conclude that for such f and f' the inequality

$$\mathrm{Re}\,((\,Hf\;|f')) \ge ((\;|f|\;|\;K^*|f'|\;)) \quad \text{is satisfied.}$$

It is the main point of the following theorem that this condition characterizes the domination relation among contraction semigroups infinitesimally.For technical reasons it is frequently desirable to work with a weaker condition which demands such an inequality only for suitable cores of the generators. For this and the proof of this theorem we refer the reader to [4].

THEOREM 3.4: Let $\mathcal{K}$ dominate $\mathcal{H}$ and assume given two contraction semigroups as above.The following conditions are equivalent:

(i) T(t) dominates S(t) for t > 0.

(ii) $(\lambda+K)^{-1}$ dominates $(\lambda+H)^{-1}$ for $\lambda > 0$.

(iii) T(t) is positivity preserving for t >0 and for all f in D(H), f' in D(K*) with f,f' absolutely paired <u>Kato's inequality</u> holds:

(3.7) $$\mathrm{Re}\,((\,Hf\;|f')) \ge ((|f|\;|\;K^*|f'|\;)).$$

4.APPLICATIONS. Returning now to the setting of section 2 we note that a proof similar to the original selfadjointness proof of Kato involving his inequality (see [7]) establishes that the closure H of $-{}^{V}\!\Delta$ on $C^\infty(V)$ is nonnegative selfadjoint.Analogously the closure K of $-\Delta$on $C^\infty(M\times R)$ is nonnegative selfadjoint.These operators therefore generate (selfadjoint) contraction semigroups $S(t)=\exp(t\,{}^{V}\!\Delta)$ and $T(t)=\exp(t\,\Delta)$ on $\mathcal{H} = L^2(V)$ and $\mathcal{K} = L^2(M\times R)$ respectively.It is well known (and also transparent from the probabilistic significance of $\exp(t\,\Delta)$) that $T(t)$ is positivity preserving.In view of equ. (2.12) and condition (iii) of Theorem 3.4 we may conclude that $\exp(t\,\Delta)$ dominates $\exp(t\,{}^V\Delta)$ for all $t>0$.

THEOREM 4.1: For all $t>0$ we have

$$(\dim V)^{-1}.\mathrm{trace}(\exp(t\,{}^V\Delta)) \leq \mathrm{trace}(\exp(t\,\Delta)). \tag{4.1}$$

PROOF: The general regularity theory of elliptic operators on compact manifolds yields the conclusion that $\exp(t\,{}^V\Delta)$ and $\exp(t\,\Delta)$ are integral operators with smooth kernels.Let $\exp(t\,{}^V\Delta)(x,y)$ be the kernel operator in $\mathrm{Hom}(V_x,V_y)$ corresponding to the points x,y in M.We write $||\exp(t\,{}^V\Delta)(x,y)||$ for the operator norm in this space.As the domination condition holds pointwise we arrive at the estimate

$$||\exp(t\,{}^V\Delta)(x,y)|| \leq \exp(t\,\Delta)(x,y). \tag{4.2}$$

Consequently $\mathrm{trace}(\exp(t\,{}^V\Delta)) = \int \mathrm{trace}_{V_x}(\exp(t\,{}^V\Delta)(x,x))\,dm(x) \leq$

$$\dim V\;.\int ||\exp(t\,{}^V\Delta)(x,x)||\,dm(x) \leq$$

$$\dim V\;.\int \exp(t\Delta)(x,x)\,dm(x) = \dim V\cdot\mathrm{trace}(\exp(t\,\Delta)).$$ QED

REMARKS:

(i) It follows from this that the lowest eigenvalue of $-{}^V\Delta$ is not lower than the lowest eigenvalue of $-\Delta$ .It seems remarkable that this holds no matter how the bundle and the connection ${}^V\nabla$ are chosen.

(ii) Addition of zero order terms to ${}^V\Delta$ and Δ are dealt with by use of the Trotter product formula (see [11]) for contraction semigroups.In particular suitably semibounded potential terms can easily be incorporated. Alternatively,via computations involving generalized Weitzenboeck type formulae one can in this way deal with the de Rham Laplacian on V-valued differential forms.(See [5]).

(iii) If M admits a spin structure one defines a suitable Dirac operator whose square is called the <u>spinor Laplacian</u>.This latter again differs from a Bochner laplacian only by zero order curvature type terms and analogous inequalities can be deduced.(See [5]).
Physically interesting related work is found in Hogreve-Schrader-Seiler [6] and Avron-Seiler [1] .

(iv) Analogous inequalities hold involving the traces of arbitrary powers of the resolvents.It would be interesting to know what spectral information can be gained from the "off-diagonal" inequalities ($x \neq y$) in the equ.(4.2).

BIBLIOGRAPHY

1. J.E.Avron and R.Seiler,"Paramagnetism for non-relativistic electrons and Euclidean massless Dirac particles",preprint,October 1978.

2. J.E.Avron,I.Herbst and B.Simon,"Schroedinger operators in magnetic fields",Duke Math.J. (to appear) and Phys.Rev.Lett.,39(1977),1068-1070.

3. J.M.Combes,R.Schrader and R.Seiler,"Classical bounds and limits for energy distributions of Hamilton operators in electromagnetic fields", Ann.Phys.

4. H.Hess,R.Schrader and D.A.Uhlenbrock,"Domination of semigroups and generalization of Kato's inequality",Duke Math.J.,44(1977),893-904.

5. H.Hess,R.Schrader and D.A.Uhlenbrock,"Kato's inequality and the spectral distribution of Laplacians on compact Riemannian manifolds", J.Diff.Geom.(to appear).

6.H.Hogreve,R.Schrader and R.Seiler,"A conjecture on the spinor functional dterminant",Nucl.Phys.B.(to appear).

7. T.Kato,"Schroedinger operators with singular potentials",Israel J.Math.,13(1972),135-148.

8. R.Schrader,"Towards a constructive approach of a gauge invariant massive $P(\phi)_2$ theory",Comm.Math.Phys.

9. B.Simon,"An abstract Kato's inequality for generators of positivity preserving semigroups",Indiana Math.J.

10. B.Simon,"Kato's inequality and the comparison of semigroups", J.Func.Anal.

11.H.F.Trotter,"On the product of semigroups of operators",Proc.A.M.S. 10(1959),545-551.

DEPARTMENT OF MATHEMATICS
UNIVERSITY OF WISCONSIN
MADISON,WI,53706,USA

Proceedings of Symposia in Pure Mathematics
Volume 36, 1980

NONLINEAR STABILIZATION OF QUASI-MODES

Alan Weinstein[1]

ABSTRACT. If Γ is an elliptic closed geodesic on a compact riemannian manifold X, it is known that the wave equation $-\psi_{tt} = -\Delta\psi+V\psi$ has solutions which remain concentrated near Γ for a long time; in the generic case, these solutions eventually drift away from Γ. In this paper, it is suggested that the nonlinear equation $-\psi_{tt} = -\Delta\psi+V\psi-|\psi|^2\psi$ may have periodic solutions remaining concentrated near Γ for all time. This is shown to be the case when X is the standard 2-sphere and the potential function V is odd. The proof uses bifurcation theory for critical manifolds, estimates on the spectrum of $-\Delta+V$, and a new result on spherical harmonics.

1. INTRODUCTION

If X is a riemannian manifold with laplacian Δ, and V is a real-valued function on X, the equation

$$-\frac{\partial^2\psi}{\partial t^2} = -\Delta\psi + V\psi \tag{1.1}$$

for a complex-valued function $\psi(x,t)$ on $X \times \mathbb{R}$ is called the (linear) wave equation on X with potential V. The <u>principle of geometric optics</u>, stated roughly in this context, is that there is a close relation between "high-frequency" solutions of (1.1) (waves) and geodesics on X (rays).

Closely related to (1.1) is the (linear) Schrödinger equation

$$i\hbar\frac{\partial\psi}{\partial t} = -\frac{\hbar^2}{2}\Delta\psi + V\psi \tag{1.2}$$

in which $\hbar$ is a positive constant. Here, the <u>correspondence principle</u> asserts that solutions of (1.2) for small $\hbar$ (quantum probability waves) are closely related to solutions of Newton's equation

$$\frac{D^2x}{Dt^2} = -\operatorname{grad} V \tag{1.3}$$

(classical particle trajectories).

1980 Mathematics Subject Classification 58C40.
[1]Supported by the National Science Foundation.

In recent years, the principles stated in the two preceding paragraphs have been validated in the form of precise theorems on the propagation of singularities. (See [G-S] for a thorough study of these results, based on the geometric calculi of Maslov [M] and Hörmander [H].) These theorems have in turn led to results on the relation between eigenvalues and eigenfunctions of $-\Delta+V$ for large eigenvalues and closed geodesics(or eigenvalues and eigenfunctions of $-\frac{1}{2}\hbar^2\Delta+V$ for small $\hbar$ and periodic solutions of (1.3)).

A limitation on the correspondence between rays and waves has been stressed by Arnol'd [AR1][AR2]. It has been proven that, for each elliptic closed geodesic Γ on X, there exist solutions of (1.1) which remain concentrated near Γ over arbitrarily long periods of time; nevertheless, Arnol'd has given arguments based upon a general position hypothesis which would imply that one cannot expect to find solutions of (1.1) which reamin concentrated near Γ for all time. In the language of optics, diffraction causes any beam to disperse eventually. A similar observation was made earlier in quantum mechanics, where diffraction is replaced by the "tunnel effect," which allows a particle to escape from a region in which classical mechanics would have it permanently trapped.

I would like to suggest in this paper that the connection between waves and rays or classical trajectories may become closer if the linear wave equations (1.1) and (1.2) are replaced by the nonlinear wave equation

$$-\frac{\partial^2\psi}{\partial t^2} = -\Delta\psi + \Delta\psi - |\psi|^2\psi \tag{1.4}$$

and nonlinear Schrodinger equation

$$i\hbar\frac{\partial\psi}{\partial t} = -\frac{\hbar^2}{2}\Delta\psi + V\psi - |\psi|^2\psi \tag{1.5}$$

These equations have been proposed to describe various physical phenomena, including the self-focusing of laser beams and the stability of elementary particles. In either case, one may imagine that an initially concentrated wave ψ can "dig its own potential well" and so protect itself against the effects of diffraction or tunneling.

In particular, I will outline a proof that (1.4) has periodic solutions concentrated along closed geodesics when X is the standard 2-sphere and V is any odd function. One ingredient in the proof is an estimate on the L^4-norm of spherical harmonics; an improvement of this estimate would enable us to drop the assumption that V be odd.

Conversations with many mathematicians have been helpful to me in the pursuit of this research, but I would like to single out P. Rabinowitz, R. Stanton, and M. Taylor for special thanks.

2. APPROXIMATE EIGENFUNCTIONS AND QUASI-MODES

We may look for periodic solutions of (1.1), (1.2), (1.4), or (1.5) by separating space from time variables. In fact, the method applies for the general forms

$$-\frac{\partial^2\psi}{\partial t^2} = -\Delta\psi + W(x,|\psi|^2)\psi \tag{2.1}$$

and

$$i\hbar\frac{\partial\psi}{\partial t} = -\frac{\hbar^2}{2}\Delta\psi + W(x,|\psi|^2)\psi \tag{2.2}$$

where W is any function on $X \times \mathbb{R}^+$. Let us try to solve (2.1) with a <u>standing wave</u>

$$\psi(x,t) = e^{i\omega t}u(x). \tag{2.3}$$

The function u will be called the <u>waveform</u> of the standing wave solution. It is easy to see that u needs to satisfy the nonlinear eigenvalue equation

$$\omega^2 u = -\Delta u + W(x,|u|^2)u. \tag{2.4}$$

Similarly, substituting the <u>stationary state</u>

$$\psi(x,t) = e^{-\frac{i}{\hbar}\lambda t}u(x) \tag{2.5}$$

in (2.2) gives

$$\lambda u = -\frac{\hbar^2}{2}\Delta u + W(x,|u|^2)u\ . \tag{2.6}$$

REMARK. By rewriting (2.1) and (2.2) as infinite dimensional hamiltonian systems, one can show that what makes possible the solution by separation of variables is that the hamiltonian is invariant under the one-parameter group $\alpha\cdot u = e^{i\alpha}u$. The standing wave solutions are then relative equilibria ([AR2][SM]). In the present case, the substitution of an equilibrium problem for a periodic orbit problem is especially useful because it eliminates small divisor problems by "isolating" a given frequency from its multiples.

For the linear wave equation (1.1), the associated eigenvalue problem is the linear problem

$$\lambda^2 u = -\Delta u + Vu. \tag{2.7}$$

The rest of this section shall be a review of some results on approximate solutions of (2.7) which are concentrated along closed geodesics.

Let Γ be a closed geodesic of length L on the compact d-dimensional riemannian manifold X. Suppose that Γ is <u>elliptic</u> in the sense that the linearized Poincaré map for the geodesic flow around Γ splits into the direct sum of two-dimensional rotations. Suppose further that the rotation angles $\theta_1,\cdots,\theta_{d-1}$ are distinct and unequal to zero; we can normalize them to be between 0 and 2π. Finally let μ be the Morse index of Γ as a critical point for the energy functional on the loop space. Then for each (d-1)-tuple of non-negative integers $(n_1,n_2,\cdots,n_{d-1})$ there exists a sequence u_n of functions on X such that $\|u_n\| = 1$ and

$$\|- \Delta u_n + Vu_n - \tau_n^2 u_n\| = O(n^{1/2}), \tag{2.7}$$

where $\| \ \|$ is the L^2 norm and

$$\tau_n = L^{-1}(2\pi n + \Sigma(n_i + \tfrac{1}{2})\theta_i + \tfrac{\pi}{2}(\mu-1)) \tag{2.8}$$

Incidentally, the constant implied by the "O" on the right hand side of (2.7) will generally depend upon $(n_1,\cdots,n_{d-1})$. Also, we remark that the potential V is irrelevant until we try to attempt to improve the error estimate to negative powers of n (see [P] or [RAL2]). The sequence u_n is concentrated along Γ in the sense that, if $\{\alpha_n\}$ is any numerical sequence with polynomial growth, then the series $\Sigma\alpha_n u_n$ converges uniformly outside any neighborhood of Γ to a smooth function.

The functions u_n can be taken to be simple oscillating exponentials along Γ, and gaussian functions multiplied by polynomials along each slice transverse to Γ. In fact, waves of this type have been known for some time in the field of laser optics, where they are known as gaussian beams (see [AN][SG]). The formula (2.8) for the approximate eigenvalues was obtained by Babich[BA] and Pyshkina [P]. Other proofs of the result above were given by Guillemin and the author [G1][G-W], and by Ralston [RAL1].

We now take our approximate eigenfunctions and eigenvalues (u_n,τ_n) back to the wave equation (1.1) and see what we can do with them. A simple argument using the minimax characterization of eigenvalues shows that there is a sequence ω_n of <u>exact</u> eigenvalues for $-\Delta+V$ such that

$$\omega_n = \tau_n + O(n^{-1/2}). \tag{2.9}$$

This estimate looks somewhat more interesting if we use (2.8) to recast it in the form

$$\omega_n = \tau_n + O(n^{-1/2}). \tag{2.10}$$

Thus, the wave equation has solutions whose frequencies approximate the arithmetic progression τ_n; the error goes to zero like $n^{-1/2}$.

Now for the punch line of this section. It is <u>not</u> the case that a sequence of exact eigen<u>functions</u> can be found anywhere near the u_n's. As explained by Arnol'd [AR1][AR2], it is possible and even likely that each

u_n is a linear combination of several eigenfunctions belonging to distinct eigenvalues which are near to λ_n^2. (Genericity results in the direction suggested by Arnol'd have been obtained by Uhlenbeck [U] and Driscoll [D].) That is, we might take u_n to be $a_1v_1+\cdots+a_kv_k$, where $(-\Delta+V)v_k = \tau_k^2 v_k$, and so u_n will be the initial value for the solution

$$\sum_{j=1}^{k} a_j e^{i\tau_j t} v_j \tag{2.11}$$

of (1.1). If the τ_j's are near λ_k but distinct then (2.11) will oscillate near u_n for a while but will eventually drift away. If the τ_j's happen to be integer multiples of a single number τ_o, then (2.11) will eventually return to u_n, but only after the time $\frac{2\pi}{\tau_o}$. In the more likely case that the τ_j's are linearly independent over the integers, the solution (2.11) will be quasi-periodic and will eventually pass arbitrarily close to every linear combination of the v_j's. In particular, if there is another closed geodesic Γ' which has the same length, Poincaré rotation angles, and index as Γ, then the beam which started along Γ may later appear along Γ'. (A related phenomenon is sometimes used for the transfer of information between optical waveguides [Y].) Because of their quasi-periodic behavior, Arnol'd has dubbed the solutions coming from approximate eigenfunctions quasi-modes.

A similar situation is quite well known in quantum mechanics. A symmetric double potential well on $\mathbb{R}$ (with potential $x^4 - x^2$, say) admits two equilibrium positions for Newton's equation, and one can find approximate eigenfunctions of $-\frac{1}{2}\hbar^2\Delta+V$ concentrated near each of these positions, with the concentration increasing as $\hbar \to 0$. But the exact eigenfunctions must be even or odd, and the eigenvalues λ_E and λ_0 corresponding to the even and odd functions turn out to be slightly different, so the solution to Schrödinger's equation which starts out concentrated in one well will eventually "tunnel" through to appear in the other well; it will oscillate between the wells at a frequency equal to $\hbar|\lambda_E - \lambda_0|$ (see [L-L], p.176). These oscillations occur, for instance, in the ammonia molecule, which performs "inversion oscillations" between mirror-image configurations (see [A-V], pp.64-69).

3. EFFECT OF THE NONLINEAR TERM

There are several reasons to believe that the addition of a nonlinear term like $-|\psi|^2\psi$ may eliminate the "drift" of a quasi-mode away from the closed geodesic around which it is initially concentrated.

If we write (1.4) in the form

$$-\frac{\partial^2\psi}{\partial t^2} = -\Delta\psi + (V - |\psi|^2)\psi \tag{3.1}$$

we may think of $-|\psi|^2$ as part of the potential. Thus, the wave ψ tends to deepen its potential well and "dig in" against the effect of diffraction.

When a laser beam is passed through certain materials, the beam changes the index of refraction of the medium through which it passes, so that an initially uniform medium becomes a "lens" which focuses the beam. The mathematical description of this "self-focusing," as it is called in optics, involves multiplying Δ by a function of $|\psi|^2$ and so leads to a quasilinear equation which is much more difficult to deal with than the semilinear equation (1.4); however, an approximation leads to a nonlinear Schrödinger equation like (1.2), where the coordinate in the direction of the beam plays the role of the variable t (see [K-K][WH]).

Finally, we note that concentrated solutions (solitons) have been found in recent years for a wide variety of nonlinear wave equations. These solitons have been found explicitly, mostly for equations in one space dimension [KK][WH]. In higher dimensions, one may have to use methods from functional analysis to prove the existence of soliton solutions (as in [BR] and [STR]). The program to be described in this paper is in that spirit.

4. THE VARIATIONAL PROBLEM

Solutions u of the eigenvalue problem (2.4) are easily seen to be extremals for the functional

$$F(u) = \int_X \left[\frac{1}{2}\|\nabla u\|^2 + K(x,|u|^2)\right]dx \tag{4.1}$$

subject to the constraint

$$\frac{1}{2}\|u\|^2 = \text{constant}. \tag{4.2}$$

Here, $K(x,\theta) = \frac{1}{2}\int_0^\theta W(x,s)ds$, and $\|\ \ \|$ is the L^2 norm as usual. The eigenvalue ω^2 in (2.4) is just the Lagrange multiplier. For the special case $K(x,|u|^2) = V(x) - |u|^2$ arising from (1.4), the functional is given by

$$F(u) = \int_X \left[\frac{1}{2}\langle\nabla u,\nabla u\rangle + \frac{1}{2}V|u|^2 - \frac{1}{4}|u|^4\right]dx. \tag{4.3}$$

Our approach in this paper will be to use topological and functional-analytic methods to find solutions for the variational problem just described. Global methods have been used successfully by Rabinowitz [RAB] and others, especially in 1-dimensional problems, but with little resulting information about the nature of the solutions. To get such information, we will use less global methods involving bifurcation theory.

Since $K(x,0) = 0$, we may write $K(x,\theta) = \frac{1}{2} V(x)\theta + \frac{1}{4} U(x,\theta)\theta^2$ for suitably chosen functions V and U. (This is consistent with the previous use of the letter V.) Now we define the "dilated" functional $F_\varepsilon(u)$ to be $\frac{1}{\varepsilon^2} F(\varepsilon u)$; we find

$$F_\varepsilon(u) = \int_X \left[\frac{1}{2} \langle \Delta u, \Delta u \rangle + \frac{1}{2} V|u|^2\right] dx + \varepsilon^2 \int_X \frac{1}{4} U(x,\varepsilon^2|u|^2)|u|^4 dx. \qquad (4.4)$$

We note that v is a critical point for $F(v)$ subject to $\|v\| = \varepsilon$ if and only if $u = v/\varepsilon$ is a critical point for $F_\varepsilon(u)$ subject to $\|u\| = 1$. Thus, we can study $F(u)$ near the origin by adding a small perturbation to the quadratic functional

$$F_0(u) = \int_X \left[\frac{1}{2} \langle \nabla u, \nabla u \rangle + \frac{1}{2} V|u|^2\right] dx. \qquad (4.5)$$

At this point, it is possible to use the bifurcation theory of nondegenerate critical manifolds as described in [RE] or [WE2]. This theory, in which the implicit function theorem is used to reduce the infinite-dimensional variational problem to a finite-dimensional one, is often called the Liapunov-Schmidt method. First, one must fix the appropriate function space; in this case it is the Sobolev space $H_1(X)$. The constraint $\|u\| = 1$ defines a smooth submanifold $S \subseteq H_1(X)$ (the unit sphere with respect to the L^2 norm). The quadratic functional F_0 restricts to a smooth function $F_0 : S \to \mathbb{R}$. The critical points of F_0 are just the normalized (with respect to the L^2 norm) eigenfunctions of the elliptic differential operator $-\Delta+V$. For each eigenvalue ω^2 of $-\Delta+V$, the intersection of the eigenspace with S is a nondegenerate critical manifold S_{ω^2} for F_0. (If V takes negative values, some ω's may have to be imaginary.)

The bifurcation theory should now tell us something about the critical points of F_ε (the restriction of F_ε to S) for sufficiently small ε. First of all, we must be sure that the nonquadratic term $\int_X K(x,|u|^2)dx$ is defined and smooth on $H_1(X)$. For this, it is sufficient to assume that K is smooth and that $K(x,\theta) = O(\theta^s)$ as $\theta \to \infty$, where $s < n/n-2$ and n is the dimension of X.

Then it follows from Theorem 7.1 of [RE] that, for sufficiently small ε (depending upon ω), there is a function $G_\varepsilon : S_{\omega^2} \to \mathbb{R}$ such that each critical point of G_ε corresponds to a nearby critical point of F_ε. Now S_{ω^2} is the unit sphere in a complex vector space of dimension equal to the multiplicity of ω^2. Everything in sight is invariant under multiplication by the circle group $\mathbb{T} = \{z \in \mathbb{C} \mid |z| = 1\}$, so G_ε is the lift of a function on $\mathbb{C}P^{n-1}$, which must have at least n critical points. Thus we arrive at the following conclusion.

THEOREM 4.1. Let ω^2 be a positive eigenvalue of multiplicity m for $-\Delta+V$. Then there is a number $\varepsilon_\omega > 0$ (depending on ω) such that for all $\varepsilon \in (0,\varepsilon_\omega)$ the nonlinear wave equation (2.1) has at least m standing wave solutions with periods near $\frac{2\pi}{\omega}$ and waveforms u having L^2 norm ε. As $\varepsilon \to 0$, the normalized functions u/ε approach the ω^2-eigenspace of $-\Delta+V$.

REMARK. The solutions found in Theorem 4.1 above are <u>a priori</u> only in H_1. But a regularity theorem can be applied to show that they are indeed smooth.

If ω^2 is a simple eigenvalue, then we can tell immediately what the periodic solution of Theorem 4.1 looks like. Unfortunately, this solution is close to a periodic solution of the linear equation, which generally does not follow a closed geodesic for all time.

If ω^2 is a multiple eigenvalue, then it is important to know <u>which</u> eigenfunctions are approximated by solutions of the nonlinear problem. Information about this point is provided by the so-called <u>bifurcation functional H</u>, obtained if we differentiate F_ε with respect to ε^2 and set $\varepsilon=0$:

$$H(u) = \frac{1}{4}\int_X U(x,0)|u|^4 dx \tag{4.6}$$

COMPLEMENT TO THEOREM 4.1. Let u_o in the ω^2-eigenspace of $-\Delta+V$ be an accumulation point as $\varepsilon \to 0$ of the functions u/ε, where the u's are as in Theorem 1. Then u_o is a critical point of the restriction of H to S_{ω^2}. On the other hand, suppose that Σ is a nondegenerate critical manifold for the restriction of H to S_{ω^2}. Σ must be invariant under the action of the group $\mathbb{T}$; let r be the minimum number of critical points for a function on the quotient space $\Sigma/\mathbb{T}$. Then there is a number $\varepsilon' \in (0,\varepsilon_\omega)$ such that for all $\varepsilon \in (0,\varepsilon'_\omega)$, the wave equation (2.1) has at least r periodic solutions u as in Theorem 4.1 with the additional property that, as $\varepsilon \to 0$, u/ε approaches Σ.

In fact, the location of these special u's can be described in terms of the critical points of a higher order bifurcation function on Σ, which will have all the symmetries of the original problem. Of course, at any stage, the bifurcation function may have critical points which do not lie on nondegenerate manifolds, and then we are stuck.

5. THE 2-SPHERE WITHOUT POTENTIAL

As an example of Theorem 4.1 and its complement, let us consider the case where X is the unit sphere S^2 with its usual metric and V=0. This case is interesting because the eigenvalues of $-\Delta$ on S^2 have large multiplicities; later we shall see that, by making careful estimates, we can obtain similar results when a potential is added, because the eigenvalues will still lie in bounded "clusters" with wide gaps between them.

On S^2 we will use both the (dependent) cartesian coordinates (x,y,z) and the spherical coordinates (ϕ,θ) given by $x = \sin\theta\cos\phi$, $y = \sin\theta\sin\phi$, $z = \cos\theta$. The laplacian on S^2 is

$$\Delta = \frac{\partial^2}{\partial\phi^2} + \frac{1}{\sin^2\phi}\frac{\partial^2}{\partial\theta^2} + \cos\theta\frac{\partial}{\partial\theta}\,. \tag{5.1}$$

The space $H_1(S^2)$ is the direct sum of eigenspaces $E_\ell = \ker(\Delta + \ell(\ell+1))$; the dimension of E_ℓ is $2\ell+1$. The standard orthonormal (with respect to L^2) basis for E_ℓ consists of the simultaneous eigenfunctions of $-\Delta$ and $\frac{\partial}{\partial\theta}$; these are the associated Legendre functions $Y_\ell^m = C_\ell^m e^{im\phi}P_\ell^m(\cos\theta)$, where C_ℓ^m is a normalizing constant and

$$P_\ell^m(z) = (z^2-1)^{m/2}\frac{d^{m+\ell}}{dz^{m+\ell}}(z^2-1)^\ell \tag{5.2}$$

The integer index m runs from $-\ell$ to ℓ. The Y_ℓ^m's are determined up to a constant by the conditions:

$$-\Delta Y_\ell^m = \ell(\ell+1)Y_\ell^m \tag{5.3}$$

$$\frac{\partial}{\partial\theta}Y_\ell^m = mY_\ell^m \tag{5.4}$$

Another basis, consisting of real-valued functions, is made up of $\eta_\ell^o = Y_\ell^o$ and, for each m between 1 and ℓ, the normalized real and imaginary parts of Y_ℓ^m: $\eta_\ell^m = \sqrt{2}\,C_\ell^m \cos m\phi\, P_\ell^m(\cos\phi)$, and $\zeta_\ell^m = \sqrt{2}\,C_\ell^m \sin m\phi\, P_\ell^m(\cos\phi)$.

The group SO(3) acts on $H_1(S^2)$, with the E_ℓ's being the irreducible invariant subspaces. The rotations around the z-axis form a maximal torus in SO(3), and the Y_ℓ^m's are the weight vectors; Y_ℓ^ℓ is called the <u>highest weight</u> vector, and Y_0^0 is called the <u>spherical function</u>. We mention all these terms to suggest how the results to be described here might be generalized from S^2 to some other homogeneous spaces.

Now the linear wave equation

$$-\frac{\partial^2\psi}{\partial t^2} = -\Delta\psi \tag{5.5}$$

on S^2 <u>does</u> have periodic solutions concentrated along closed geodesics. In fact, the highest weight function Y_ℓ^ℓ is, up to a constant, $(x+iy)^\ell$; we note that $|(x+iy)^\ell| = (x^2+y^2)^{\ell/2} = (1-z^2)^{\ell/2} = \sin^\ell\theta$, which is equal to 1 when

$z = 0$ and decreases rapidly to 0 for all other values of z in $[-1,1]$. Thus, Y_ℓ^ℓ is concentrated along the equator, and therefore so is the solution $e^{i\sqrt{\ell(\ell+1)}\,t}\,Y_\ell^\ell$ of (5.5). (We could also use η_ℓ^ℓ or ζ_ℓ^ℓ in place of Y_ℓ^ℓ.) By applying rotations, we can find standing wave solutions of (5.5) concentrated along all the great circles on S^2.

Since we have found standing waves concentrated along closed geodesics for the linear wave equation on S^2, it may appear pointless to go any further. However, these waves are very unstable in the sense that they can be destroyed by the slightest perturbation. If a potential, no matter how small, is added to the laplacian, the eigenspace E_ℓ will generally "split" into $2\ell+1$ one-dimensional eigenspaces, and these new eigenfunctions could branch off from spherical harmonics which are nowhere near $Y_\ell^\ell, \eta_\ell^\ell, \zeta_\ell^\ell$, or their translates under the action of $SO(3)$. Although these orbits do carry approximate eigenfunctions and thereby play an important role in our understanding of the splitting of the multiple eigenvalues when a potential is added (see [G2][LI]), there appears to be nothing to prevent the quasi-mode phenomenon from occurring.

Passing to the nonlinear equation

$$-\frac{\partial^2\psi}{\partial t^2} = -\Delta\psi - |\psi|^2\psi \tag{5.6}$$

we may apply Theorem 4.1 and its complement. From the theorem, we find immediately that for each ℓ there exist for small ε (depending on ℓ) at least $2\ell+1$ standing wave solutions of (5.6) whose waveforms u have L^2 norm ε and are such that u/ε is close to the unit sphere in E_ℓ.

In order to find out which functions in E_ℓ are approximated by the waveforms, we must look at the bifurcation functional $H(u)$ defined in (4.6). In the present case, $U(x,0) \equiv 1$, so we are led to the following:

PROBLEM. Let N_ℓ be the function on the unit sphere $S_\ell \subseteq E_\ell$ defined by

$$N_\ell(u) = \int_{S^2} |u(x)|^4 dx = \|u\|_{L_4}^4$$

Describe the nondegenerate critical manifolds of N_ℓ.

A first result comes from using only the symmetries of N_ℓ and the following general principle: if N is any function on a manifold S, invariant under the action of a compact group G, and $s \in S$ is a point such that the linearized isotropy group at s fixes no non-zero vector, then s is a critical point of N.

PROPOSITION 5.1. Let $N: S_\ell \to \mathbb{R}$ be any function which is invariant under the group G generated by the transformations:

(a) $u \to \rho\varepsilon O(3)$

(b) $u \to -u$

(c) $u \to \bar{u}$

Then each Y_ℓ^m is a critical point of N. In addition, if m=0 or $|m| > \ell/3$, then η_ℓ^m and ζ_ℓ^m are also critical points of N.

REMARKS AND QUESTIONS.

1. Any point in the G-orbit of one of the points listed in Proposition 5.1 is critical for every G-invariant function. Are there any "universal critical points" not on these orbits? (Incidentally, η_ℓ^m and ζ_ℓ^m lie on the same orbit.)

2. If $0 < |m| \leq \ell/3$, then there exists a G-invariant function on S_ℓ which is not critical at η_ℓ^m. Is N_ℓ such a function?

3. Does the function N_ℓ have any critical points not on one of the orbits mentioned in Remark 1? (Partial answer: N_2 is <u>constant</u>.)

4. What does the global Morse theory of N_ℓ look like?

5. How do the answers to the questions above change if N_ℓ is replaced by $\int|u|^\rho$ for $\rho = 6,8,10,\cdots$?

Proposition 5.1 gives us some possibilities for the spherical harmonics from which nonlinear waveforms may bifurcate, namely the G-orbits of Y_ℓ^m for all m and η_ℓ^m for m=0 and $\ell/3 < m \leq \ell$. To see whether such bifurcation actually occurs, and to analyze the stability of the resulting standing waves, it is necessary to look at the hessian of N_ℓ in the directions transverse to an orbit.

Robert Stanton and I have obtained the following result [ST-W]:

THEOREM 5.2. For $\ell \geq 3$ the hessian of N_ℓ is negative definite on the space transverse to the orbit of η_ℓ^ℓ. Thus, for $\ell \geq 3$ the orbit of η_ℓ^ℓ is a nondegenerate critical manifold of local maxima for the function N_ℓ.

We defer discussing the proof of Theorem 5.2 until section 7 and go on now to state its immediate consequence: equation (5.6) has periodic solutions concentrated along closed geodesics. More precisely, we have:

THEOREM 5.3. Given any $\ell \geq 3$ and any $\delta > 0$, there exists a number $\varepsilon_o > 0$ (depending on δ and ℓ) such that, for every $\varepsilon \in (0,\varepsilon_o)$, there is a standing wave solution of the nonlinear equation (5.6) on S^2 which has frequency within δ of $\sqrt{\ell(\ell+1)}$ and whose waveform u satisfies:

(a) $||u|| = \varepsilon$

(b) $||u/\varepsilon - \eta_\ell^\ell|| < \delta$.

Of course, since the equation (5.6) is rotationally invariant, there also exist standing wave solutions with waveforms near any point in the orbit of η_ℓ^ℓ, i.e., concentrated near any given closed geodesic. The proof of Theorem 5.3 actually gives more: the family of waveforms with $||u|| = \varepsilon$ for a fixed ℓ is acted upon transitively by $O(3)$ and forms a nondegenerate critical manifold for the variational problem; furthermore, the linearized (infinite-dimensional) Poincaré map around the periodic standing wave associated with each such waveform is elliptic, so that the wave is infinitesimally stable.

As we already remarked, the existence of standing waves concentrated along closed geodesics of S^2 should come as no surprise, since such solutions exist already for the linear equation. However, the nondegeneracy alluded to in the preceding paragraph implies that these families of standing waves cannot disappear entirely when the symmetry is broken by a perturbation. In fact, the bifurcation theory of critical manifolds tells us that, for a sufficiently small perturbation of the metric or the addition of a sufficiently small potential, at least three standing waves concentrated near closed geodesics remain for each value of ε and ℓ. The allowable size of the perturbation goes to zero as $\varepsilon \to 0$ (as it should, since we need "sufficient nonlinearity" to overcome the quasi-mode effect), but unfortunately it also goes to zero as $\ell \to \infty$. In the next section, we shall explain how the latter deficiency can be remedied for certain perturbations.

6. THE 2-SPHERE WITH POTENTIAL

Consider now the full nonlinear equation with potential on S^2:

$$-\frac{\partial^2\psi}{\partial t^2} = -\Delta\psi + V\psi - |\psi|^2\psi . \tag{6.1}$$

If the amplitude of ψ is small, and if ψ is now an eigenspace E_ℓ for large ℓ, then both $V\psi$ and $-|\psi|^2\psi$ will be small in comparison with $\Delta\psi$. Unfortunately, it does not seem to be possible to keep track of both of these perturbations by a simple scaling argument as in section 4. Instead, when we apply the Liapunov-Schmidt bifurcation theory, it is necessary to use estimates of the size of perturbation for which the implicit function theorem will apply.

Roughly speaking, we have two competing effects: the term $V\psi$ tends to disperse the waves concentrated along closed geodesics, and the term $-|\psi|^2\psi$ tends to make them more concentrated. There is a range of amplitudes in which these two effects balance out to produce stable waveforms.

In order to carry out the required estimates, we need two kinds of information beyond that needed for the case $V=0$.

(a) We need information about the behavior of the function N_ℓ in some finite neighborhood of the orbit of η_ℓ^ℓ, not just the infinitesimal neighborhood as was used above. For this, it suffices to have a global upper bound for N_ℓ.

(b) We need estimates for the size of the clusters into which the multiple eigenvalues of $-\Delta$ are split by the addition of the potential V. We can also use information on the structure of these clusters.

The best result which is available to us for part (a) is the following fact which can be proven quite easily from well known properties of spherical harmonics (see [ST-W]).

THEOREM 6.1. $\sup\{N_\ell(u) \mid u \in S^\ell\} = O(\ell)$.

This theorem is just enough to give us a theorem on nonlinear waves, if we have a very sharp estimate in part (b) above. Such an estimate is available if the potential V is odd (i.e., $V(x) = -V(-x)$).

THEOREM 6.2. (Guillemin [G3]) If V is an odd function on S^2, then the ℓ'th eigenvalue cluster of $-\Delta+V$ lies in an interval of the form $[\ell(\ell+1) - a_\ell, \ell(\ell+1) + a_\ell]$, where $a_\ell = O(\ell^{-2})$.

Plugging Theorems 6.1 and 6.2 into our general bifurcation machine, we can prove the following theorem about nonlinear waves. (A more precise statement and a proof will appear in a paper now in preparation.)

THEOREM 6.3. Let V be an odd function on S^2. Given any ℓ sufficiently large (this depends upon V) there exist positive numbers $\varepsilon_1 < \varepsilon_2$ (depending on ℓ) such that, for any $\varepsilon \in (\varepsilon_1, \varepsilon_2)$, there is a standing wave solution of (6.1) which has frequency near $\sqrt{\ell(\ell+1)}$ and whose waveform u satisfies:

(a) $\|u\| = \varepsilon$

(b) u/ε is L^2 close to the orbit of η_ℓ^ℓ.

Since the equation is no longer O(3)-symmetric, only certain closed geodesics will "carry" waveforms for standing waves. We suspect that these will be the critical points of Guillemin's "order -3 symbol" in [G3].

We would like to extend Theorem 6.3 to an arbitrary potential, but this is much more delicate. In this case, the eigenvalue clusters have a width which is bounded from below (see [G1]), and there is no hope of improving Theorem 6.1 enough to compensate for this entirely. However, we know enough about the structure of these clusters [C][G2][G3][LI][WE1][WE2] to conclude that somewhere in the orbit of η_ℓ^ℓ there is an appropriate eigenfunction for $-\Delta+V$ whose "approximate eigenvalue" lies within $O(\ell^{-1})$ of the lower end of the

ℓ'th cluster. Now we may on the fact that η_ℓ^ℓ is a maximum for N_ℓ on S_ℓ, and not just a critical point, in order to obtain "cooperation" between the linear and nonlinear effects. But we still need some improvement of Theorem 6.1. If the answer to the following question is "yes," then Theorem 6.3 can be extended to arbitrary potentials.

QUESTION 6.4. Is $\sup\{N_\ell(u) \mid u \in S^\ell\} = O(\ell)$?

This problem, to estimate the L^4 norm of spherical harmonics, does not seem to be solved in the literature. We remark that $N_\ell(\eta_\ell^\ell)$ grows like $\ell^{1/2}$, so that if these local maxima are actually global maxima, Theorem 6.1 can actually be improved quite a lot. Numerical calculations suggest that $N_\ell(\eta_\ell^m) = O(\ell^{1/2})$ as well, uniformly for $|m| \leq \ell$, but this is probably not such strong evidence, since the η_ℓ^m's are eigenfunctions of the operator $\frac{\partial}{\partial\psi}$ as well as $-\Delta$, so that they might be expected to satisfy estimates characteristic of one variable rather than two. Finally, we remark that if $\int u^4$ is replaced by $\int u^p$ for p large enough (probably $p > 6$), then it appears that the maxima is attained at η_ℓ^0 instead of η_ℓ^ℓ. (This is definitely true for $p = \infty$, and is due to the peak of the spherical function at the poles of the sphere.)

7. ON THE PROOF THAT THE HESSIAN IS NEGATIVE

The proof outlined in this section is the result of work with Robert Stanton; details will be published in a joint paper [ST-W].

The hessian of N_ℓ at η_ℓ^ℓ is a quadratic form on the tangent space to N_ℓ at η_ℓ^ℓ; this tangent space may be identified with the orthogonal space to η_ℓ^ℓ in E_ℓ. Simple calculations and symmetry arguments show that we can restrict our attention to calculating the hessian in the directions of $\eta_\ell^0, \eta_\ell^1, \cdots, \eta_\ell^{\ell-2}$. (The hessian in imaginary directions is even less, the η_ℓ^m's are "hessian-orthogonal," and $\eta_\ell^{\ell-1}$ is in the direction of the orbit of η_ℓ^ℓ.) Up to a constant independent of ℓ, the hessian of N_ℓ at η_ℓ^ℓ in the direction of η_ℓ^m turns out to be given by

$$2\int_{-1}^{1} \hat{P}_\ell^\ell(z)^2\, \hat{P}_\ell^m(z)^2 dz - \int_{-1}^{1} \hat{P}_\ell^\ell(z)^4 dz \tag{7.1}$$

where $\hat{P}_\ell^m(z)$ is the multiple of

$$P^m(z) = \frac{(z^2-1)^{m/2}}{2^\ell \ell!} \frac{d^{\ell+m}}{dz^{\ell+m}} (z^2-1)^\ell$$

for which $\int_{-1}^{1} \hat{P}_\ell^m(z)^2 dz = 1$. Theorem 5.2 says, then, that (7.1) is negative when $0 \leq m \leq \ell-2$ (except when $m=0$, $\ell=2$).

The second integral in (7.1) can be easily evaluated. It is equal to

$I_{2\ell}/I_\ell^2$, where $I_\ell = \int_{-1}^{1} (1-z^2)^\ell d\ell = \frac{2^{2\ell+1}\ell!^2}{(2\ell+1)!}$, which is asymptotic to $\sqrt{\frac{\pi}{\ell}}$. The first integral in (7.1) can be expressed in terms of Clebsch-Gordan coefficients, but this expression, although explicit, is a rather horrendous alternating sum which does not lend itself to being estimated. Stanton has found a much simpler expression in terms of generalized hypergeometric functions; while it is still useless for estimation, this expression is quite convenient for numerical evaluation.

A useful parameter in understanding (7.1) is the difference $r = \ell-m$. For each value of r, the hessian (7.1) can be written as the product of a factor which is obviously negative times a polynomial $H_r(\ell)$. It is necessary to show that, for $r \geq 2$ $H_r(\ell)$ is positive for all $\ell \geq r$ (except $\ell=r=2$). Using a computer, one can verify this for as many values of r as desired. Stanton and I carried this out for all $r \leq 11$, verifying in each case that H_r and all its nontrivial derivatives are positive when evaluated at $\ell=r$.

Of course, one cannot prove the entire theorem by such calculations. For large r, one has another method, namely that of asymptotic approximation to the Legendre functions. The function $u(\theta) = \sin^{1/2}\theta\, P_\ell^m(\cos\theta)$ satisfies Legendre's differential equation

$$\frac{d^2u}{d\theta^2} = -\left[(\ell+\tfrac{1}{2})^2 - \frac{m^2-\frac{1}{4}}{\sin^2\theta}\right] u ,$$

and one can compute the value $u(\pi/2)$ without much difficulty. Using this as a starting point, one can use the W.K.B. method to get an approximation to u which is good in a neighborhood of $\theta = \pi/2$. This gives an approximation to $\hat{P}_\ell^m(z)^2$ which is good in a neighborhood of z=0. Outside this neighborhood, $\hat{P}_\ell^\ell(z)^2$ is relatively small. Now the first integral in (7.1) can be estimated in two parts -- near z=0 and away from z=0, where the dividing point approaches zero like $\ell^{-1/2}$. Of course, one needs precise error bounds for the W.K.B. approximation; those are available in the form we need in the work of F. Olver [O]. The estimation remains delicate, and it only seems to work if r is large enough, say for $r \geq 10$. The computer calculation just fills the gap, and Theorem 5.2 is proven.

8. OTHER MANIFOLDS

The proofs of our results on S^2 depend heavily upon the fact that the eigenvalues of $-\Delta+V$ are clustered around widely spaced points. This suggests immediately that the results might be extended to other manifolds, all of whose geodesics are closed (see [BS] for a study of such manifolds). For the other symmetric spaces of rank 1, this would require an extension of the hessian estimate to other special functions -- a first step in this direction would be to simplify out proof for Legendre functions. For Zoll surfaces (diffeomorphic to S^2), it could be useful to understand the behavior of certain Fourier integral operators with respect to the L^4 norm.

None of the examples just mentioned have periodic geodesics which are elliptic in the sense that the Poincaré rotation angles are unequal to zero (though S^2 with a potential may have elliptic periodic solutions of Newton's equation with high velocity, and these may in fact carry the concentrated nonlinear waves). For the elliptic case, the spectrum is never clustered, so we must work much harder. A vague idea which I have for this case is that, microlocally near an elliptic periodic orbit in T^*X, the spectrum is in some sense clustered around the arithmetic progressions discussed in Section 2. Here I see two immediate difficulties? First of all, microlocal analysis with respect to L^p norms ($p \neq 2$) is not very well developed. Secondly, there are arithmetic progressions corresponding to different values of $(\eta_1, \cdots, \eta_{d-1})$, and interactions between the various modes must be considered. This corresponds to the problem of excitement of higher transverse modes by gaussian beams in laser optics -- the optics literature (e.g. [AN][SG]) might be profitably consulted in this regard.

BIBLIOGRAPHY

[A-V] M. Alonso and H. Valk, Quantum Mechanics: Principles and Applications, Addison-Wesley, Reading, 1973.

[AN] J. A. Arnaud, Beam and Fiber Optics, Academic Press, New York, 1976.

[AR1] V. I. Arnol'd, Modes and quasimodes, Funct. Anal. Appl., 6 (1972), 94-101.

[AR2] V. I. Arnol'd, Mathematical Methods of Classical Mechanics, Graduate Texts in Math. Vol 60, Springer-Verlag, New York (1978).

[BA] V. M. Babich, Eigenfunctions concentrated in a neighborhood of a closed geodesic, Mathematical Problems of Wave Propagation Theory, Seminars in Mathematics, V. A. Steklov Mathematical Institute, 9 (1968). English translation by Consultants Bureau (1970).

[BR] M. S. Berger, On the existence and structure of stationary states for a nonlinear Klein-Gordon equation, J. Functional Anal., 9 (1972), 249-261.

[BS] A. L. Besse, Manifolds all of whose geodesics are closed, Ergebnisse der Math. und Grenzgebiete, 93, Springer-Verlag, 1978.

[C] Y. Colin de Verdière, Spectre conjoint d'opérateurs pseudo-différentielles qui commutent, I. Duke Math. J., 46 (1979), 169-182.

[D] B. H. Driscoll, The multiplicity of the eigenvalues of a symmetric drum, Ph.D. Thesis, Northwestern University, Evanston, 1978.

[Gl] V. Guillemin, Symplectic spinors and partial differential equations, Géométrie Symplectique et Physique Mathématique, Colloques Internationaux du C.N.R.S., 237 (1975), 217-252.

[G2] V. Guillemin, Addenda to: Some spectral results on rank one symmetric spaces, Advances in Math., 28 (1978).

[G3] V. Guillemin, Some spectral results for the Laplace operator with potential on the n-sphere, Advances in Math., 27 (1978), 273-286.

[G-S] V. W. Guillemin and S. Sternberg, Geometric Asymptotics, Amer. Math. Soc., 1976.

[G-W] V. Guillemin and A. Weinstein, Eigenvalues associated with a closed geodesic, Bull. A.M.S., 82 (1976), 92-94.

[H] L. Hörmander, Fourier integral operators. I, Acta Math., 127 (1971), 79-183.

[K-K] B. B. Kadomtsev and V. I. Karpman, Nonlinear waves, Soviet Physics Uspekhi, 14 (1971), 40-60.

[L-L] L. D. Landau and E. M. Lifshitz, Quantum Mechanics, Non-relativistic Theory, Addison-Wesley, Reading, 1958.

[LI] W. Lichtenstein, Qualitative behavior of special functions on compact symmetric spaces, Ph.D. Thesis, University of California, Berkeley, 1978.

[M] V. P. Maslov, Theory of Perturbations and Asymptotic Methods (in Russian), Moscow State Univ., Moscow, 1965; translated as Théorie des Perturbations et Méthodes Asymptotiques, Dunod, Gauthier-Villars, Paris, 1972.

[O] F.W.J. Olver, Error bounds for the Liouville-Green (or WKB) approximation, Proc. Camb. Phil. Soc., 57 (1961), 790-810.

[P] M. F. Pyshkina, Asymptotic behavior of eigenfunctions of the Helmholz equation concentrated near a closed geodesic, Mathematical Problems of Wave Propagation Theory II, Seminars in Mathematics, V. A. Steklov, Mathematical Institute, 15 (1969). English translation by Consultants Bureau(1971).

[RAB] P. Rabinowitz, Free vibrations for a semilinear wave equation, Comm. Pure Appl. Math., 31 (1978), 31-68.

[RAL1] J. V. Ralston, On the construction of quasimodes associated with stable periodic orbits, Comm. Math. Phys., 51 (1976), 219-242.

[RAL2] J. V. Ralston, Approximate eigenfunctions of the laplacian, J. Diff. Geom., 12 (1977), 87-100.

[RE] M. Reeken, Stability of critical points under small perturbations. Part II: analytic theory, Manuscripta Math., 8 (1973), 69-92.

[SG] A. Siegman, An Introduction to Lasers and Masers, McGraw Hill, New York, 1971.

[SM] S. Smale, Topology and mechanics. I,II., Inventiones Math., 10 (1970), 305-331; 11 (1970), 45-64.

[ST-W] R. Stanton and A. Weinstein, On the L^4 norm of spherical harmonics (to appear).

[STR] W. A. Strauss, Existence of solitary waves in higher dimensions, Comm. in Math. Phys., 55 (1977).

[U] K. Uhlenbeck, Generic properties of eigenfunctions, Amer. J. Math., 98 (1976), 1059-1078.

[WE1] A. Weinstein, Asymptotics of eigenvalue clusters for the laplacian plus a potential, Duke Math.J., 44 (1977), 883-892.

[WE2] A. Weinstein, Bifurcations and Hamilton's principle, Math. Z., 159 (1978), 235-248.

[WH] G. B. Whitham, Linear and Nonlinear Waves, Wiley-Interscience, New York, 1974.

[WI] H. Widom, Eigenvalue distribution theorems in certain homogeneous spaces, J. Funct. Anal. (to appear).

[Y] A. Yariv, Guided-wave optics, Scientific American, 240 (1979), 64-72.

DEPARTMENT OF MATHEMATICS
RICE UNIVERSITY
HOUSTON, TEXAS 77001

Current Address:
Department of Mathematics
University of California
Berkeley, California 94720

Proceedings of Symposia in Pure Mathematics
Volume 36, 1980

ASYMPTOTICS OF COMPRESSIONS TO SPECTRAL SUBSPACES OF THE LAPLACIAN

Harold Widom[1]

ABSTRACT. The classical Szegö theorem on the asymptotics of the spectra of Toeplitz matrices is generalized to pseudodifferential operators on two-point homogeneous spaces. A corollary is Weinstein's theorem on the asymptotics of the eigenvalue clusters of the Schrödinger operator on such spaces.

We shall be concerned with operators of the form $P_n A P_n$ where $\{P_n\}$ is a sequence of spectral projections for the positive Laplace operator Δ on a compact two-point homogeneous space (= rank 1 homogeneous space) M and A is an order zero pseudodifferential operator on M. The main result is a "Szegö theorem" describing the asymptotic behavior of the spectra of these operators under certain conditions. An application is another proof of a theorem of A. Weinstein [2] on the asymptotics of the eigenvalues of the operator $\Delta + M_q$ where M_q denotes multiplication by the smooth function q.

It was T. Spencer who first suggested the existence of a Szegö theorem for pseudodifferential operators which would yield the conclusion of Weinstein's theorem. The present work was the result of that suggestion.

With one exception (the proof of the special case of the conjecture at the end of the paper) the results stated here are proved in [3], perhaps somewhat differently. We shall therefore not always supply detailed proofs.

We begin with a simple lemma from functional analysis. The setting is a compact space X with a compact group G acting transitively on X. There is then a unique G-invariant measure μ on X normalized so that $\mu(X) = 1$.

LEMMA. Let E be a normed linear space, Φ a bounded linear mapping from $C(X)$ to E satisfying $\Phi(\sigma \circ g) = \Phi(\sigma)$ for all $g \in G$, $\sigma \in C(X)$. Then

$$\Phi(\sigma) = \int \sigma \, d\mu \cdot \Phi(1).$$

Proof. In case E is the scalar field the Reisz representation theorem tells us that

---

1970 Mathematics Subject Classification 58G15

[1]Supported by the National Science Foundation

$$\Phi(\sigma) = \int \sigma d\nu$$

for some measure ν, and the G-invariance of Φ implies that

$$\nu = \mu \cdot \Phi(1).$$

For general E, take any $\lambda \in E^*$. Then $\lambda \circ \Phi$ is a G-invariant linear functional so the assertion holds for it. Since $\lambda \in E^*$ is arbitrary the Hahn-Banach theorem implies that the assertion holds for Φ itself.

In our application of the lemma X will be the cosphere bundle S^*M and G the isometry group of M lifted to S^*M. Observe that G acts transitively on X precisely because M is a two-point homogeneous space. The space E will be ℓ^∞/c_o, the bounded sequences modulo the sequences converging to zero.

The mapping Φ is obtained as follows. Let $\{P_n\}$ be any fixed sequence of finite rank spectral projections for Δ such that

(1) $$r_n = \text{rank } P_n \to \infty .$$

Take a function $\sigma \in C^\infty(X)$ and denote by $\text{op}\,\sigma$ any order zero pseudodifferential operator with principal symbol σ. Then $\Phi(\sigma)$ is defined to be the image in ℓ^∞/c_o of the sequence

(2) $$\{r_n^{-1} \text{ tr } P_n \text{ op}\,\sigma\, P_n\} \in \ell^\infty .$$

If in (2) the operator $\text{op}\,\sigma$ were replaced by an arbitrary compact operator the sequence would belong to c_o. It therefore follows from the identity

$$\inf\{\| \text{op}\sigma + K\| : K \text{ compact}\} = \max\{|\sigma(v)| : v \in X\}$$

that $\Phi(\sigma)$, the representative of the sequence (2) in ℓ^∞/c_o, is independent of the choice of $\text{op}\sigma$ and that

$$\|\Phi(\sigma)\| \le \max|\sigma| .$$

In particular Φ extends to all of $C(X)$. To see that Φ is G-invariant let g be an isometry of M and $\bar{g}$ the corresponding diffeomorphism of X. If $U_g : L_2(M) \to L_2(M)$ is defined by $U_g f = f \circ g$ then

$$\text{tr } P_n \text{ op}\,\sigma\, P_n = \text{tr } U_g^* P_n \text{ op}\,\sigma\, P_n U_g = \text{tr } P_n U_g^* \text{ op}\,\sigma\, U_g P_n$$

since U_g commutes with Δ and so with its spectral projections. But

$$U_g^* \text{ op}\,\sigma\, U_g = \text{op}\,\sigma \circ \bar{g}$$

(this is just the formula for change of symbol under a diffeomorphism), so Φ is G-invariant.

Thus the lemma applies. The conclusion is that (2) is, modulo c_o, the constant sequence $\{\int \sigma d\mu\}$. Thus:

THEOREM 1. Let A be an order zero pseudodifferential operator on M with principal symbol σ. Then for any sequence of spectral projections $\{P_n\}$ satisfying (1) we have

$$\lim_{n\to\infty} r_n^{-1} \operatorname{tr} P_n A P_n = \int_{S^*M} \sigma d\mu .$$

Suppose now that A is self-adjoint and denote by μ_n the measure on R which assigns measure r_n^{-1} to each eigenvalue of $P_n A P_n$. Denote by μ_σ the push-forward to R, via the real-valued function σ, of the measure μ on S^*M:

$$\mu_\sigma(E) = \mu(\sigma^{-1}(E)).$$

The conclusion of Theorem 1 may be stated

$$\int x \, d\mu_n(x) \to \int x \, d\mu_\sigma(x).$$

Now we would like to assert that $\mu_n \to \mu_\sigma$ (weakly) and this amounts to showing that the above relation holds with x replaced by an arbitrary power x^k. This in turn may be restated

$$r_n^{-1} \operatorname{tr}(P_n A P_n)^k \to \int \sigma^k d\mu$$

which, in view of Theorem 1 applied to $A^k = \operatorname{op} \sigma^k$, is equivalent to

$$\operatorname{tr}\{P_n A^k P_n - (P_n A P_n)^k\} = o(r_n).$$

For $k = 2$ this may be written (with the subscript indicating Hilbert-Schmidt norm)

$$\text{(3)} \qquad \| P_n A(I-P_n) \|_2^2 = o(r_n)$$

and this implies that for each k the trace norm of $P_n A^k P_n - (P_n A P_n)^k$ is $o(r_n)$. Thus:

THEOREM 2. The measure μ_n converges to μ_σ if an only if (3) holds.

Note that (3) with o replaced by O always holds, but (3) itself need not. Two conditions of quite different kinds, each of which is sufficient, are the following.

(a) If P_n corresponds to the set Λ_n of $\operatorname{spec} \Delta^{1/2}$ and $P_{n,\delta}$ corresponds to the set

$$\{\lambda \in \operatorname{spec} \Delta^{1/2} : \lambda \notin \Lambda_n, \ d(\lambda, \Lambda_n) < \delta\}$$

then for each $\delta > 0$

$$\operatorname{rank} P_{n,\delta} = o(r_n)$$

as $n \to \infty$.

(b) The commutator $[A, \Delta^{1/2}]$ is compact. Equivalently σ is invariant under the geodesic flow on S^*M.

Proofs may be found in [3]. From (a) it follows that $\mu_n \to \mu_\sigma$ if $\{\Lambda_n\}$ is a sequence of intervals whose lengths tend to ∞. In particular we recover Szegö's theorem for Toeplitz matrices [1, §5.2] (albeit for continuous symbols only) which is the case $M = S^1$, $\Lambda_n = [0,n]$.

Weinstein's theorem may be obtained from (b) as follows. For $\dim M > 1$ the eigenvalues λ_n of Δ have multiplicities $r_n \to \infty$. The eigenvalues of $\Delta + M_q$ may be written

$$\lambda_n + \alpha_{n,k} \qquad (k=1,\ldots,r_n)$$

and the question is that of the behavior of the $\alpha_{n,k}$ as $n \to \infty$. If ν_n is the measure on R which assigns measure r_n^{-1} to each $\alpha_{n,k}$ and P_n is the spectral projection of Δ corresponding to the eigenvalue λ_n then it is easy to show that

$$\nu_n - \mu_n \to 0$$

where μ_n is the measure associated with the spectrum of $P_n M_q P_n$. If T is the common period of the closed geodesics of M and we define

$$A = T^{-1} \int_0^T e^{it\Delta^{1/2}} M_q e^{-it\Delta^{1/2}} dt$$

then $P_n A P_n = P_n M_q P_n$. Moreover by Egorov's theorem A is an order zero pseudodifferential operator whose symbol at a cotangent vector v is $\hat{q}(v)$, the average of q around the closed geodesic associated with v. Since $\hat{q}$ is clearly invariant under the geodesic flow (b) is applicable and we find $\mu_n \to \mu_{\hat{q}}$. Thus

$$\nu_n \to \mu_{\hat{q}},$$

which is Weinstein's result. (Weinstein actually proves more: ν_n has an asymptotic expansion of which $\mu_{\hat{q}}$ is the first term.)

The method of proof of Theorem 2 also gives information on the joint behavior of the eigenvalues and eigenfunctions of $P_n A P_n$ whenever (3) holds. In particular it is easy to deduce that if $k_n(x,y)$ is the kernel of $r_n^{-1} P_n A P_n$ with respect to the volume element dx on M and if one defines for $x \in M$

$$\bar{\sigma}(x) = \int_{\pi^{-1}(x)} \sigma \frac{d\mu}{dx},$$

where π is the projection from S^*M to M, then

$$(4) \qquad k_n(x,x) \to \bar{\sigma}(x) \text{ weakly}$$

in the sense that $k_n(x,x)dx \to \bar{\sigma}(x)dx$ weakly as measures.

CONJECTURE. $k_n(x,x) \to \bar{\sigma}(x)$ in $C^\infty(M)$.

Here is a proof of the conjecture in the one-dimensional case, $M = S^1$. Look at the operator

$$r_n^{-1}P_n[d/dx,A]P_n. \tag{5}$$

Since P_n commutes with d/dx this is the same as

$$[d/dx,\ r_n^{-1}P_nAP_n]$$

whose kernel is just

$$\left(\frac{\partial}{\partial x} + \frac{\partial}{\partial y}\right) k_n(x,y) \ . \tag{6}$$

Now it is true of any self-adjoint operator of finite rank that the L_1 norm of the restriction to the diagonal of its kernel is at most the norm of the operator times its rank. (This follows immediately from the eigenfunction expansion of the kernel.) The restriction of (6) to the diagonal is just

$$\frac{d}{dx} k_n(x,x) \ .$$

The operators (5) have rank at most r_n and, since $[d/dx,A]$ is bounded, norm at most r_n^{-1}. We deduce

$$\int \left|\frac{d}{dx} k_n(x,x)\right| dx = O(1) \ .$$

In particular $\{k_n(x,x)\}$ is an equicontinuous family and this, together with (4), implies that $k_n(x,x) \to \bar{\sigma}(x)$ uniformly. Further commutation with d/dx yields C^∞ convergence.

BIBLIOGRAPHY

1. V. Grenander and G. Szegö, Toeplitz forms and their application, Univ. of Cal. Press, Berkeley, 1958.

2. A. Weinstein, "Asymptotics of eigenvalue clusters for the laplacian plus a potential", Duke Math. J., 44 (1977), 883-892.

3. H. Widom, "Eigenvalue distribution theorems for certain homogeneous spaces", J. Functional Anal., 32 (1979).

DIVISION OF NATURAL SCIENCES
UNIVERSITY OF CALIFORNIA
SANTA CRUZ, CALIFORNIA 95064